Conducting Polymers
Bioinspired Intelligent Materials and Devices

RSC Smart Materials

Series Editors:
Professor Hans-Jörg Schneider, *Saarland University, Germany*
Professor Mohsen Shahinpoor, *University of Maine, USA*

Titles in this Series:
1: Janus Particle Synthesis, Self-Assembly and Applications
2: Smart Materials for Drug Delivery: Volume 1
3: Smart Materials for Drug Delivery: Volume 2
4: Materials Design Inspired by Nature
5: Responsive Photonic Nanostructures: Smart Nanoscale Optical Materials
6: Magnetorheology: Advances and Applications
7: Functional Nanometer-Sized Clusters of Transition Metals: Synthesis, Properties and Applications
8: Mechanochromic Fluorescent Materials: Phenomena, Materials and Applications
9: Cell Surface Engineering: Fabrication of Functional Nanoshells
10: Biointerfaces: Where Material Meets Biology
11: Semiconductor Nanowires: From Next-Generation Electronics to Sustainable Energy
12: Supramolecular Materials for Opto-Electronics
13: Photocured Materials
14: Chemoresponsive Materials: Stimulation by Chemical and Biological Signals
15: Functional Metallosupramolecular Materials
16: Bio-Synthetic Hybrid Materials and Bionanoparticles: A Biological Chemical Approach Towards Material Science
17: Ionic Polymer Metal Composites (IPMCs): Smart Multi-Functional Materials and Artificial Muscles, Volume 1
18: Ionic Polymer Metal Composites (IPMCs): Smart Multi-Functional Materials and Artificial Muscles, Volume 2
19: Conducting Polymers: Bioinspired Intelligent Materials and Devices

How to obtain future titles on publication:
A standing order plan is available for this series. A standing order will bring delivery of each new volume immediately on publication.

For further information please contact:
Book Sales Department, Royal Society of Chemistry, Thomas Graham House, Science Park, Milton Road, Cambridge, CB4 0WF, UK
Telephone: +44 (0)1223 420066, Fax: +44 (0)1223 420247
Email: booksales@rsc.org
Visit our website at www.rsc.org/books

Conducting Polymers
Bioinspired Intelligent Materials and Devices

Toribio Fernández Otero
Technical University of Cartagena, Spain
Email: toribio.fotero@upct.es

THE QUEEN'S AWARDS
FOR ENTERPRISE:
INTERNATIONAL TRADE
2013

RSC Smart Materials No. 19

Print ISBN: 978-1-78262-315-1
PDF eISBN: 978-1-78262-374-8
ISSN: 2046-0066

A catalogue record for this book is available from the British Library

Published by the Royal Society of Chemistry,
Thomas Graham House, Science Park, Milton Road,
Cambridge CB4 0WF, UK

Registered Charity Number 207890

For further information see our web site at www.rsc.org

Printed in the United Kingdom by CPI Group (UK) Ltd, Croydon, CR0 4YY, UK

The proximity paradox: human scientific models predicted the existence of subatomic particles and astronomical structures beyond the borders of the known universe years before the construction of the required equipment for their observation, but they cannot describe life and life functions that constitute our everyday experience.

TF Otero

Preface

Human scientific knowledge has advanced to different borders. One of these limits is shown around us and every day we observe it; in fact it takes part of ourselves. Life, life functions and malfunctions are outside of the descriptions or predictions of any scientific model.

Present physical models were able to predict the existence of subatomic particles or astronomical structures beyond the borders of the explored universe long before we designed and constructed devices required for their observation. But they cannot describe or predict health or illnesses that constitute our every experience. Life is chemistry. Biochemical reactions include biopolymers and macromolecules as reactants. The reactions induce conformational movements, but current chemical models do not include quantification of any conformational, allosteric or structural reaction-driven changes.

Up until the 1980s, chemists did not have at their disposal dense reactive gels that could mimic the intracellular matrix (ICM) of living cells, which constitutive polymeric chains may participate as reactants in chemical or electrochemical reactions. From them, a plethora of different materials has emerged, giving electroactive films in liquid electrolytes, including conducting polymers, redox polymers, fullerenes, carbon nanotubes, graphenes, phthalocyanines and so on.

In this book the electrochemistry of conducting polymers as reactive gels that mimic, in its simplest expression (reactive chains, reaction-driven conformational movements, ions and water), the composition of the ICM in living cells is described.

Electrochemical reactions (oxidation or reduction) from the different families of conducting polymers (Chapter 4) drive conformational movements of the constitutive polymeric chains and structural macroscopic changes of the film, such as conformational relaxation, swelling, shrinking and

RSC Smart Materials No. 19
Conducting Polymers: Bioinspired Intelligent Materials and Devices
By Toribio Fernández Otero

Published by the Royal Society of Chemistry, www.rsc.org

conformational compaction from different energetic states of trapped ions. Chapter 6 describes how these reaction-driven structural changes that mimic biochemical reactions are identified, controlled and quantified.

Such reactive gels can be used as material models of the ICM to obtain the empirical kinetics of the reaction. The new aspect, from a chemical kinetic point of view, is the possibility of repeating the full, selected kinetic procedure using different shrunken or conformationally packed initial states. Surprising results have been attained, such as the activation energy (E_a), the reaction coefficient (k) and the reaction order being the three of them, which are functions of the initial energetic state of packed conformations. The kinetic magnitudes include conformational and structural quantitative information: the *chemical kinetics* become *structural chemical kinetics* (Chapter 7).

The electrochemically stimulated conformational relaxation (ESCR) model and the consequential structural chemical kinetic (SCK) model here presented describe the obtained empirical results. Variation of the chain conformational energy during a reaction reaches a quantitative magnitude. The final structural equations for E_a, k and the reaction orders are adapted to those biochemical reactions taking place in the absence of electric fields, like enzymatic reactions, and allosteric and folding/misfolding effects.

The faradaic ionic exchange shifts the gel composition by several orders of magnitude, as functional reactions do in organs. The values of any composition-dependent properties of the material (volume, color, charge storage, ionic storage, ionic conductivity and so on) will change under the control of the driving current through several orders of magnitude. Chapter 5 is devoted to these composition-dependent properties, each mimicking a biological function. Each of these biomimetic properties allows the development of a biomimetic device that works in a way driven by the reaction of the device's constitutive material, as biological organs do. In this regard, Chapter 8 describes artificial muscles, smart membranes, artificial glands, decontaminating systems, artificial chemical synapses, chemo-ionic-conformational memories, smart surfaces, electrochromic devices, organic batteries, biosensors and mechanical sensors. However, the reaction shifts the value of each of the composition-dependent properties simultaneously. This fact opens an unexplored and unexpected scientific and technological world of multi-tool devices, where several tools work simultaneously in a physically uniform device driven by the same reaction. Chapter 9 is devoted to the exploration of sensing motors, inspired by haptic muscles. Haptic muscles and the brain give rise to proprioception. Based on electrochemical, polymeric and mechanical basic principles, a theoretical model is presented describing artificial proprioception and the attained experimental results.

A short review of some of the basic electrochemical cells and methods is given (Chapter 2). Chapter 3 is devoted to the complex mechanism of the electrochemical synthesis of conducting polymers whose control allows the synthesis of tailored materials and may help new researchers to understand the new concepts and principles presented here.

Some final consideration and remarks related to reactions and structures, materials and biological structural processes are collected in the final chapter. Challenges in the quantification and prediction of biological reactions and functions are presented, while the door is left open for a quantitative description of electro-chemo-conformational memories and brain memory.

Toribio Fernández Otero

Acknowledgements

I have received immeasurable support and encouragement from my parents and brothers, friends, and co-workers. I could never have completed this book without them. I thank God for sending them to support me on this journey. This acknowledgements section could never be complete without expressing my sincere gratitude to these individuals.

I will always be in permanent debt to each and every one of my cultural cowherd (brañeiros y vaqueiros) brothers from the Cantabric Mountains Range in the northern part of Spain, and in particular to those from my hometown, Palacios del Sil, where I spent my childhood. I am also in debt to my brañeiros ancestors that kept this isolated and hardy but efficient civilization frozen in time for centuries. They helped that eager boy to mature, while he enjoyed plants, flowers, animals, water springs, rain, snow, sunny days, storms, stars, and the sky and land. Breathing fresh and clean air from the mountain tops and listening every night to old stories around a fire helped me fall in love with nature forever. That civilization of transhumance shepherds has dissolved into the wind over the last 30 years but my childhood probably helped me grow as an independent researcher, keeping in mind that the scientific literature is just focused on a small fraction of the truth sought. We should always be ready to think outside of the boxes that others have built.

Finding a way through a fog-filled forest under a severe storm is a difficult task if you don't have solid reference points. I was blessed when I found my way to Don José Escudero, my primary school teacher. Had it not been for him, I would never have become a researcher and would most likely have spent my life as a coal miner, as many of my elementary school fellows did. His incessant pressure on my parents and his help teaching me as a free student for three years (exams were taken annually during a day in June at the secondary school located 50 km from my town) gave me the possibility to combine my student and shepherd duties. I was quite fortunate to find

RSC Smart Materials No. 19
Conducting Polymers: Bioinspired Intelligent Materials and Devices
By Toribio Fernández Otero

Published by the Royal Society of Chemistry, www.rsc.org

some good teachers in my life, but Don José Escudero remains my master and milestone.

Some days with grazing cows and goats were long when I was growing up. The temptation to climb the surrounding mountains was invariably there and I learned that behind a mountain you will always find new mountains and, in between them, you will find new forests, rivers, lakes and rocks. I learned that exploring is good—a good lesson for a future scientist.

I am both scientifically and personally indebted to many people that I have met on my journey. Dr M. Sánchez Cruz from the Rocasolano Institute of the CSIC was my thesis supervisor, who revealed to me the fascinating and surprising ways of electrochemical kinetics. Prof. Parsons, the editor for many years of the *Journal of Electroanalytical Chemistry*, encouraged me by correcting several of my first papers and giving me useful experimental advice. Without the heartening effort of this personally involved editor, I may not have pursued my career as a scientist. After finishing my Ph.D. I moved to the newly founded Faculty of Chemistry in San Sebastian (Basque Country), which was to become devoted to the development of polymer science and polymer technologies. After several hard years organizing and delivering teaching activities I started a new research group, with the help of tenacious students, with the aim of combining my electrochemical background with polymers. This was at the beginning of the 1980s, when the first papers from the MacDiarmid, Diaz and Osada groups were published. By repeating the experimental results from Prof. Diaz's papers on electropolymerization and the electrochemical characterization of the generated films, this fascinating journey was initiated. Without Prof. MacDiarmid's and Osada's encouraging comments, which they shared any time I had the chance to meet them (at least once per year), I probably would have moved to another polymeric field. Since then, tens of Ph.D. and Masters students have devoted their time and effort to develop most of the ideas here presented: my permanent gratitude is with all of them. Dr Melling has reviewed the original manuscript giving a lot of useful advice and comments, thanks Dan.

Finally, I must be thankful for the deep economic crisis that has affected Spain over the last five years: the number of students in my laboratory has dropped to one. The upside of this is that I finally got the time to write this book.

Without the unwavering emotional and spiritual support of my wife Ana, my sons Miguel and Juan, and my daughters-in-law, I would never have embarked on this journey, let alone complete it. Their constant support is the cornerstone of my life and I can never thank them enough.

Contents

RSC Smart Materials No. 19
Conducting Polymers: Bioinspired Intelligent Materials and Devices
By Toribio Fernández Otero

Published by the Royal Society of Chemistry, www.rsc.org

CHAPTER 1

Life, Bioinspiration, Chemo-Biomimesis and Intelligent Materials

1.1 Introduction

Life is chemistry, but current chemical models, developed from reactions taking place in the gas phase or dilute solutions, are unable to describe life and life functions. Scientists are concerned with the development of theoretical models for the description of life functions, predicting health and diseases, and advancing the different ways for health restoration, even before the emergence of new diseases. In parallel, the progressive development of a new technological world constituted by devices and tools working, as biological organs do, driven by the chemical reactions of the device's constitutive materials should be expected.[1]

1.2 Basic Hypotheses

- Life and life functions originate from chemical reactions taking place in the dense gel of the intracellular matrix (ICM) in living cells.
- Most of the reactions sustaining both life and life functions (such as walking, memory, thinking or consciousness) involve complex organic molecules (enzymes, proteins, nucleic acids and so on) as reactants.
- Biochemical reactions induce specific and convoluted conformational movements (allosteric effects, folding or unfolding) of the organic reactants and, when required for charge and osmotic balance, the exchange of water and ions with the surroundings.

RSC Smart Materials No. 19
Conducting Polymers: Bioinspired Intelligent Materials and Devices
By Toribio Fernández Otero

Published by the Royal Society of Chemistry, www.rsc.org

- Current physical or chemical kinetic models do not include any conformational or structural movements induced by reactions.
- Current chemical models were developed from reactions taking place in the gas phase at low pressure, or in solutions using very dilute reactants.
- Gas phase and dilute solutions are far away from the dense reactive gel structure constituting the ICM of living cells.
- Any attempt to describe life and life functions requires advanced chemical models attained from systems where at least one of the reactants is a complex molecule or carbonaceous structure making up part of the dense gel.
- The gel reaction must drive conformational movements of the reactant polymer or macromolecule, and the simultaneous exchange of ions and water.

1.3 Bioinspiration, Biomimesis, Chemo-Biomimesis, Intelligent Materials and Systems

Evolution may be considered the most powerful engineering tool or form of designer machinery that has worked for billions of years to create a plethora of efficient molecules, reactions, structures, cells, organs, functions, systems and beings. These evolutionary molecules have been, and will continue to be, the inspiration of the human species to develop tools, devices, structures, arts, science and technology.

In this context, terms such as *bioinspiration*, *biomimicry* or *biomimesis*, *chemo-biomimesis*, *intelligent materials* and *intelligent systems* appear with rising frequency in scientific papers. Their widespread use has resulted in different meanings when used by different authors and speakers. At this stage, the best possible clarification for some of these concepts may be given by a reference defining the characteristics of the inspiring top-level biological organs or functions and those attained by the new biomimetic material, device or structure.

Bioinspiration: learning from nature's macroscopic, microscopic or molecular structures and being inspired by them to try to adapt physical or mechanical structural efficiency using different materials and scales to solve human problems.

Biomimicry or biomimesis: construction of new tools, devices or robots mimicking some biological physical functions from the extracellular matrix (ECM) of living cells.

Chemo-biomimesis, chemo-biomimicry, electro-chemo-biomimicry or electro-chemo-biomimesis: construction of new chemical- or electrochemical-driven tools, devices or robots mimicking biological functions (from walking or proprioception to consciousness or brain memory) generated by chemical or electrochemical reactions in the ICM of living cells.

Intelligent materials: the most intelligent materials and systems come from nature and are, simultaneously, actuators, sensors and self-healing, *e.g.*, haptic muscles. Artificial intelligent materials are as intelligent as they fit most of these biological characteristics. For those covering the same number of characteristics the most intelligent are those getting the highest efficiency.

In this context the present book is mainly concerned with the exploration of the incipient electro-chemo-biomimetic and chemo-biomimetic scientific and technological space. The emerging scenery is based on new reactive dense gel materials that can mimic, in its simplest expression, the contents of the ICM of living cells from biological organs, including reactive macromolecules, reaction-driven conformational movements, ions and water (or solvent).

A basic, important and differential point related to metals and inorganic materials is that, despite deep change of the material composition (polymer/ion ratio) during reactions, a relatively low variation of the mechanical consistency is observed, thus maintaining the material's integrity.

Similar composition variations by several orders of magnitude can only be observed inside the functional cells of biological organs when they pass from rest to work states. Reactions involving these artificial dense gels promote variation of the gel (polymer/ion) composition. These material properties, the values of which change with the material composition, are named *composition-dependent properties*. Parallel variation during reactions involving the composition-dependent properties mimics biological functions. The development of a new technological field giving new chemo-biomimetic and electro-chemo-biomimetic devices and envisaging new tools and robots based on those biomimetic properties is emerging. The state of the art will be presented here.

In parallel, such reactive gels, the reactions of which drive the conformational movement of the reacting polymer chains and structural macroscopic changes, can be taken as a new system and reaction model. New theoretical tools will allow exploration of unknown fields beyond the borders of chemical kinetic models discussed in current chemical, biochemical and biological textbooks. The quantitative inclusion of these reaction-driven conformational and structural changes in present chemical models should result in more advanced structural chemical kinetic models. The theoretical description and quantification of reaction-driven conformational changes can allow the subsequent theoretical description and quantification of biochemical reactions, life, life functions, health and diseases.

1.4 Available Reactive Materials

During the last 30 years a plethora of new redox organic molecules and carbon-based structures has been discovered: conducting polymers (CPs), redox polymers, viologens, porphyrins, phthalocyanines, fullerene derivatives, carbon nanotubes, graphenes and so on. Most of them form electronic conducting films, or can be supported by polymers forming electronic conducting films. These conducting films can be used as self-supported film electrodes in liquid electrolytes for the study of their redox reactions.

Regarding a general description of the available reactive materials, this book focuses on CPs.[2–13] However, concepts, reactions, properties, devices and theoretical models here presented can be translated to any of the different reactive materials mentioned above and those that have yet to be discovered, as well as biological reactions.

1.5 Intrinsic CPs

Monomeric units linked by σ bonds (insulating chain; Figure 1.1a) constituting any neutral chain (*i.e.*, not charged) can be named an intrinsic CP. By oxidation or reduction, polaronic (radical anions or radical cations, using chemical terminology) π-conjugated structures, involving several monomeric units (Figure 1.1b), are formed along every chain. These conjugated structures allow the flow of electrons across the material under potential gradients. In partially oxidized or partially reduced films of CPs, electrons can jump between neighbor polaronic structures from the same chain (intra-chain jumping) or take part in different chains (inter-chain jumping). Both individual chains and films become electronic conductors. The electronic conductivity increases under the control of the chain oxidation or reduction state, which determines the number of polarons per chain and the polaronic concentration (carrier concentration) in the film. More polarons per chain and a higher polaronic concentration mean shorter distances for electronic jumps, lower electrical resistance and higher conductivity. The electronic conductivity is an intrinsic property of the polymer chain, being controlled by its oxidized/reduced state.

Obtaining the neutral state of the polymer chains in films becomes, as will be explained in this book, a difficult task. This means that obtaining insulating films becomes almost impossible, hence the name *conducting polymers*.

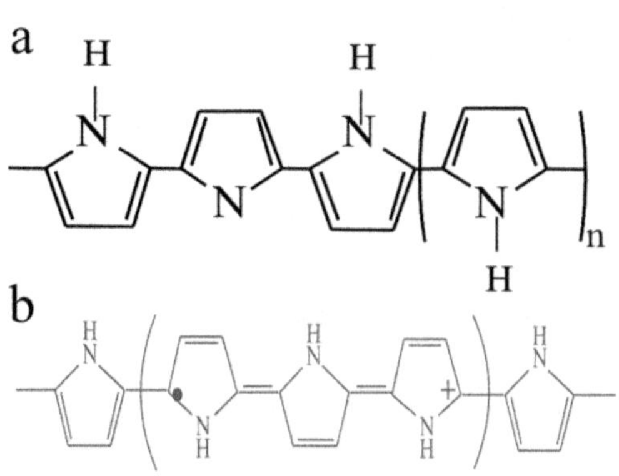

Figure 1.1 (a) Neutral polypyrrole chain. (b) Conjugated polaronic (radical cation) structure involving several monomeric units generated by extraction of one electron from the neutral chain.

1.5.1 Available Material Families

The literature presents thousands of CPs, which represent only a tiny fraction of the possible CPs that will be synthesized during the coming decades. To try to clarify the field of this intricate forest, a basic monomeric unit can be taken as a reference. Every monomeric unit can give a family of polymer–ion compounds of the basic CP attained by polymerization and seven derived families of intrinsically CP materials. Thus, each basic monomer can give eight different families of intrinsically conducting materials.[11,14,15]

- *Basic CP compounds* may be obtained by polymerization of each of the simplest monomers in the presence of different salts: polyacetylene, polyaniline, polypyrrole, polyindole, polyfurane, polycarbazole and so on. The synthesis generates a salt polymer anion. Different polymeric compounds are generated by polymerization of the same monomer in the presence of different anions.
- *Substituted CPs* are any basic monomer including several hydrogen atoms, each linked to a carbon atom, with substitution of one, two or three of those hydrogen atoms (two must always remain unsubstituted in order to allow the subsequent polymerization) by an organic group. The substituted monomer gives, by polymerization, a substituted CP.[16–20] Long organic substituents induce steric polymerization limitations giving short or very short oligomers.

The substituents can be selected (electron donors, electron acceptors, long and flexible, long and rigid *etc.*) in order to modify, related to the basic polymer, the physical (optical, electrical, mechanical) or chemical (electrochemical potentials, chemical stabilization) properties of the resulting polymer.

- *Self-doped CPs*[21–25] appear when the substituent is an organic salt with the organic anion covalently linked to the monomer balanced by a small cation; the resulting polymer is an intrinsically conducting polyelectrolyte, where each monomeric unit is a salt of an organic monomeric anion and a cation. By electrochemical oxidation of the polymer, the generated positive charges in the chain are compensated by its own anions, forcing the exit of the cations (cation-exchange materials).
- *Copolymers*[26–37] can be made by a combination of monomers (substituted or not), dimers or trimers, each including two or three different monomeric units. Their polymerization gives random or alternative copolymers with some specific properties, such as faradaic electrodissolution. Dimers or trimers attained by the combination of basic and substituted monomers can allow broad control of the polymer's physical and chemical properties.
- *CP–organic macroion blends*[38–49] require the synthesis of CP films in the presence of large organic counterions (organic salts or organic acids, polyelectrolytes, ionic liquids) to give blend materials of the CP with

the organic macroion. The electro-generated, positively charged chains of the CP are compensated by the negative charges of the organic macroion that are incorporated and trapped inside the growing polymer material. The trapped macro-anion will force the exchange of balancing cations during the subsequent oxidation/reduction of the material.

The large macroion can be selected to provide the resulting CP with some specific physical or chemical properties. When the CP is mixed with commodity polymers or other organic molecules the result is also known as a CP blend.

- *CP–inorganic macroion hybrids*[50–55] requires the electrosynthesis of CPs in the presence of large inorganic macroions, such as polyoxometallates, giving CP–macroion hybrid materials. The macroion is selected to provide the CP with some magnetic, electrical or electrochemical properties. Most of the polyoxometallates are also electroactive.
- *Composites*[56–59] involve mixing CPs with particles or nanoparticles of different materials (carbon, carbon nanotubes, fullerenes, graphenes, commodity plastics, semiconductors, biological materials, metals and so on) to give conducting composite materials.
- *CP salts*. Each of the above families can be oxidized (or reduced, see Chapter 4) in the presence of electrolytes, generating a new polymer salt:

$$\mathrm{Pol} + n\mathrm{A}^- + m\mathrm{H_2O} \rightleftarrows [\mathrm{Pol}^{n+}(\mathrm{A}^-)_n(\mathrm{H_2O})_m]_{\mathrm{gel}} + ne^-, \quad (1.1)$$

where Pol represents any polymeric chain in the film and A^- is the anion solved in water. Being a reversible reaction, when the polymer is translated into a different electrolyte with a different anion, after some consecutive oxidation/reduction cycles, the first anion can be completely exchanged by the new anion and the oxidized material is a new CP–anion compound.

The procedure can be repeated as many times as new compounds are required. A similar procedure can be applied with different cations for CPs exchanging cations (Section 4.3) with the electrolyte. In a similar way to metal atoms that can give different inorganic and organic salts, each constituting a different material, every CP can form hundreds of different compounds through very easy, soft and clean electrochemical procedures by substituting the charge balance anion or cation by a new one.

Figure 1.2 shows the basic molecular structure corresponding to seven of the eight polypyrrole families.

1.6 Biomimetic Reactive Gels

Films of CPs can be used as freestanding electrodes or as film coatings for metal electrodes in liquid or solid electrolytes. The flow of anodic or cathodic currents promotes the oxidation or reduction, respectively, of the film [eqn (1.1)]. By oxidation/reduction, electrons are extracted or injected from

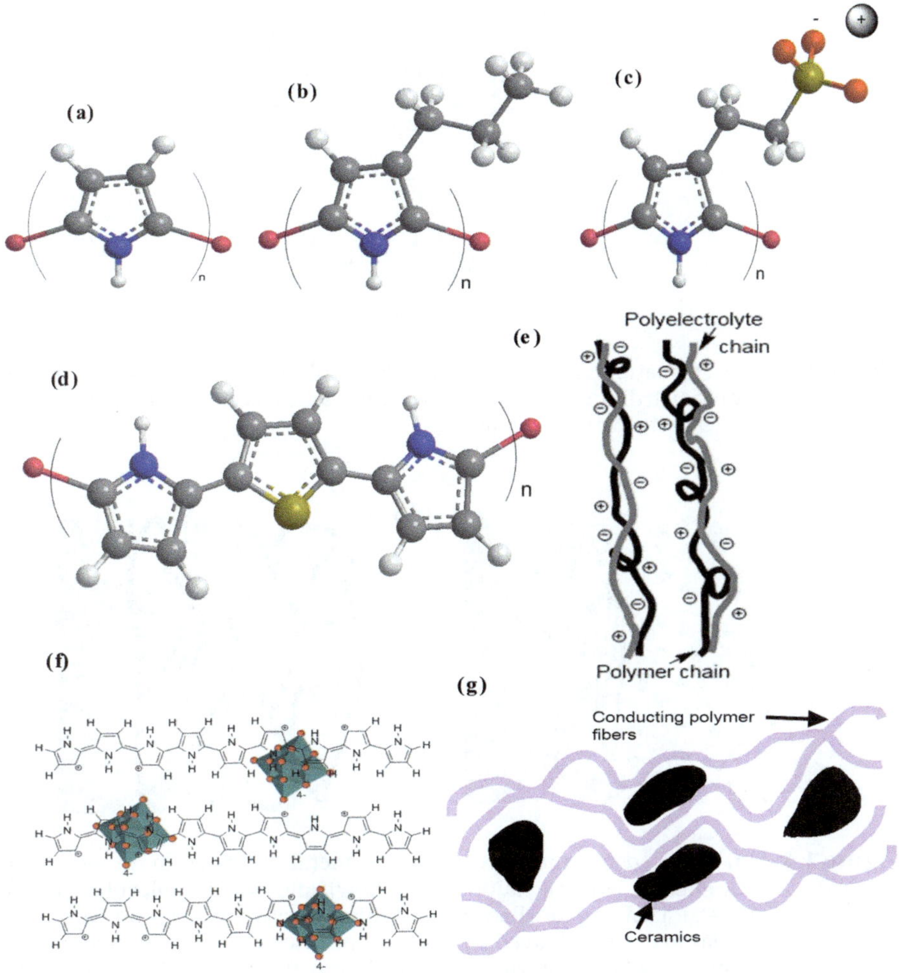

Figure 1.2 Seven different polypyrrole-based material families: (a) basic polypyrrole, (b) substituted polypyrrole, (c) self-doped polypyrrole, (d) poly(pyrrole-thiophene) copolymer, (e) polypyrrole–organic macroanion blend, (f) hybrid polypyrrole–inorganic macroanion, and (g) polypyrrole composite. With kind permission from Springer Science and Business Media, adapted from ref. 15.

or towards, respectively, each polymer chain constituting the film, generating positive or negative charges. The reaction of the material only occurs under charge balance. Emerging positive (oxidation) or negative (reduction) charges on any film chain forces the simultaneous entrance of ions from the electrolyte to keep the film's charge balance: the material becomes a polymer–ion (polyelectrolyte) compound (Figure 1.3). The presence of charged chains and balancing counterions in the film forces the exchange of water for osmotic balance: the film becomes a dense gel. In order to lodge balancing ions and solvent, the volume of the film increases (swelling), driven by the reaction.

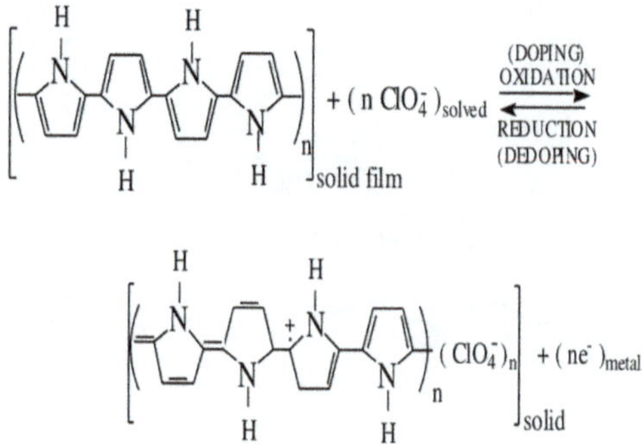

Figure 1.3 The oxidation/reduction of a polypyrrole chain changes the double bond distribution, forming polaronic (radical cation) conjugated structures in the oxidized state.

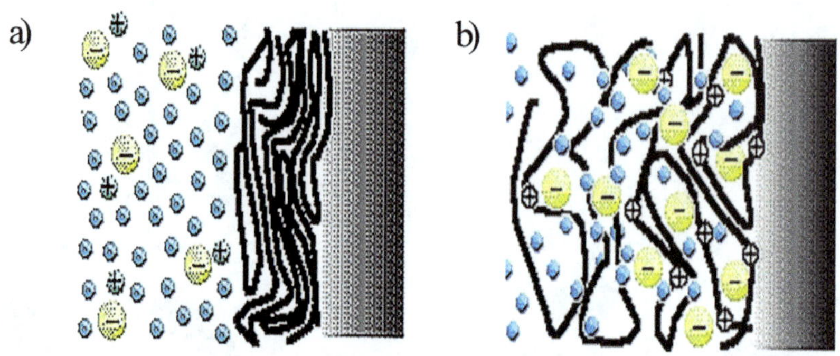

Figure 1.4 (a) Reduced polymer film coating a metal electrode in an electrolyte [cations (+, green), anions (−, yellow) and water molecules (blue)]. (b) Oxidized and swollen polymer film: positive charges on the chains (polarons) are balanced by anions and water molecules to guarantee osmotic balance. With kind permission from John Wiley and Sons, adapted from ref. 60.

In summary, under the reaction conditions the film becomes a swelling/shrinking dense electroactive gel (Figure 1.4),[60] the of composition of which mimics, in its simplest expression (reactive macromolecules, ions and water), the ICM of living cells. These dense gels usually include low water (or solvent) percentages (5–20%) compared with the usual polyacrylamide gels, which can have water contents of greater than 95%.

References

1. *Structural Biological Materials: Design and Structure-Property Relationships*, ed. M. Elices, Pergamon, Oxford, 2000.
2. *Handbook of Conducting Polymers*, ed. T. A. Skotheim and J. Reynolds, CRC Press, New York, 3rd edn, 2006.

3. *Electropolymerization: Concepts, Materials and Applications*, ed. S. Cosnier and A. Karyakin, Wiley Vch Verlag Gmbh, Weinheim, 1st edn, 2010.
4. *Conjugated Polymers: A Practical Guide to Synthesis*, ed. K. Müllen, J. R. Reynolds and T. Masuda, Royal Society of Chemistry, Cambridge, 2013.
5. G. G. Wallace, G. M. Spinks and L. A. P. Kane-Maguire, *Conductive Electroactive Polymers: Intelligent Polymer Systems*, CRC Press, London, 3rd edn, 2008.
6. F. Carpi and E. Smela, *Biomedical Applications of Electroactive Polymer Actuators*, John Wiley & Sons, Chichester, 1st edn, 2009.
7. N. Gospodinova and L. Terlemezyan, *Prog. Polym. Sci.*, 1998, **23**, 1443–1484.
8. G. Inzelt, M. Pineri, J. Schultze and M. Vorotyntsev, *Electrochim. Acta*, 2000, **45**, 2403–2421.
9. J. Heinze, B. A. Frontana-Uribe and S. Ludwigs, *Chem. Rev.*, 2010, **110**, 4724–4771.
10. A. Pron and P. Rannou, *Prog. Polym. Sci.*, 2002, **27**, 135–190.
11. T. F. Otero, *Polym. Rev.*, 2013, **53**, 311–351.
12. T. F. Otero, J. G. Martinez and J. Arias-Pardilla, *Electrochim. Acta*, 2012, **84**, 112–128.
13. T. F. Otero, *Mod. Aspects Electrochem.*, 1999, **33**, 307–434.
14. T. F. Otero and J. Rodriguez, in *Intrinsically Conducting Polymers: An Emerging Technology*, ed. M. Aldissi, Kluwer Academic Publishers, Dordrecht, 1993, vol. 246, pp. 179–190.
15. T. F. Otero, in *Modern Aspects of Electrochemistry*, ed. R. E. White, J. O. Bockris and B. E. Conway, Springer US, New York, 1999, pp. 307–434.
16. G. Zotti, M. Gallazzi, G. Zerbi and S. Meille, *Synth. Met.*, 1995, **73**, 217–225.
17. J. Reynolds, J. Ruiz, A. Child, K. Nayak and D. Marynick, *Macromolecules*, 1991, **24**, 678–687.
18. F. Garnier, G. Tourillon, M. Gazard and J. Dubois, *J. Electroanal. Chem.*, 1983, **148**, 299–303.
19. I. Rabias, I. Hamerton, B. J. Howlin and P. J. S. Foot, *Comput. Theor. Polym. Sci.*, 1998, **8**, 265–271.
20. H. A. M. van Mullekom, J. Vekemans, E. E. Havinga and E. W. Meijer, *Mater. Sci. Eng., R*, 2001, **32**, 1–40.
21. Y. Ikenoue, N. Uotani, A. Patil, F. Wudl and A. Heeger, *Synth. Met.*, 1989, **30**, 305–319.
22. M. S. Freund and B. A. Deore, *Self-Doped Conducting Polymers*, John Wiley & Sons, 2007.
23. H. R. Ghenaatian, M. F. Mousavi and M. S. Rahmanifar, *Electrochim. Acta*, 2012, **78**, 212–222.
24. A. Patil, Y. Ikenoue, N. Basescu, N. Colaneri, J. Chen, F. Wudl and A. Heeger, *Synth. Met.*, 1987, **20**, 151–159.
25. E. Havinga, L. Vanhorssen, W. Tenhoeve, H. Wynberg and E. Meijer, *Polym. Bull.*, 1987, **18**, 277–281.
26. E. Yildiz, P. Camurlu, C. Tanyeli, I. Akhmedov and L. Toppare, *J. Electroanal. Chem.*, 2008, **612**, 247–256.

27. M. C. Iovu, C. R. Craley, M. Jeffries-EL, A. B. Krankowski, R. Zhang, T. Kowalewski and R. D. McCullough, *Macromolecules*, 2007, **40**, 4733–4735.
28. R. Mazeikiene and A. Malinauskas, *Synth. Met.*, 1998, **92**, 259–263.
29. O. Turkarslan, M. Ak, C. Tanyeli, I. M. Akhmedov and L. Toppare, *J. Polym. Sci., Part A: Polym. Chem.*, 2007, **45**, 4496–4503.
30. S. Martwiset, R. C. Woudenberg, S. Granados-Focil, O. Yavuzcetin, M. T. Tuominen and E. B. Coughlin, *Solid State Ionics*, 2007, **178**, 1398–1403.
31. K. Li, H. Guo, Z. Q. Liang, P. Thiyagarajan and Q. Wang, *J. Polym. Sci., Part A: Polym. Chem.*, 2005, **43**, 6007–6019.
32. B. C. Roy, M. D. Gupta, L. Bhowmik and J. K. Ray, *J. Appl. Polym. Sci.*, 2002, **86**, 2662–2669.
33. S. Ogi, K. Sugiyasu and M. Takeuchi, *ACS Macro Lett.*, 2012, **1**, 1199–1203.
34. M. Nguyen, P. Kasai, J. Miller and A. Diaz, *Macromolecules*, 1994, **27**, 3625–3631.
35. S. Hong and D. Marynick, *Macromolecules*, 1992, **25**, 3591–3595.
36. J. Carrasco, A. Figueras, T. Otero and E. Brillas, *Synth. Met.*, 1993, **61**, 253–258.
37. T. Otero, J. Carrasco, A. Figueras and E. Brillas, *J. Electroanal. Chem.*, 1994, **370**, 231–239.
38. T. F. Otero, P. Herrasti, P. Ocon and C. R. Alves, *Electrochim. Acta*, 1998, **43**, 1089–1100.
39. T. F. Otero and V. Olazabal, *Electrochim. Acta*, 1996, **41**, 213–220.
40. L. Buckley, D. Roylance and G. Wnek, *J. Polym. Sci., Part B: Polym. Phys.*, 1987, **25**, 2179–2188.
41. S. P. Armes and M. Aldissi, *Synth. Met.*, 1990, **37**, 137–144.
42. W. Wernet, M. Monkenbusch and G. Wegner, *Mol. Cryst. Liq. Cryst.*, 1985, **118**, 193–197.
43. D. S. Maddison and J. Unsworth, *Synth. Met.*, 1989, **30**, 47–55.
44. R. Bull, F. Fan and A. Bard, *J. Electrochem. Soc.*, 1984, **131**, 687–690.
45. J. Laakso, J.-E. Österholm, P. Nyholm, H. Stubb and E. Punkka, *Synth. Met.*, 1990, **37**, 145–150.
46. H. Weiss, O. Pfefferkorn, G. Kotora and B. Humphrey, *J. Electrochem. Soc.*, 1989, **136**, 3711–3714.
47. M. Depaoli, R. Waltman, A. Diaz and J. Bargon, *J. Chem. Soc., Chem. Commun.*, 1984, 1015–1016.
48. S. E. Lindsey and G. B. Street, *Synth. Met.*, 1984, **10**, 67–69.
49. W. Wernet, M. Monkenbusch and G. Wegner, *Makromol. Chem., Rapid Commun.*, 1984, **5**, 157–164.
50. M. O. Wolf, *Adv. Mater.*, 2001, **13**, 545–553.
51. P. Gomez-Romero, *Adv. Mater.*, 2001, **13**, 163–174.
52. T. F. Otero, S. A. Cheng, E. Coronado, E. M. Ferrero and C. J. Gomez-Garcia, *ChemPhysChem*, 2002, **3**, 808–811.
53. G. Yu, X. Xie, L. Pan, Z. Bao and Y. Cui, *Nano Energy*, 2013, **2**, 213–234.
54. A. Walcarius, *Chem. Mater.*, 2001, **13**, 3351–3372.
55. Y. Zhou, M. Eck and M. Krueger, *Energy Environ. Sci.*, 2010, **3**, 1851–1864.
56. M. Helgesen, R. Sondergaard and F. C. Krebs, *J. Mater. Chem.*, 2010, **20**, 36–60.

57. C. Peng, S. Zhang, D. Jewell and G. Z. Chen, *Prog. Nat. Sci.*, 2008, **18**, 777–788.
58. G. A. Snook, P. Kao and A. S. Best, *J. Power Sources*, 2011, **196**, 1–12.
59. R. Gangopadhyay and A. De, *Chem. Mater.*, 2000, **12**, 608–622.
60. T. F. Otero and J. G. Martinez, *Adv. Funct. Mater.*, 2013, **23**, 404–416.

CHAPTER 2

Electrochemical Methods

2.1 Introduction

This book treats different aspects of the unexplored scientific borders between chemistry, electrochemistry, polymer science, biochemistry, biology, the behavioral sciences and mechanics. Conducting polymers (CPs) may be synthesized by electrochemical methods, which will be discussed here. The chemo-biomimetic properties, functions and devices, as well as the mimicked biological functions and organs, studied and described here include electrical (electronic or ionic) pulses linked to chemical reactions in CPs. Electrochemical textbooks cover links between electrical signals and chemical reactions.

In this context, basic electrochemical methodologies are expected to play a central role to attain our aims. The most common methodologies used for the study of the electrosynthesis, the material's oxidation and reduction reactions, and for the actuation of the final electrochemical devices are linear potential sweeps or linear potential cycles, potential steps and consecutive square potential waves, the flow of constant currents and consecutive square current steps. A brief description of both methods and the electrochemical cells used will be presented here. For a deeper study, some excellent electrochemical textbooks (on methods and kinetics) may be consulted.[1–4]

2.2 Two Electrode Electrochemical Cells

The electropolymerization or electrosynthesis of CP films, the study of the electrochemical properties of the attained materials, and the characterization of electrochemical devices based on these properties and their materials are performed in electrochemical cells, including two, three (Figure 2.1) or

RSC Smart Materials No. 19
Conducting Polymers: Bioinspired Intelligent Materials and Devices
By Toribio Fernández Otero

Published by the Royal Society of Chemistry, www.rsc.org

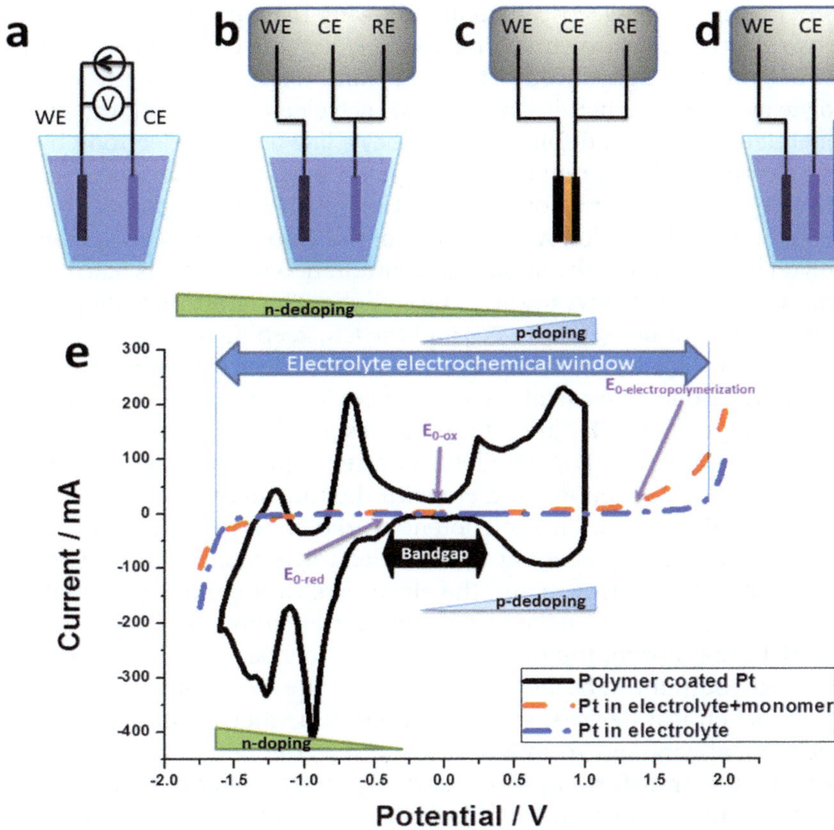

Figure 2.1 (a) A two electrode electrochemical cell connected to a current generator and voltmeter (V), with a working electrode (WE) and a counter electrode (CE). (b) Electrode connections using a conventional potentiostat/galvanostat (RE, reference electrode). (c) Solid state (two electrode configuration: CP/electrolyte/CP) electrochemical device. (d) Three electrode electrochemical cell. (e) Electrolyte potential window (dot-dash, blue line), monomeric oxidation polymerization (dashed, red line) and voltammetric response of an n-doped and p-doped CP. Reprinted with permission from ref. 14: *J. Phys. Chem. B*, 2010, **114**, 12777–12784. Copyright (2010) American Chemical Society.

four electrode cells. The macroscopic aspect of these cells can be the usual glass cells used in electrochemical laboratories (Figure 2.1a, b and d) or those required for laminar devices with a "solid" laminar electrolyte (Figure 2.1c).

The two electrode electrochemical cell includes (Figure 2.1a) a working electrode (WE), a counter electrode (CE) and the liquid electrolyte: a solvent with a solved *and dissociated* salt, acid or base. Electronic insulating polymer films including dissociated salts with high ionic mobility (casted from solutions including the polymer and a salt followed by solvent evaporation)[5–8] of one of the constituting ions[9,10] are the most usual solid electrolytes (Figure 2.1c). The electrolyte can also be a pure ionic liquid or any molten salt.[11–13]

In order to close the electrical circuit, allowing the current to flow, at each of the two electrode/electrolyte interfaces the electronic charge flowing by the metal wires must be transduced to the same ionic charge flowing through the electrolyte by an electrochemical reaction at each electrode: *n* monovalent ions are generated or destroyed following Faraday's law on each electrode by the flow of *n* electrons. The current flows through the electrolyte (electronic insulator) driven by the dissociated ions (carriers) and pushed by the potential gradient between both electrodes. Anions (each of the anion species present in the electrolyte) move towards the anode and, simultaneously, cations (each of the present cation species) move towards the cathode. The ionic movements close the electrical circuit between the WE and the CE, keeping the charge balance around each of the two electrodes during the two electrochemical reactions required to transduce the metal electronic flow into ionic flow in the electrolyte. A current generator (Figure 2.1a) produces the flow of a constant current between the WE and the CE. A parallel-connected voltmeter, V, follows the evolution of the WE/CE potential gradient with the time of the current flow, *i.e.*, the cell potential gradient. This cell potential gradient is a result of three components: the potential gradient at the WE/electrolyte interface, the potential gradient across the electrolyte and the electrolyte/CE interface potential gradient. A potentiostat/galvanostat also can be used (Figure 2.1b) as the current or potential generator for the two electrode electrochemical cell; its reference electrode (RE) output will be short-circuited with the CE output.

In electrochemical devices, such as batteries, artificial muscles or smart electrochemical windows to mention just a few, the device actuation originates from the transformation of each electrode material, occurring by an electrochemical reaction. The electrolyte (a solid or wet membrane) is required to avoid short-circuits between both electrodes, to close the electrical circuit guaranteeing the ionic conductivity between both electrodes and to supply the counterions required for charge balance in every electrode material during its reaction. Consequently, the electrolyte must occupy a minimum volume constituting the minimum weight fraction of the device. The most suitable design is a laminar structure of the full electrochemical cell components (Figure 2.1c), the electrolyte being as thin as possible. When the ion expelled from one of the electrode reactions is incorporated for charge balance during the second electrode reaction, it's called a rocking chair: the thin electrolyte supports the flow of the ion in one direction during direct actuation (*e.g.*, discharge in batteries) and in the opposite direction during reverse actuation (*e.g.*, charge in batteries).

In the two electrode cell configuration the evolution of the potential gradient across the cell can be followed at any time during current flow, but the potential gradient at each electrode/electrolyte interface cannot be determined. This electrode/electrolyte potential gradient is the electrochemical motor of each electrode reaction: it modifies the reaction activation energy [eqn (7.7) and (9.2)] of the reaction.

The potential gradient across the electrolyte is required to obtain the high mobility of every constituent ion. The electrolyte acts simultaneously as a

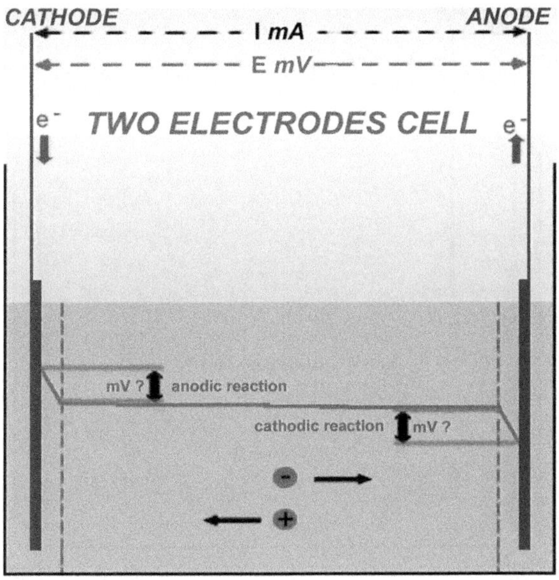

Figure 2.2 Two electrode electrochemical cell (*I*, current; *E*, potential).

resistance to the current flow (ionic movement) and consumes energy by heating the solution (the Joule effect), so electrolyte concentrations higher than 10^{-2} M L^{-1} of a fully dissociated salt minimize the electrolyte potential gradient.

In summary, using a two electrode cell structure the full potential gradient across the cell can be measured at any time of current flow, but none of its three components (Figure 2.2) can be determined separately.

2.3 Three Electrode Electrochemical Cells

The three electrode cell configuration includes a WE, CE and RE. A RE is an ideal non-polarizable electrode/electrolyte interface; the electrode equilibrium reactions (forwards and backwards) are fast enough (with large reaction coefficients) to support the flow of low currents, maintaining the electrode equilibrium potential. By contrast, in ideal polarizable electrodes any very low current flow separates the electrode/electrolyte potential gradient from its equilibrium potential by several tens or hundreds of millivolts. In a circuit with a high RE entrance impedance, the RE constitutes a stable potential that is taken as the potential reference level to follow or control the evolution of the WE/electrolyte potential gradient with or without current flow between the WE and the CE. Used as the reference potential level, a potentiostat can subject the WE/electrolyte potential gradient to potential steps and linear or sinusoidal variations under total control (*i.e.*, potentiostatic experiments). During galvanostatic experiments (*i.e.*, the flow of a constant current or square current waves between the WE and CE), an RE allows the

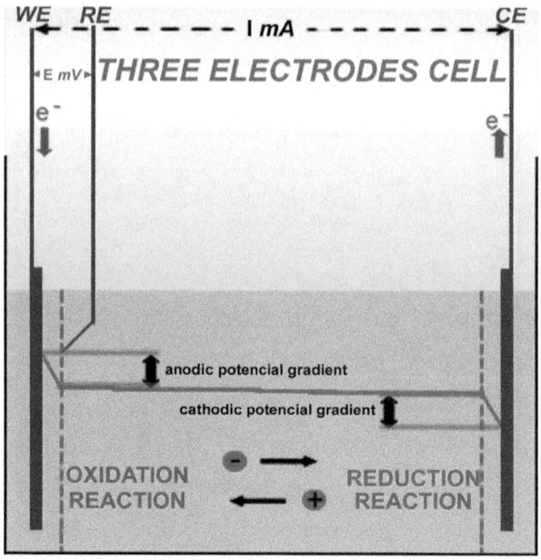

Figure 2.3 Three electrode electrochemical cell.

evolution of the WE/electrolyte potential gradient to be followed during the experiment. As mentioned in Section 2.2, the control of the WE/electrolyte potential gradient and its evolution allows control of the activation energy of the WE reaction [eqn (7.7) and (9.2)], including the reduction of metal salts that conventionally require oven temperatures of over 1000 °C to take place (metal synthesis), but may be performed under ambient laboratory conditions from the solutions by electrochemical methodologies (Ni, Cr, Au *etc.* electroplating) with the energy supplied by a common battery.

The CE can be designed with a large surface area in order to consume (during its electrode reaction) a minimum amount of energy during current flow (*i.e.*, the minimum potential gradient and minimum reaction resistance). When required to obtain uniform material films (with uniform surface reactions) on the WE, a uniform electric field must be generated: parallel WE and CE plates—two CE plates fully parallel and at the same distance from the WE with one on each side, or a cylindrical CE around a WE wire located at the cylinder axis.

A computer running electrochemical software allows control of the potentiostat/galvanostat to apply any selected electrochemical methodology to the WE (Figure 2.3).

2.4 Four Electrode Electrochemical Cells

For the electrochemical characterization of membranes, a four electrode electrochemical cell (Figure 2.4) is required. Two REs, one on each side of the membrane, allow the four electrode potentiostat to establish a selected potential gradient across the membrane. The two auxiliary electrodes (AE) or CEs allow the flow of the ionic current across the membrane under the

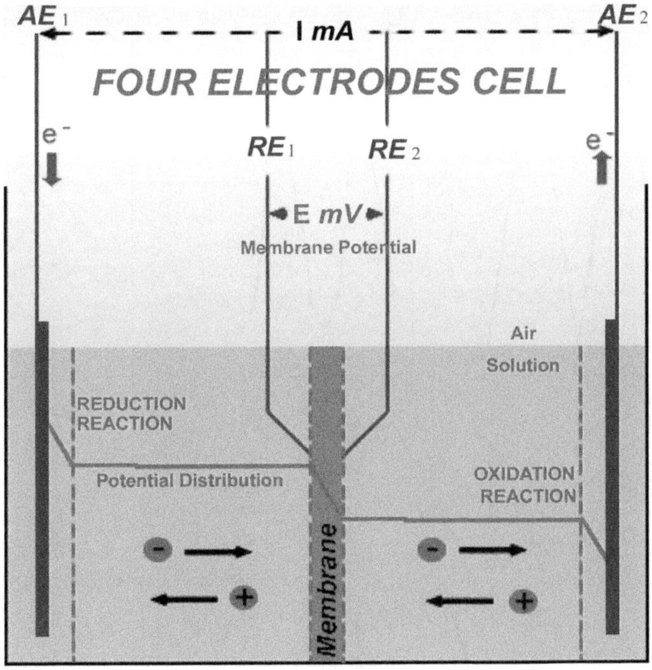

Figure 2.4 Four electrode electrochemical cell.

selected membrane potential gradient, supporting the concomitant oxidation and reduction reaction rates. The ionic current flowing across the membrane equals the electronic current flow by the external circuit. In order to identify which ions, the type of ion and the current fraction carried by each ion (*i.e.*, the transport number) across the membrane, a different salt (*i.e.*, a different anion and a different cation) must be present on each side of the membrane. After the flow of a defined charge Q (in C), the ion species and concentrations present on each side can be identified and quantified.

2.5 Cyclic Voltammetry

Using a three electrode cell configuration, the potential gradient at the WE/electrolyte interface is forced to change linearly in relation to the constant potential of the RE with time at a constant sweep rate, v (mV s^{-1}), from an initial potential, E_1 (mV), to a final potential, E_2 (mV), and back to the initial potential E_1 (Figure 2.5a). The electrical generator (the potentiostat) guarantees the flow of the required current (between the WE and the CE) to charge the electrical double layer capacitor and to support the required reaction rates at every imposed potential, giving a voltammetric (current/potential) response.

The procedure is parallel to those used for any spectroscopic analysis when the frequency (*i.e.*, photon energy) of the light flowing across the sample is shifted between two limits at a constant rate. The absorption or transmission spectra give information about the atomic and molecular electronic

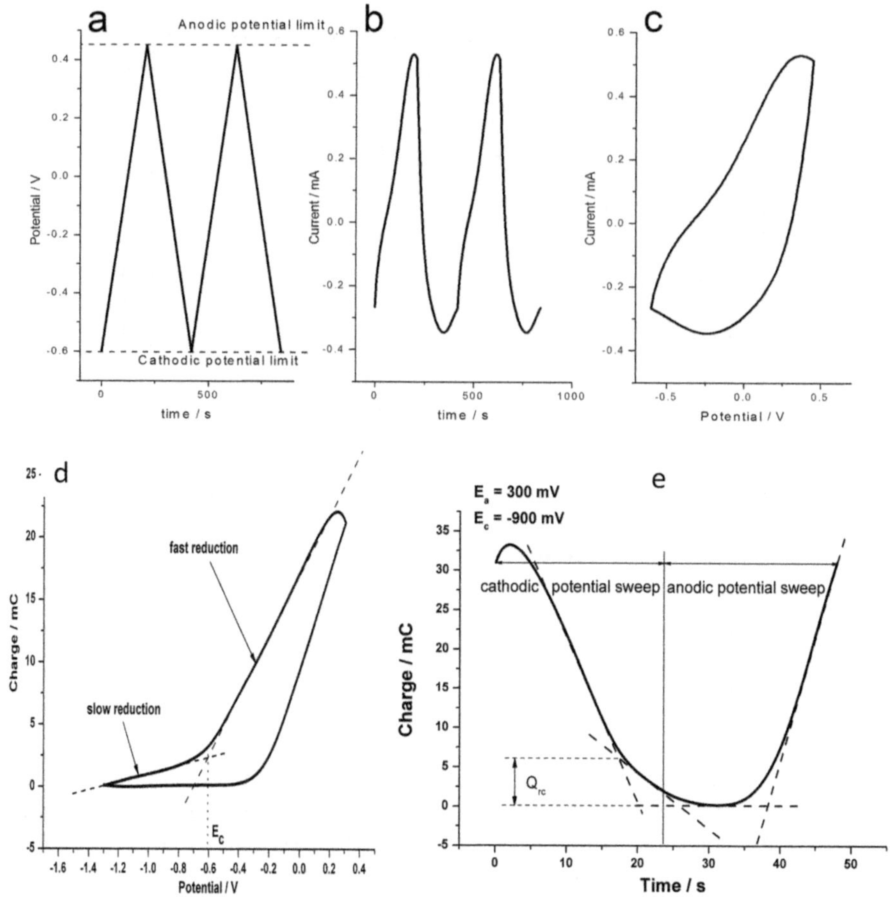

Figure 2.5 (a) Consecutive potential sweeps between a cathodic potential limit and an anodic potential limit at a constant sweep rate (slope). (b) Voltammetric (current–time) response. (c) Cyclic voltammetric (current–potential) response. (d) Cyclic coulovoltammetric (charge–potential) response. (e) Chronocoulometric (charge–time) response. With kind permission from John Wiley and Sons, adapted from: *Adv. Funct. Mater.*, 2013, **23**, 3929–3940. Copyright (2014) John Wiley and Sons.

levels. Voltammetric responses give information about atomic and molecular ionization (oxidation by electron extraction) or electroaffinity (reduction by electron injection), and the electronic levels of atoms, molecules and ions present during the electrochemical experiment.

The voltammetric (current/potential) response from a clean electrolyte allows definition of the electrolyte potential window (Figure 2.1e, blue dot-dash line) from the reduction potential of the electrolyte component (the solvent molecule, anion or cation) with the higher electroaffinity until the oxidation potentials of its components with the lower ionization potentials.[14]

By adding the monomer to the electrolyte, the first voltammetric response gives the monomer oxidation–polymerization potential [assuming that the electrolyte oxidation potential is higher (*i.e.*, more anodic) than the monomer oxidation; Figure 2.1e, red dashed line].

By using a self-supported CP film as the WE in a fresh electrolyte, the first oxidation potential, the first reduction potential, the polymer band-gap, the charge consumed for the film oxidation and for the film reduction during p-doping/p-dedoping or during n-doping/n-dedoping can be obtained (Figure 2.1e).

2.5.1 Voltammetric and Coulovoltammetric Responses

The applied potential sweep described in Section 2.5 shifts the Fermi level of the metal WE (the energetic level between occupied and unoccupied electronic band levels). Oxidation (electron loss) starts when the unoccupied levels are shifted by the anodic potential sweep until the lowest highest occupied molecular orbital (HOMO) level of the atoms, molecules or ions present at the WE/electrolyte interface. Reduction (electron injection) reactions start during the cathodic potential sweep, when the Fermi level of the metal WE is moved above the lowest unoccupied molecular orbital (LUMO) level of the atoms, molecules or ions present at the WE/electrolyte interface. The system responds with the flow of a current between the WE and the CE (provided by the potentiostat), which is required to adjust the oxidation or reduction reaction rates of any atom, ion or molecule (electroactive species) according to its oxidation or reduction overpotential: voltammetric (I/E) response (Figure 2.5c). The electroactive species can take part of the electrode itself (metal corrosion, CPs, trapped analytes) or be present in the electrolyte (solved molecules, solvent, ions). The electrodic reaction supporting the current flow eliminates or generates ions, inducing the simultaneous movement (migration and diffusion) of the electrolyte ions (each of the ionic species) to guarantee both current flow through the electrolyte and charge balance at any point of the cell.

Unfortunately for students and non-specialists, two different criteria of potential scales have persisted, based on the protection of specific economic interests. Positive potential increments and positive currents are considered anodic (exploring oxidation processes of increasing oxidation energies at increasing anodic potential), while negative potential increments and negative currents are considered cathodic (exploring reduction processes of increasing reduction energies at increasing cathodic potential) for most international journals. Other countries follow the opposite criteria with regards to journal publications, hardware and industrial processes: *i.e.*, positive potentials and currents for reductive (cathodic) processes and negative potentials and currents for oxidative (anodic) processes. Thus, readers are warned that they may find voltammetric responses in journal and books akin to those presented in Figures 2.1e and 2.5c or central symmetrical ones (the magnitude of specific economic interests overlapped with human stupidity surpassing, by orders of magnitude, scientific and general human welfare).

The potential voltammetric axis becomes time by using the potential sweep v (in mV s^{-1}) as the transformation factor: E/v (mV/mV s^{-1} = s), transforming voltammetric responses into chronoampero–voltammetric responses (Figure 2.5b).

By integration, the anodic or cathodic voltammetric areas between two voltammetric points gives the consumed anodic or cathodic charges (Q), respectively:

$$\int I\mathrm{d}t = Q. \tag{2.1}$$

By its integration, any voltammetric response becomes a coulovoltammetric (Q/E) response (Figure 2.5d). Nowadays, most potentiostat/galvanostat devices give directly and simultaneously both the voltammetric and coulovoltammetric responses. They also allow a direct change from "European" to "American" plotting and calculation criteria. Following the "European" criteria, Q/E responses used to be presented with positive charge increments related to electrodic oxidation processes and negative charge increments related to electrodic reduction processes. Q/E responses, practically ignored for decades, will constitute a great support for the identification and quantification of the different reaction-driven conformational and structural processes discussed in this book (see Section 6.3).

Coulovoltammetric responses are also being rediscovered as powerful tools to reveal the simultaneous presence of reversible and irreversible electrodic reactions, allowing a quantitative separation of the reversible and irreversible consumed charges[15] (see Section 4.5).

By direct translation of E (in mV) into t (in s), the close coulovoltammetric response becomes an open chronocoulometric (Q/t) response (Figure 2.5e).

2.5.2 Electrolyte Potential Window

Any WE (*i.e.*, an electronic conductor material, for example, platinum) placed in a deaerated electrolyte (for example, H_2O + Na_2SO_4) results in potential cycles between the potential limits E_1 and E_2. After two consecutive cycles the anodic potential limit E_2 is shifted to a more anodic value. The process is repeated for consecutive anodic potential limits. When the potential limit E_2 attains the lowest oxidation potential required for the oxidation of the atom, molecule or ion present at the electrode/electrolyte interface (*i.e.*, Pt, H_2O, SO_4^{2-} or Na^+ for our example) with the lower HOMO, electrons are extracted from the HOMO giving rising anodic currents for the increasing electrodic potentials; so, the chemical transformation of the concomitant species begins (oxygen evolution from H_2O molecules for our example here). The potential where the increasing current flow starts is the anodic potential limit (E_2') of the studied electrolyte.

Now the procedure is repeated for a constant anodic potential limit lower than E_2', shifting the cathodic potential limit E_1 each time. When this limit overcomes the potential (E_1') required to reduce the atom, molecule or ion with the highest electronegativity, the reduction starts at the highest LUMO and increasing cathodic currents flow for increasing overpotentials. With aqueous electrolytes, hydrogen evolution is the most usual reaction. The potential E_1' is the cathodic potential limit of the studied electrolyte.

The potential range between E_1' and E_2' is the potential window of the studied electrolyte. Inside this potential range, only the current required to charge the electrical double layer at the electrode/electrolyte interface and for the formation/destruction of some monolayers (metal oxides on some noble metals, or ion adsorption) occurs.

Thus, with regards to monomers and CPs, voltammetric experiments can provide: (a) the electrolyte potential window, (b) the monomer oxidation and reduction potentials inside that potential window, (c) the monomer oxidation–polymerization potential range and (d) the potential where the monomer oxidation–polymerization overlaps with the electrolyte discharge potential (see Sections 3.1, 3.2, 3.3 and 3.3.1). After electrogeneration, the voltammetric characterization of the generated polymer will provide: (a) the potential range (inside the electrolyte potential window) of the polymer oxidation/reduction (p-doping/p-dedoping) and reduction/oxidation (n-doping/n-dedoping), (b) the involved charges and (c) the potentials where these reactions overlap with other irreversible reactions. These are key points to obtain fully reproducible, long-term and durable electrochemical devices. The studied reactions must occur inside the electrolyte potential window in a potential range as far away as possible from both electrolyte discharge limits. The potential window of organic solvents can attain very high cathodic potential limits: in order to avoid the water discharge, both components (the organic solvent and the salt) must be free of water. Moreover, the selected salt must dissociate in the organic solvent: the electrolyte concentration in electrochemistry refers to the concentration of dissociated ions, the only current carriers. Note that handbooks and books on organic electrochemistry used to give a list of the potential windows for different electrolytes and ionic liquids.

When the material presents different oxidation or reduction processes, each of them will show a voltammetric maximum in a different potential range and always inside the electrolyte potential window. A voltammetric response is equivalent to an optical spectrum, revealing the different oxidation and reduction processes, the potential domain and the required energy for each process.

2.6 Square Potential Steps: Chronoamperometric, Chronocoulometric and Reaction Kinetic Responses

Once identified from the stationary voltammetric response of a material and the potential domain for every oxidation and reduction process by cyclic voltammetry, the use of square potential waves (Figure 2.6a) will allow the separate kinetic study of each oxidation or reduction reaction.[16] For one of the oxidation processes the selected initial (reduction) and final (oxidation) potentials must include a fraction or the full potential range of the concomitant voltammetric process. The WE is subjected to consecutive potential steps from the initial potential E_1 applied for time t_1 to the final potential E_2 applied for time t_2 and back to E_1 and so on. Both potentials and applied times must guarantee full reduction of the material at E_1 and its oxidation at E_2. Potentials E_1 and E_2 must be as far away as possible from both the

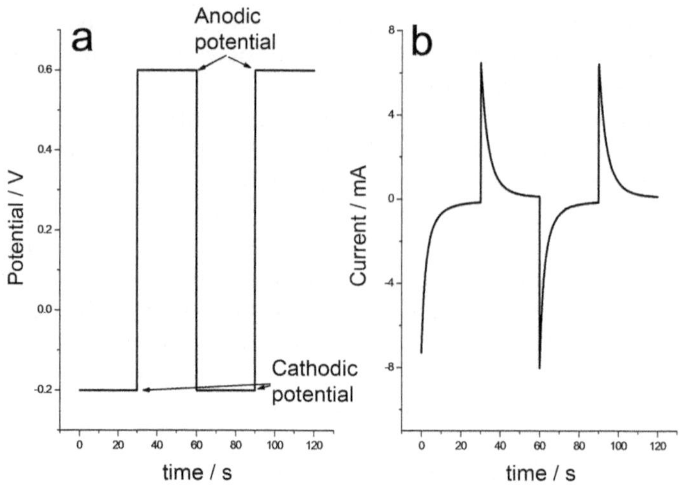

Figure 2.6 (a) Consecutive square potential waves between a constant cathodic potential (E_1) applied to the WE for time t_1 and an anodic potential E_2 applied for time t_2; (b) chronoamperometric responses from the WE. Adapted with permission from ref. 16: *ChemPhysChem*, 2012, **13**, 2108–2114. Copyright (2012) John Wiley and Sons.

potential range for the oxidation/reduction processes of other materials and the electrolyte potential limits in order to avoid overlapping of the currents and charges related to those interfering reactions.

The potential E_2 (measured *vs.* the RE) determines the oxidation overpotential ($\eta = E_2 - E_0$), where E_0 is the formal oxidation potential of the material, molecule or ion. After each potential step the electrode chronoamperometric (I/t) responses show a current peak (Figure 2.6b). The anodic chronoamperometric response shows the evolution of the current flow between the WE and the CE electrode during the time required to complete the WE material oxidation. The cathodic chronoamperometric response shows the current flow required to reduce, at the reduction overpotential, the WE material. By integration, each chronoamperometric response gives the concomitant chronocoulometric (Q/t) response.

If the volume V (in L) or the mass w (in g) of the reacting material film are known, the Q/t response becomes the evolution of the volumetric or specific concentration c (in M L^{-1} or M g^{-1}) of the oxidized or reduced species with time (c/t) in the electrodic material: this is the electrode oxidation or reduction kinetics. The charge (in C) consumed at any time after the oxidation or reduction potential step gives the variation of the oxidized or reduced species concentration (n or n'):

$$\Delta[\text{Pol}*] = \frac{\Delta Q}{FV} = n\left(\text{in Eq L}^{-1}\right); \tag{2.2}$$

$$\Delta[\text{Pol}*] = \frac{\Delta Q}{Fw} = n'\left(\text{in Eq g}^{-1}\right), \tag{2.3}$$

where F is Faraday's constant (96 500 C Eq^{-1}) and [Pol*] refers to the concentration of active sites in the CP film. Thus, the slope of the Q/t response gives the oxidation or the reduction reaction rate:

$$\frac{\Delta Q}{\Delta t} = \frac{\Delta n}{\Delta t}\left(\text{in Eq L}^{-1}\text{ s}^{-1}\text{ or Eq g}^{-1}\text{ s}^{-1}\right). \tag{2.4}$$

This reaction rate corresponds to the specific current (i; per unit of weight of the electroactive material) or to the current per unit of electroactive material volume flowing through the electroactive electrode (chronoamperometric response) at the oxidation time after the potential step:

$$\frac{\Delta Q}{\Delta t} = i\left(\text{in A g}^{-1}\text{ s}^{-1}\text{ or A L}^{-1}\text{ s}^{-1}\right). \tag{2.5}$$

2.7 Galvanostatic Methodologies: Chronopotentiometric Responses

The WE is subjected to the flow of a constant current. Anodic currents promote the anodic shift of the electrode potential (chronopotentiometric response) to meet the oxidation potential of the atom, ion or molecule with the lower HOMO and that which has the concentration that is able to support the oxidation rate, fitting the imposed current [eqn (2.4) or (2.5)]. When the rate of electron extraction from the concomitant species equals the electronic flow imposed by the current through the electrochemical cell, the electrode potential evolution attains a plateau. After oxidation of that species has been completed, the potential steps again to find the ionization potential of the next electroactive species from the electrode material or from the electrolyte.

The flow of a constant cathodic current promotes the cathodic shift of the electrode potential to meet the reduction potential of the ion or molecule with the higher LUMO. When the reduction rate equals the current flow (the electrons injected at the electrode/electrolyte interface), the chronopotentiometric response shows a constant potential plateau.

Some experiments, such as charge/discharge materials for battery electrodes or coloring/bleaching electrochromic electrodes, require subjecting the electrode material to consecutive square current steps (Figure 2.7a), giving consecutive chronopotentiometric responses (Figure 2.7b). After the initial reduction by flow of cathodic current I_1 for time t_1 to obtain the same initial state, a constant anodic current (I_2 in A) is imposed to the electrode for time t_2, followed by the cathodic current I_1 for t_1 and so on. The potential of the electrode responds by adapting its evolution to the reaction rate imposed by the current flow [eqn (2.4) or (2.5)], *i.e.*, the flowing charge controls the reaction rate. Anodic and cathodic currents and the times of current flow are selected to try to keep the potential evolution inside the charge consumed by the process, as defined by integration of the

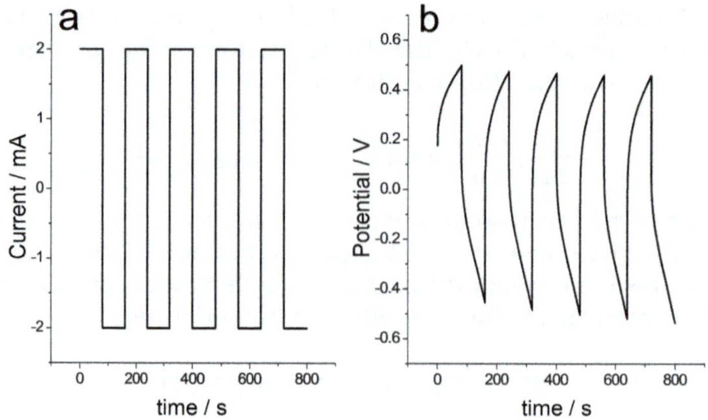

Figure 2.7 (a) Consecutive square current waves applied to the WE. (b) Concomitant consecutive chronopotentiometric responses.

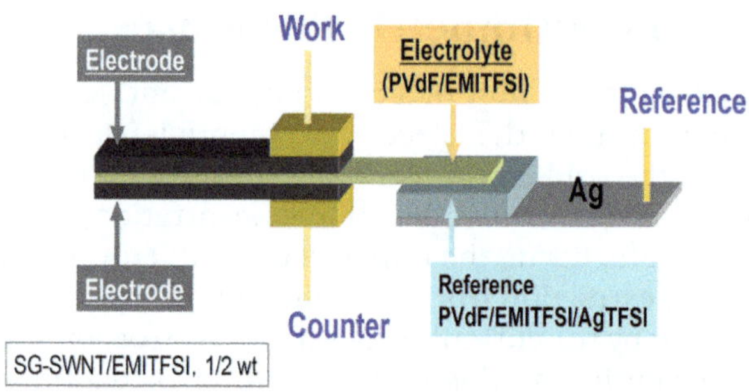

Figure 2.8 Three electrode system for the control of a planar electrochemical device (actuator or artificial muscle). *Work* means the WE and *counter* means the CE; the electrolyte is a polymeric film (polyvinylidene fluoride; PVdF) with a solved ionic liquid (1-ethyl-3-methylimidazolium bis(trifluoromethyl-sulfonyl)imide, EMITFSI), where SG-SWNT refers to second generation single-walled carbon nanotubes and AgTFSI refers to silver bis(trifluoromethane-sulfonyl)imide. Reprinted with permission from ref. 17: *Electrochim. Acta*, 2012, **60**, 177–183. Copyright (2012) Elsevier.

concomitant peak in the voltammetric response, and far from the electrolyte potential limits.

2.8 Electrochemical Cells and Methods Using Solid State Electrolytes

When control of the WE potential is required, the design of the three electrode solid state electrochemical cell can be described by Figure 2.8.[17] Once constructed, any of the previously described electrochemical methodologies can be applied in solid state electrochemical cells.

References

1. A. J. Bard and L. R. Faulkner, *Electrochemical Methods: Fundamentals and Applications*, Wiley, New York, 1980.
2. A. C. Michael and L. Borland, *Electrochemical Methods for Neuroscience*, CRC Press, 2006.
3. E. Gileadi, *Electrode Kinetics for Chemists, Chemical Engineers, and Materials Scientists*, VCH, 1993.
4. K. Vetter, *Electrochemical Kinetics: Theoretical and Experimental Aspects*, Academic Press, New York, NY, 1967.
5. M. Armand, *Solid State Ionics*, 1983, **9–10**, 745–754.
6. J. Y. Song, Y. Y. Wang and C. C. Wan, *J. Power Sources*, 1999, **77**, 183–197.
7. K. M. Abraham, *J. Electrochem. Soc.*, 1996, **143**, 1.
8. M. Rikukawa and K. Sanui, *Prog. Polym. Sci.*, 2000, **25**, 1463–1502.
9. J. W. Fergus, *J. Power Sources*, 2010, **195**, 4554–4569.
10. P. Knauth, *Solid State Ionics*, 2009, **180**, 911–916.
11. T. Welton, *Chem. Rev.*, 1999, **99**, 2071–2084.
12. M. C. Buzzeo, R. G. Evans and R. G. Compton, *ChemPhysChem*, 2004, **5**, 1106–1120.
13. M. Armand, F. Endres, D. R. MacFarlane, H. Ohno and B. Scrosati, *Nat. Mater.*, 2009, **8**, 621–629.
14. J. Arias-Pardilla, W. Walker, F. Wudl and T. F. Otero, *J. Phys. Chem. B*, 2010, **114**, 12777–12784.
15. T. F. Otero, M. Alfaro, V. Martinez, M. A. Perez and J. G. Martinez, *Adv. Funct. Mater.*, 2014, **24**, 1265–1274.
16. J. G. Martinez, T. Sugino, K. Asaka and T. F. Otero, *ChemPhysChem*, 2012, **13**, 2108–2114.
17. P. Gimenez, K. Mukai, K. Asaka, K. Hata, H. Oike and T. F. Otero, *Electrochim. Acta*, 2012, **60**, 177–183.

CHAPTER 3

Electrosynthesis of Conducting Polymers

3.1 Introduction

Polymeric chains, whether they are conducting polymers or commodity polymers, are generated from monomers through a polymerization mechanism. Most of the accepted polymerization mechanisms described by polymer textbooks include three consecutive steps to obtain the polymer chain: initiation, propagation and termination.

In conducting polymers, the chain formation is initiated from the monomer by its physical, chemical or electrochemical attack to produce a monomeric radical cation. Since this book is mainly devoted to electrochemical methodologies, here I will describe how these methodologies can be used to generate films of conducting polymers from monomeric solutions through the electropolymerization or the electrogeneration process.[1–6] After generation, electrochemical methodologies combined with gravimetric determinations will be used to check how the synthesis conditions influence the specific electroactivity (charge stored per unit of polymer mass) of the electrogenerated material.[1,7] The evolution of this electroactivity with different values for the variables during synthesis can reveal the presence of new reactions during the polymerization mechanism, thus influencing the electroactivity and the concomitant electrochemical properties of the synthesized material.[8,9] A deeper knowledge of how the synthesis conditions influence the electrochemical properties of the generated material allows selection of the best synthesis conditions to produce improved materials for the construction of electrochemical devices.

A three electrode electrochemical cell contains the electrolyte (a solvent, and a solved and dissociated salt) and the solved monomer. The dissociated

RSC Smart Materials No. 19
Conducting Polymers: Bioinspired Intelligent Materials and Devices
By Toribio Fernández Otero

Published by the Royal Society of Chemistry, www.rsc.org

salt provides the ions acting as charge carriers for current flow through the solution during the electrochemical reactions. Only dissociated ions participate in the current flow. If the salt is not dissociated, the solution is an insulator, the current does not flow and the monomeric reaction does not happen. In order to avoid interferences of the reaction intermediates and products with solved oxygen during reactions, the cell must be closed and degassed under pressure by bubbling an inert gas through the monomeric solution for 20 min before electropolymerization using a bubbling probe. A bubbler for the gas exit guarantees some overpressure in the cell, which is required to avoid air entrance. Before arriving at the cell, the inert gas must be cleaned and saturated by bubbling through a flask with pure solvent in order to avoid solvent evaporation from the cell and a subsequent concentration increase of both ions and the monomer during degassing.

3.2 Linear Potential Sweep: Monomer Oxidation Potential

The working electrode (WE) is subjected to an anodic potential sweep: the metal Fermi level separating the occupied and unoccupied electronic levels is shifted to a lower energetic value. When this level is pushed below the monomeric highest occupied molecular orbital (HOMO) level, the transfer of electrons from the monomer molecules to the metal's unoccupied electronic levels begins (Figure 3.1, solid line), before arriving at the anodic limit of the electrolyte potential window (Figure 3.1, dotted line).[10–12] The monomer oxidation begins at E_0, the standard oxidation potential of the monomer. From there, rising oxidation currents flow for the increasing overpotentials, $\eta = E - E_0$ (Figure 3.1). Each monomeric molecule (M) arriving from the solution produces at the electrode–electrolyte interface a monomeric radical cation ($M^{\bullet+}$) on the electrode:

$$M_{solv} \rightarrow M^{\bullet+}_{ad} + e^{-}_{metal}, \tag{3.1}$$

where the sub-indexes solv and ad mean solvated (in solution) and adsorbed on the electrode, respectively. Eqn (3.1) corresponds to the electropolymerization initiation reaction. In order to avoid the simultaneous electrolyte discharge, which would give decreasing electropolymerization efficiencies of the consumed charge, the reaction shown in eqn (3.1) must start at a lower potential than the anodic limit of the potential window for the background electrolyte (Figure 3.1).[13]

3.3 Electropolymerization by Consecutive Potential Sweeps

Starting from a clean cell, clean electrodes and using a fresh monomeric solution, the WE potential (*vs.* the reference electrode, RE) is linearly swept at a constant potential sweep rate, v (in mV s^{-1}), from an initial potential

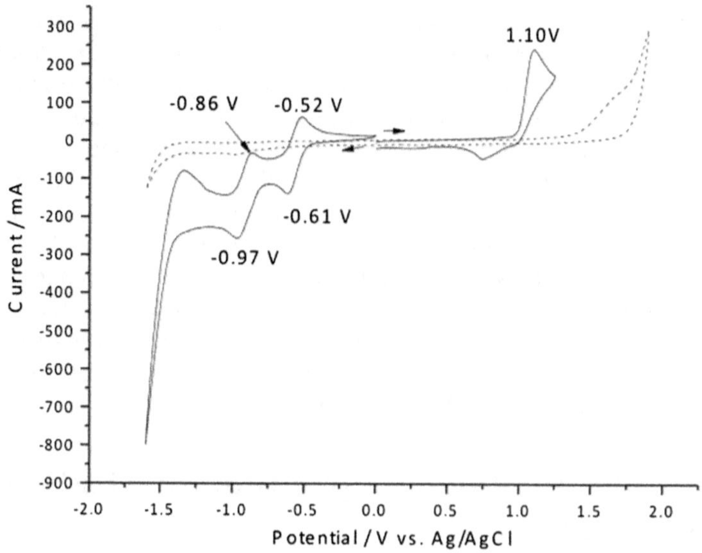

Figure 3.1 Cyclic voltammogram obtained in 0.1 M tetrabutylammonium hexafluorophosphate ($TBAPF_6$) in a dichloromethane solution (dotted line): the potential window ranges from −1.5 to 1.3 V. Voltammetric responses in 1 mM naphthalenediimide-terthiophene (NDI-3T) and 0.1 M $TBAPF_6$ in a dichloromethane solution (solid lines): the monomer oxidation maximum at 1.1 V and the first monomer reduction maximum at −0.61 V are indicated. The working electrode was a 1 cm^2 Pt plate; scan rate, 50 mV s^{-1}; temperature, 20 °C. Reprinted with the permission of the Royal Society of Chemistry from ref. 13: *Phys. Chem. Chem. Phys.*, 2011, **13**, 16513–16515.

located between the monomer oxidation potential and the cathodic limit of the electrolyte potential window. The voltammetric response shows (Figure 3.2) the flow of very low constant anodic currents up to the monomer oxidation potential (E_0). The concomitant charge is consumed to charge the electrical double layer (EDL). Beyond the monomer oxidation potential (E_0), rising anodic currents start to flow through the cell for increasing anodic potentials.

During the first sweep and simultaneous with the flow of the increasing currents, the color of the electrode changes due to the formation and growth of the polymer film. When the potential sweep is reversed ($-v$ in mV s^{-1}) the monomer oxidation occurs at the beginning of the cathodic potential sweep, giving more anodic currents (Figure 3.2, inset) than those observed during the anodic potential sweep: the oxidation–polymerization current gives a loop characteristic of a nucleation process. The polymer formation is initiated only at some points on the electrode surface, forming nuclei. These nuclei expand and grow with the subsequent increase of the polymer–electrolyte surface area. For a constant polymer oxidation per surface unit, the increasing polymer–electrolyte interface area means more monomer oxidation per unit of time, giving rise to increasing oxidation currents and

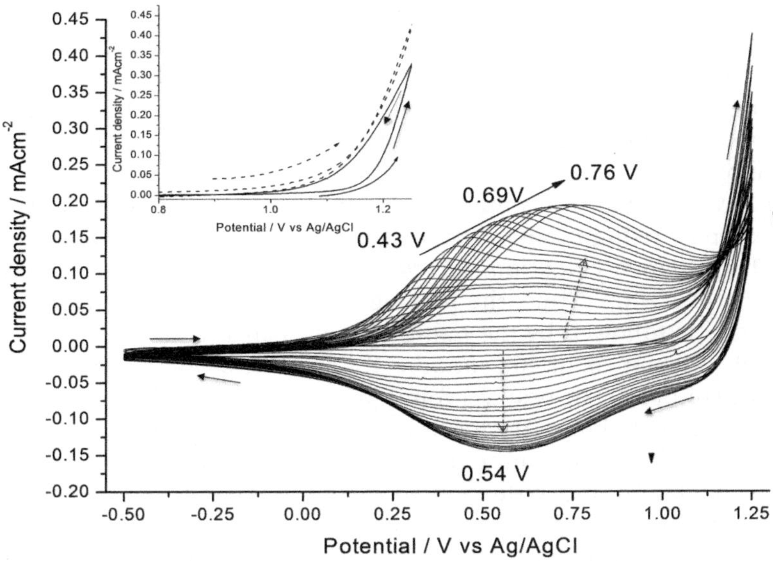

Figure 3.2 Consecutive cyclic voltammograms (25 cycles) between −0.50 and 1.25 V in a 1 mM bis-ethylenedioxythiophene (bis-EDOT) and 0.1 M $TBAPF_6$ dichloromethane solution using a 1 cm^2 Pt electrode at a scan rate of 20 mV s^{-1} and a temperature of 20 °C. Inset: enlargement of the monomer oxidation during the first (continuous line showing a nucleation loop) and second (dotted line) potential cycles. Figure reproduced from ref. 15 with the permission of the Royal Society of Chemistry: *J. Mater. Chem.*, 2012, **22**, 4944–4952.

easier (more catalytic) monomeric oxidation on the polymer related to the metal. By integration of the anodic charge below the two branches of the oxidation–polymerization loop (below the anodic branch plus that of the anodic currents flowing at the beginning of the cathodic branch), the consumed oxidation–polymerization charge is attained.

When the nucleation loop closes, the current becomes cathodic and shows a large and low cathodic current maximum. This means that the polymer was generated in its oxidized state and is now being reduced: the charge under the maximum quantifies the polymer reduction charge.

During the subsequent anodic potential sweep, the polymer oxidizes at low potentials, giving a large anodic maximum. After the maximum, the current increases due to the monomer oxidation–polymerization. Related to the voltammetric response from the first sweep, the monomer oxidation potential (E_0') is lower and the current flow is higher: the oxidized polymer coating, as a fraction of the metal surface, is a better catalyst for the monomer oxidation than the clean metal. As a partial conclusion, the electropolymerization process takes place at more anodic potentials than E_0 through two well-differentiated stages: an initial monomeric oxidation–polymerization–nucleation on the clean metal followed, once the metal has been coated, by the monomer oxidation–polymerization on the polymer, with E_0' being the new monomer oxidation potential.

The next cathodic branch of the second sweep shows a higher (relative to the first sweep) polymer reduction maximum: the polymer generated during the second sweep accumulates with that generated during the first sweep, requiring the flow of higher currents (higher charge) to be reduced. The next anodic branch confirms the formation of a higher polymer oxidation maximum.

Thus, by consecutive potential sweeps, monomer oxidation–polymerization is observed by rising monomeric oxidation–nucleation–polymerization currents. The polymer oxidation/reduction occurs at lower potentials than the monomer oxidation. Thus, the polymer growth during each of the consecutive cycles is followed by the rising current, which is depicted by the concomitant polymer oxidation (anodic voltammetric branch) and reduction (cathodic voltammetric branch) voltammetric maxima.[10,14,15,16]

The consumed charge during the consecutive potential sweeps (anodic minus cathodic) is the monomer oxidation–polymerization charge. The polymer oxidation and reduction charges (anodic and cathodic maxima) compensate for each other during the consecutive sweeps. The generated polymer mass is attained by the mass difference between that of the clean electrode before the polymerization process and that of the coated, reduced, rinsed and dried electrode after polymerization.

The polymerization process can be repeated under different experimental conditions, varying the potential sweep rate, anodic potential limit (taking care to avoid the possible parallel discharge of the electrolyte), number of voltammetric cycles, solved salt (inorganic, organic, polyelectrolytes, polyoxometallates, ionic liquids), salt or monomer concentration, and temperature.

3.3.1 Electropolymerization and Polymer Passivation (Degradation)

In order to get a faster polymerization growth, the first temptation is to increase the anodic potential limit: very high currents will flow through the electrode giving (in a supposed faradaic process) higher polymer masses per cycle. In order to explore the idea at the maximum level a very high anodic potential limit can be chosen. One voltammetric cycle is then performed in the background electrolyte (Figure 3.3, dashed line) and after adding the monomer (Figure 3.3, solid line). The voltammetric response in Figure 3.3 shows that the current increase that begins at the monomer oxidation potential goes on through an inflexion point and then through a maximum. The electrode shows a black color, indicating the formation of a thick polymeric film. When the potential sweep is reversed after this anodic maximum, the expected high polymer reduction maximum does not appear because the electrogenerated polymer film is not electroactive.[15]

The high anodic maximum present in the voltammetric response is equivalent to the formation of oxide passivation maxima in electroactive metals. Here, the inflexion point present in the rising current branch of the maximum overlaps with the anodic limit of the electrolyte potential window

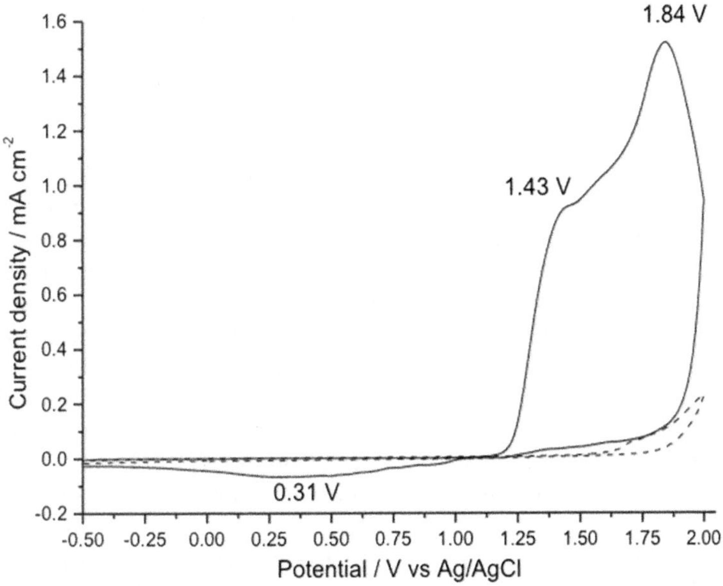

Figure 3.3 First voltammetric cycle in 1 mM bis-EDOT and 0.1 M $TBAPF_6$ in dichloromethane (solid line) and in the background electrolyte (dashed line) using a 1 cm^2 clean Pt electrode each time: initial potential, −0.5 V; scan rate, 20 mV s^{-1}; temperature, 25 °C. Figure reproduced from ref. 15 with the permission of the Royal Society of Chemistry: *J. Mater. Chem.*, 2012, **22**, 4944–4952.

(Figure 3.3, dashed line): the maximum can be due to conjugation degradation in the chains of the conducting polymer film, originated from reaction with the electrolyte oxidation intermediates and products. Thus, in aqueous solution and for this potential range, monomeric oxidation gives monomeric radical cations and the oxidized state of the polymer presents polaronic (radical cation) species in the polymer chains. The water electrolysis at the polymer–metal interface produces OH radicals, with oxygen evolution and H^+ liberation inside the growing conducting polymer film. The recombination of the radicals ($^{\bullet}OH$ and polarons) and the strong acid media inside the film promotes film degradation with loss of both the chain's conjugation and the film's electroactivity (reversible oxidation/reduction). Thus, in an initial approach we can consider the anodic potential limit of the electrolyte potential window as the potential where the polymer degradation starts: E_d. This degradation process is also known as polymeric overoxidation.

3.4 Electropolymerization at a Constant Potential (Potentiostatic)

For the studied system (electrode, solvent, salt, monomer), the voltammetric response from the monomeric solution gives the monomer oxidation–polymerization potential on the metal (E_0) and the potential for the beginning of the polymer degradation (E_d). A faster polymer growth can be obtained by monomeric oxidation–polymerization under a constant electrodic potential beyond the monomer oxidation potential.[11,12] The polymerization potential must be higher than the monomer oxidation potential (E_0) but lower than the beginning of the polymer degradation potential (E_d).

Using a clean cell, including a clean WE, counter electrode (CE) and RE, and a fresh monomeric solution, the WE is subjected to a potential step from an initial potential E_1, which is lower than that of the monomer oxidation, and kept for a short time (1 s) before moving to a final potential, E_2, which is more anodic than the monomer oxidation potential E_0. The amount of electrogenerated polymer increases with the oxidation–polymerization time. By repeating the experiment using a clean cell, the same clean electrodes located at the same distance and a fresh solution each time, as well as the same oxidation–polymerization potential and polarization time, the same polymer weight (±2%) is electrogenerated as many times as required: the electrogeneration of conducting polymers is a very reproducible method of synthesis.

The chronoamperometric (I/t, where I is current and t is time) response from the WE shows a sharp initial maximum related to the charge of the EDL at the electrode–electrolyte interface (Figure 3.4). Then, a minimum is followed by a fast increase of the current through an inflexion point towards a constant current. The consecutive events are characteristic by the formation, growth and coalescence of polymeric nuclei.[16–22] After nuclei coalescence, a constant polymer–solution surface area gives a constant monomeric diffusion layer, providing a constant amount of monomeric molecules per surface unit and a constant oxidation–polymerization rate with flow of a constant current. By integration of the chronoamperometric response, the charge consumed during the polymerization process is attained. The generated

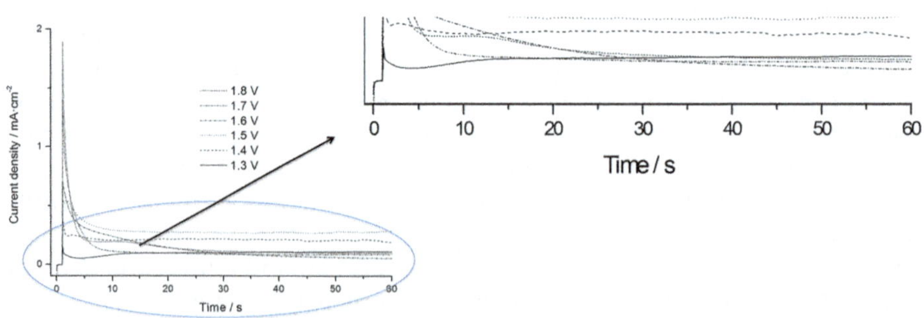

Figure 3.4 Chronoamperograms recorded from 1 mM anthraquinone-functionalized EDOT (AQ-EDOT) and 0.1 M TBAPF$_6$ in a dichloromethane solution (with a fresh solution every time) using a Pt electrode at constant potentials of 1.3, 1.4, 1.5, 1.6, 1.7 or 1.8 V. Reprinted from ref. 85 with the permission from Elsevier: *Electrochim. Acta*, 2010, **55**, 1535–1542.

polymer mass can be attained by the mass difference between the rinsed and dry-coated electrode and the clean electrode.

The electropolymerization kinetics under constant chemical conditions can be studied by repeating the procedure using a clean cell and electrodes each time, as well as a fresh solution for the different polymerization times.[23] The electropolymerization empirical kinetics are attained by repeating the kinetics under different experimental conditions: different polymerization times, higher oxidation–polymerization potentials (faster polymerization rates but possible simultaneous electrolyte discharge), different monomer or electrolyte concentrations, different electrolytes (inorganic, organic, polyelectrolytes, polyoxometallates, ionic liquid) or different temperatures.[1]

High anodic potentials give chronoamperometric responses showing a maximum. The high conductive film generated at the beginning of the current flow becomes an increasingly resistive film by polymeric degradation, giving (according to Ohm's law) decreasing currents at increased polymerization times. These high potentials overlap with the discharge of the background electrolyte, which can promote the parallel degradation of the electrogenerated polymer.

3.5 Electropolymerization by Consecutive Square Potential Waves

As a conclusion from the two previous points, the electropolymerization of conducting polymers starts by monomeric oxidation with formation of polymer nuclei. As a consequence, quite irregular cauliflower or dendritic polymeric surfaces are attained: the monomer diffusion from the solution is faster towards the dendrite peaks (the shortest distance from the diffusion layer border) than towards the valleys between the dendrites. In order to destroy the diffusion layer, thus obtaining a uniform monomeric concentration distribution at the growing surface, the polymerization process can be performed by consecutive potential steps. During the short duration of the anodic potential step the monomer oxidation–polymerization starts by nucleation. The monomer concentration drops at the electrode interface, developing a concentration gradient from the solution to the electrode surface; this is the monomer diffusion layer. The preferential diffusion of the monomer towards the top of the nuclei begins. The cathodic potential step promotes the reduction of the polymer nuclei, the monomer oxidation stops and the uniform monomeric concentration at the electrode–electrolyte interface is restored with destruction of the monomer diffusion layer. At the beginning of the next anodic potential step, the old nuclei start to grow again, but new monomer is oxidized at the inter-nuclei uncoated metal spaces, generating new nuclei and giving a more uniform and shiny surface to the electrogenerated polymer film. Suitable square potential waves allow the electrogeneration of smooth and shiny films of conducting polymers.[24–26]

The electrogeneration of organic polymeric blends with large organic ions, amphiphilic ions or polyelectrolytes gives very uniform and shiny films by polymerization at a constant potential. The large organic ions are preferentially adsorbed at the tops of the starting dendrites, inducing a low passivation and shifting the electric fields towards the valleys, in a similar way to additive actuation during the electrodeposition of metals.

3.6 Electropolymerization by Flow of a Constant Current (Galvanostatic)

The first potential sweep using a clean electrode (Figure 3.3) provides information about the relationship between the flowing current density and the overpotential from E_0 until the beginning of the degradation process at E_d. The polymer film can be generated now under the flow of a constant anodic current density higher than zero and lower than that of the inflexion point. The flow of the selected constant anodic current through a clean cell with clean electrodes and a fresh monomeric solution gives a chronopotentiometric (E/t) response during the film electrogeneration, showing the evolution of the electrode potential (related to the RE) during the polymerization time (Figure 3.5). After a progressive potential increase related to the polymer nucleation, nuclei growth and nuclei coalescence, a constant polymerization

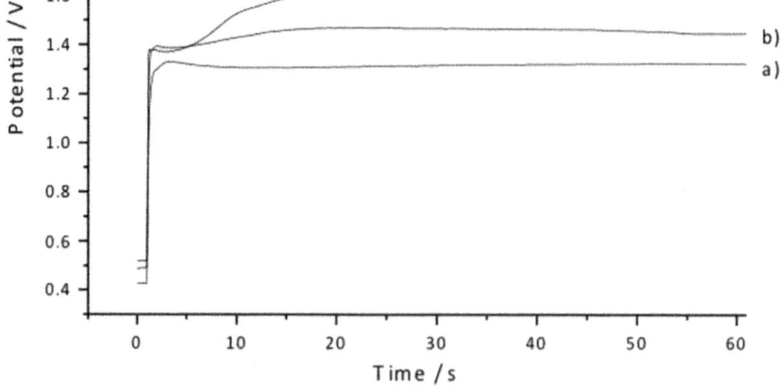

Figure 3.5 Chronopotentiograms obtained from 1 mM AQ-EDOT and 0.1 M $TBAPF_6$ in a dichloromethane solution on Pt at a constant anodic current density of (a) 0.1 mA cm^{-2}, (b) 0.2 mA cm^{-2} or (c) 0.3 mA cm^{-2}. Temperature, 20 °C.

potential is attained (Figure 3.5, line a): the conductivity of the attained material is high. Looking at the voltammetric response (Figure 3.3), this constant potential correlates to some value between the oxidation–polymerization potential E_0 and the polymer degradation potential for the used electrolyte, E_d. The plateau of the constant potential from the chronopotentiometric response evolves at higher potentials (Figure 3.5, line b) for rising polymerization currents, resulting in faster polymer growth. When the constant current used approaches a value around the inflexion voltammetric point in Figure 3.3, the chronopotentiometric response shows rising potentials with the polymerization time (Figure 3.5, line c), thus not attaining a constant plateau. After a fast initial polymerization rate, the polymer film becomes an increasingly resistive material due to its simultaneous degradation.

3.7 Tafel Slope Mechanism Using Clean Metal Electrodes

In organic electrochemical kinetics the oxidation rate (i, in A) and the oxidation overpotential ($\eta = E - E_0$, in V) were first empirically related by the Tafel expression [eqn (3.2)]:

$$\eta = A\ln\frac{i}{i_0}, \tag{3.2}$$

where $A = RT/\alpha nF$ with T (in °K) being the experimental temperature, α the charge transfer electrochemical symmetry factor, the value of which must be between 0 and 1 (usually 0.5), n the number of electrons transferred per monomeric molecule and F being Faraday's constant.[27]

When conducting polymers were first investigated, the oxidation overpotential from the voltammetric responses, which appears after the monomer oxidation potential, was represented *versus* the rising current logarithm and the attained Tafel slope corresponded to the exchange of one electron ($n = 1$) per monomeric unit.

3.8 Electropolymerization Mechanism

Following the Tafel results, a very simple electropolymerization mechanism was proposed by Diaz:[28–30]

Initiation: by monomeric oxidation, generating a monomer radical cation; see eqn (3.1).

Propagation: by condensation of two radical cations with elimination of two protons and consecutive oligomeric oxidation.

$$M^{\bullet+} + M^{\bullet+} \rightarrow M_2 + 2H^+ \tag{3.3}$$

$$M_2 \rightarrow M_2^{+\bullet} + e^- \tag{3.4}$$

$$M_2^{+} + M^{\bullet+} \rightarrow M_3 + 2H \quad (3.5)$$

$$M_3 \rightarrow M_3^{+\bullet} + e^- \quad (3.6)$$

...and so on.

Termination: by combination of oligomeric or polymeric radical cations and simultaneous interruption of the current flow to stop the formation of new radical cations (monomeric, oligomeric or polymeric).

$$M_n^{\bullet+} + M^{\bullet+} \rightarrow M_{n+1} + 2H^+ \quad (3.7)$$

According to the Tafel empirical results from eqn (3.1) to (3.7), every new monomeric unit incorporated into the growing chain loses one electron. Eqn (3.1) and (3.3) to (3.7) constitute the radical cation polycondensation mechanism that is generally accepted in the literature.[31–45] This mechanism is comprised of faradaic processes and the electrogenerated polymer weight must be proportional (Faraday's law) to the consumed charge.

This simple mechanism supposes that the same polymer electrogenerated in different laboratories under "*the same*" experimental conditions or in the same laboratory at different times must have the same physical and chemical properties.

Experimental results have contradicted these assumptions: the electronic conductivity of oxidized polythiophene films electrogenerated under similar conditions (monomer, acetonitrile solvent and salt using the electrochemical method) in different laboratories during the 1980s covered several orders of magnitude (from 10^{-5} to 10^{5} S cm^{-1}).[46–52] These results can only be explained by the simultaneous presence of some parallel reaction during the electropolymerization process that degrades the generated polymer or generates a fraction of the insulating polymer chains (Section 3.15). Under this hypothesis the final material should be a mixture of conducting and degraded (insulating) chains.

Also under the above mechanism, constant electrochemical properties of the electrogenerated material should be expected, independent of the physical and chemical conditions of the electropolymerization. Again, the experimental results have contradicted the expected ones: the charge stored per unit of electrogenerated polymer mass changed when the films were generated under polarization at a constant potential for different polymerization times or under different electrochemical (potential limits, potentials, currents) or chemical (temperature, electrolyte concentration, monomer concentration) conditions.[1,23,53–55]

Finally, the above-presented electropolymerization by consecutive potential sweeps indicated that the conducting polymer is generated in its oxidized state. Polymer oxidation [eqn (1.1) in Chapter 1] incorporates counterions inside the film to balance positive charges on the chains: the polycondensation mechanism of radical cations doesn't describe the real generation of oxidized polymeric films.

Summarizing the experimental results, it can be indicated that:

1. The electropolymerization mechanism must be more complex than the polycondensation of radical cations above-presented, which is generally accepted by the scientific community.
2. Different electropolymerization, chemical or physical conditions give materials with different redox electroactivity. Following the polycondensation mechanism and whatever the synthesis conditions, the synthesized materials should have the same properties: the polycondensation mechanism missed some actual steps.
3. In order to generate reproducible materials with the same properties every time and, if possible, the best properties for a defined technological application, detailed knowledge of the electropolymerization mechanism is required.

3.9 Electrochemical and Gravimetric Methodologies

The alternative way is to recover some of the classical and most laborious methodologies to determine the empirical polymerization kinetics, which can be re-adapted to the current electropolymerization methodology.

For the electrogeneration of conducting polymers the polymerization charge (Q_{pol}, in C) must be consumed to produce a polymer mass (w, in g). In the absence of any parallel reaction the polycondensation mechanism is a faradaic synthesis and the consumed charge (Q_{pol}) should give constant polymerization productivities, w/Q (in g C^{-1}) under different polymerization conditions. Nevertheless, results confirm the influence of experimental variables on the productivity.[8,13,15,23,53–85]

Thus, the most confident way to follow the polymerization empirical kinetics should be the classical gravimetric procedure. The mass (w, in g) of each electrogenerated polymer film is determined after different polymerization times and the procedure is repeated for different values of each polymerization variable. After every electrogeneration the electrode coated with the conducting polymer film is rinsed and dried, and the generated polymer mass is determined by *ex-situ* microgravimetry: the mass of the coated electrode minus the mass of the clean electrode.

Going one step further, the final aim of any electropolymerization in the context of this book should be the synthesis of materials with improved properties for use in the production of new and improved electrochemical devices. Taking into account the empirical evidence that the properties of the generated material change with the synthesis conditions, knowing how these conditions influence the electrochemical properties of the synthesized material should provide the required tools to obtain tailored materials for specific applications and devices.

The electrochemical methodologies also allow fast control of the electrochemical responses from each electrogenerated film, giving the charge stored during oxidation (Q_{ox}) and the charge released during reduction (Q_{red}) from every electrogenerated film. With the film mass and the stored charge, the charge stored per unit of mass (*i.e.*, the specific charge, in C g^{-1}) may be determined for every film. In this way a deep knowledge can be attained about how this electrochemical property evolves as a function of the synthesis conditions.

Thus, in order to reveal the actual polymerization mechanism, a parallel study of the empirical kinetics through the consumed charge (Q_{pol}), the generated polymer mass (w) and the specific charge (Q_{ox}/w) stored in every film electrogenerated during the kinetic study is proposed. In addition, the ratio ($w : Q_{pol}$, in g C^{-1}) gives the productivity of the charge consumed to generate each film, *i.e.*, the electropolymerization productivity. Any variation of the electropolymerization productivity or of the stored specific charge from films electrogenerated under different values of the same synthesis variables (polymerization time, electrolyte or monomer concentration, temperature or electrical conditions), with the other variables remaining constant, will indicate a change of the polymerization mechanism with the presence of parallel processes (degradation, cross-linking, adsorption) consuming the charge and giving a mixed material, the composition and electrochemical properties of which evolve with the shift of each synthesis variable.

Having a deep knowledge of the polymerization mechanism will allow the easy generation of reproducible and tailored materials for each specific application.

3.10 Gravimetric Empirical Electropolymerization Kinetics

In this context the electro-synthesis mechanism giving a conducting polymer film from a monomeric (M) solution containing the anion A^- must include the initiation and propagation steps described in Section 3.8, which is corroborated by the Tafel experiments, to produce an oxidized polymer film, $[M_m{}^{n+}(A^-)_n]$.[8,13,15,23,53–85]

The electropolymerization process can be summarized by the reaction in eqn (3.8):

$$m\text{M} + n\text{A}^- \rightarrow (\text{M}_m)^{n+}(\text{A}^-)_n + (2m+1)\text{H}^+ + (2m+n+1)\text{e}^-. \tag{3.8}$$

The empirical polymerization rate (R_p) from eqn (3.8) is:

$$R_p = k_p[\text{M}]^a[\text{A}^-]^b = k_{p0}\exp(-E_a/RT)[\text{M}]^a[\text{A}^-]^b, \tag{3.9}$$

where [M] and $[A^-]$ represent the monomer and anion concentrations, respectively, in the polymerization solution, with a and b giving the reaction orders related to the monomer and the electrolyte concentrations, respectively; k_p is the polymerization rate coefficient, E_a is the polymerization activation energy, R is the universal constant for gases and T is the working temperature. By taking logarithms, eqn (3.10) is produced:

$$\ln R_p = \ln k_{po} - (E_a/RT) + a\ln[M] + b\ln[A^-]. \quad (3.10)$$

In order to obtain the polymerization kinetic magnitudes for E_a, k_p, a and b, the polymerization process can be performed by imposing a constant potential to the WE. The polymerization rate R_p is attained now from the polymer mass electrogenerated under a constant potential for different polymerization times using a clean cell and electrodes, and a fresh monomer solution each time, keeping constant either the monomer concentration, the electrolyte concentration or the working temperature. During each electrogeneration, the chronoamperometric response is recorded and integrated, giving the charge consumed during the electropolymerization (Q_{pol}). After electrogeneration, each film is subjected to consecutive potential sweeps, with the anodic potential limit being lower than the oxidation–polymerization potential and the charge stored in the film being obtained from the stationary voltammetric response by integration of the anodic (Q_{ox}) and cathodic (Q_{red}) voltammetric branches. Then, each electrode coated with the polymer film is rinsed, dried and weighed to get the polymer mass (w, in g). By plotting the electrogenerated polymer mass (w) *vs.* the polymerization time (in s), a linear variation is attained, the slope of which represents the gravimetric polymerization rate (R_p, in g s^{-1}; Figure 3.6a).

By repeating the procedure for a different value of the monomeric concentration each time and under constant $[A^-]$ and constant temperature,

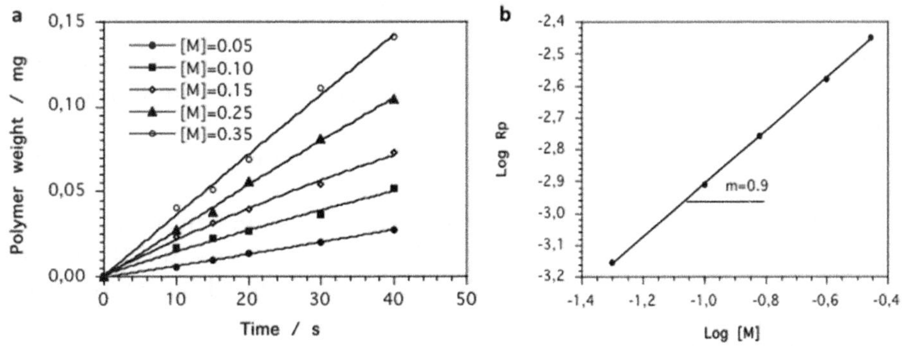

Figure 3.6 (a) Evolution of the weight of the polypyrrole films electrogenerated on a Pt electrode (1 cm^2) at 800 mV from a 0.1 M $LiClO_4$ aqueous solution for different polarization times and different pyrrole concentrations, which are indicated on the figure. Every film was generated using a clean cell, clean electrodes and a fresh solution. (b) Double logarithmic plot of the slopes (polymerization rates) from (a) *versus* the monomeric concentration: the new slope is given by eqn (3.11) and is the reaction order. With kind permission from Springer Science and Business Media, adapted from ref. 1: T. F. Otero, in *Modern Aspects of Electrochemistry*, ed. R. E. White, J. O. Bockris and B. E. Conway, Springer US, New York, 1999, pp. 307–434.

a different value of R_p at each experimental [M] value is attained (Figure 3.6a). Under these conditions, eqn (3.10) becomes:

$$\ln R_p = k' + a\ln[M] \tag{3.11}$$

with:

$$k' = \ln k_p + b\ln[A^-]. \tag{3.12}$$

The slope of the empirical double logarithmic plot of R_p *vs.* ln[M] (Figure 3.6b) is the reaction order *a* and from the origin ordinate k_p will be obtained after determination of *b*.

By repeating the procedure, now changing [A^-] under constant [M] and *T*, the attained values of R_p must follow eqn (3.13):

$$\ln R_p = k'' + b\ln[A^-], \tag{3.13}$$

where the slope is the reaction order *b*, with:

$$k'' = \ln k_p + a\ln[M]. \tag{3.14}$$

Thus, from the intercept with the origin ordinate (k''), k_p is attained using the known values of *a* and [M].

By repeating the procedure under constant concentrations and changing the polymerization temperature each time, and considering the Arrhenius expression for k_p, eqn (3.15) can be obtained:

$$\ln R_p = k''' - (E_a/RT). \tag{3.15}$$

Plotting $\ln R_p$ *vs.* the inverse temperature gives the slope of the attained linear variation, which gives the reaction activation energy (E_a/R). From the origin ordinate, the pre-exponential factor *A* is obtained.

In this way, all the gravimetric empirical polymerization kinetics in eqn (3.9) can be determined.

The full procedure can now be repeated for the polymer electrogeneration at different potentials under the flow of different currents in different solvents or in different electrolytes. If, as generally accepted in the literature, the mechanism is unique, the same values should be expected under any of these circumstances for the reaction orders, activation energy and kinetic coefficient of the empirical electropolymerization kinetics equation.

3.11 Empirical Kinetics from the Electropolymerization Charge

If we assume that the electropolymerization process is a faradaic process, only the monomeric oxidation consumes charge to produce the polymer. Under these circumstances, Faraday's law states that the electrogenerated polymer weight (W, in g^{-1}) must be proportional to the consumed electropolymerization charge (Q_p, in C):

$$W = kQ_p, \tag{3.16}$$

where k (in g C^{-1}) is the charge productivity.

The area below the chronoamperometric responses attained above during each film synthesis is the charge consumed during every electropolymerization:

$$Q_p = \int i(t)\mathrm{d}t. \tag{3.17}$$

The charge is directly given by most electrochemical equipment at the end of the experiment. Thus, Q_p is available for films generated at different polymerization times from different monomeric or electrolyte concentrations and under different temperatures.

The procedure described in Section 3.9 for the gravimetric determination of the empirical polymerization kinetics [eqn (3.9)] can now be repeated using the polymerization charge instead of the polymer weight. Under faradaic conditions the charge is only expended to produce polymer; thus, k, E_a, a and b attained from the polymerization charges must overlap those above attained from the polymer masses. Plotting the consumed charge *vs.* the polymerization time gives R_p (in C s^{-1}) for each value of an empirical variable that is required to apply eqn (3.15). The presence of any parallel reaction during the electropolymerization process should give a different set of values from gravimetric and consumed charge determinations.

A faster way to check if the polymer is generated through a faradaic process, following eqn (3.16), is from the charge productivity. For each generated film, both the consumed charge and the polymer mass are available, giving the charge productivity (k, in g C^{-1}). Under faradaic synthesis, k should be independent of the polymerization conditions. Any productivity variation indicates the presence of parallel reactions consuming charge without polymeric generation.

3.12 Electrochemical Polymerization Kinetics: Tafel Slopes from Clean Metal Electrodes

As stated in Section 3.7, the monomeric oxidation rate (i, in A) and the oxidation overpotential ($\eta = E - E_0$, in V) are linked by the Tafel expression: $i = a + b \ln \eta$.

According to a pure initiation, propagation and termination polymerization mechanism [eqn (3.1) and (3.3) to (3.7)], the monomeric oxidation rate should determine the polymerization rate. The Tafel representation is attained from the first anodic voltammetric response beyond the beginning of the monomer oxidation. Getting the Tafel response for different monomeric concentrations, different electrolyte concentrations or different temperatures gives the reaction orders, reaction kinetic coefficient or activation

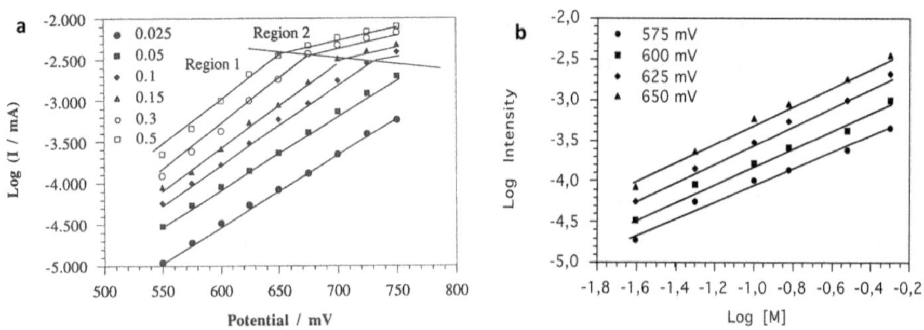

Figure 3.7 (a) Tafel plots (log*I* *vs.* *E*) obtained from the voltammetric responses of a clean Pt electrode from fresh 0.1 M $LiClO_4$ acetonitrile solutions having a different thiophene concentration each at 2 V s^{-1}. (b) Double logarithmic plots from the first voltammetric response using a clean Pt electrode and a fresh solution each time subjected to a sweep rate of 2 mV s^{-1} in 0.1 M acetonitrile solutions with different concentrations of thiophene. Reprinted with permission from ref. 27: *Synth. Met.*, 1993, **55–57**, 1436–1440. Copyright (1993) Elsevier.

energy, as in the previous kinetic methodologies. Thus, from the voltammetric results obtained from different monomeric concentrations, Figure 3.7 can be obtained, which shows the double logarithmic plot (at the same potential) of the current (polymerization rate) *versus* the monomeric concentration. The slope is the Tafel reaction order related to the monomer concentration.

3.13 Tafel Slopes from Polymer-Coated Electrodes

Both Tafel slopes from Sections 3.7 and 3.8 only account for the monomer oxidation on the clean electrode: the initiation process. But, during the electrogeneration of any polymer film thicker than a few nm, the electron transfer from the monomer takes place on a polymer electrode surface most of the time, so the nature of the material electrode has changed. By repeating the Tafel slope using a polymer-coated electrode in a clean monomeric solution, a different Tafel slope is attained: the number of electrons transferred per monomeric unit, the reaction orders, kinetic coefficient and activation energy all change. That means that the polymerization mechanism also changes, yielding reaction orders similar to those attained from the gravimetric methodology (Table 3.1).

3.14 Electropolymerization and the Properties of the Electrogenerated Films

From the electrogenerated polymer mass and the charge stored during the voltammetric control, the specific charge (Q_{ox}/w, in C g^{-1}) stored in every electrogenerated film is attained. The inverse of the specific stored charge

Table 3.1 Reaction orders of the empirical kinetics ($R_p = k[\text{Monomer}]^a[\text{electrolyte}]^b$) for the electropolymerization of different conducting polymers (Py, pyrrole; Th, thiophene; and 3MTh, 3-methyl-thiophene) in different electrolytes (ACN, acetonitrile; PC, propylene carbonate; DMF, dimethylformamide) from: (1) the charge consumed during each electropolymerization (Q_{pol}), (2) the attained polymer mass, (3) the Tafel plots using a clean Pt electrode, (4) the Tafel slopes using a polymer-coated electrode and (5) from voltammetric charges stored by the electrogenerated films. Adapted from ref. 77: *Electrochim. Acta*, 1994, **39**, 245–253, with permission from Elsevier.

Method	Py/H_2O	Py/ACN	Th/ACN	3MTh/PC	Py/ACN + 2% H_2O	Py/DMF
Polymerization charge	$a = 0.6$; $b = 0.8$	$a = 0.4$; $b = 0.5$	$a = 0.8$; $b = 1$			
Gravimetric method	$a = 0.5$; $b = 0.8$	$a = 0.5$; $b = 0.5$	$a = 1$; $b = 1$	$a = 1.8$; $b = 0.8$		$a = 0.5$; $b = 1.3$
Tafel on Pt	$a = 1$; $b = 1$	$a = 0.7$; $b = 1$	$a = 1.5$; $b = 1.8$		$a = 1$; $b = 1.4$	
Tafel on polymer	$a = 0.5$; $b = 0.7$	$a = 0.5$; $b = 0.5$	$a = 0.5$; $b = 1$		$a = 0.5$; $b = 0.5$	
Stored charge	$a = 1$; $b = 0.8$	$a = 0.4$; $b = 0.5$	$a = 0.8$; $b = 0.8$			

Table 3.2 Specific charge (in mC mg^{-1}) stored by polythiophene films electrogenerated on a Pt electrode at 1700 mV with increasing polymerization time from 0.1 M $LiClO_4$ acetonitrile solutions with different monomeric concentrations. The stored charge was obtained by voltammetric control between −500 and 1400 mV at 50 mV s^{-1} in the background electrolyte. Every electrogenerated, rinsed and dried film mass was determined by gravimetry. Adapted from ref. 77: *Electrochim. Acta*, 1994, **39**, 245–253, with permission from Elsevier.

	[M]/mol L^{-1}			
Time/s	0.05	0.10	0.15	0.25
10	562	315	272	260
20	337	280	196	205
40	212	243	194	184
60	178	215	186	149
100	174	174	140	110

is the polymer mass required to store a unit of charge (w/Q_{ox}, in g C^{-1}). Both magnitudes can be used in eqn (3.15) to get the kinetic magnitudes from the empirical chemical kinetics. In addition, if the polymerization mechanism doesn't change when the kinetic variables (polymerization time, concentration, temperature or potential) are shifted, the attained specific charge and the polymer mass required to store a unit of charge will remain constant. Table 3.2 illustrates the different, surprising results. Under constant conditions, decreasing specific charges are stored in films generated during increasing polymerization times, indicating that behind the polymerization front a degradation process advances, yielding films that should have an asymmetric transversal composition. In addition, decreasing specific charges are stored from rising monomeric concentrations, indicating the presence of a parallel process (Section 3.16) generating a non-electroactive polymeric fraction of the final film.

3.15 Analysis of the Polymerization Kinetics

Summarizing the attained results, we can conclude that the attained empirical electropolymerization kinetics and mechanism from the Tafel slope (condensation of radical cations) using a clean metal electrode that is generally accepted by the literature only represents the initiation process of a complex polymerization mechanism.

The gravimetric empirical kinetics, considering the mass of the electrogenerated, rinsed and dried polymer as the final reaction product, gives uneven reaction orders (Table 3.1) that indicate a complex electropolymerization mechanism.

When a metal electrode coated with the conducting polymer film is used in a fresh monomeric solution of the polymer, the Tafel slopes give the same kinetic magnitudes as the gravimetric empirical kinetics.

The variation of both the productivity of the consumed charge and the stored specific charge as a function of the experimental conditions also indicates that the conducting polymer is generated through a complex mechanism of parallel processes, the relative rate of which changes with the synthesis conditions.

3.16 Parallel Polymeric Degradation–Cross-Linking During Synthesis

The specific stored charge changes as a function of the electrogeneration conditions: polymerization time, electrolyte or monomer concentration, temperature or polymerization potential (Table 3.2).[86–88] Whatever the oxidation–polymerization conditions, the generated chemical environment around the electrode (strong pH variation and high concentration of radical cations) is strong enough to initiate the polymer degradation (or overoxidation) reaction.

Table 3.2 shows how the charge stored per unit weight of the electrogenerated polymer decreases as the polymerization time increases, indicating that behind the electropolymerization front there is a polymeric degradation beginning. In thick films synthesized during long polymerization times, the polymer close to the metal is degraded, pushing down the average specific charge stored by the full film. As a consequence, any electrogenerated film presents a degradation concentration gradient that produces a bending movement of the self-supported film electrode during its reversible oxidation/reduction.[88]

In fact, the degradation kinetics of a film can be followed by its polarization in the background electrolyte (in the absence of monomer) at any potential (including the same potential used for the polymerization) for a constant time. Then the electroactivity (stored charge) of the film can be checked by voltammetric control.[89] By repeating the procedure for different kinetic conditions (polarization time, degradation potential, electrolyte concentration, temperature), the degradation kinetics can be attained (Figure 3.8a) as a percentage of the loss of stored charge per unit of polarization time. From similar results, either the reaction orders, reaction coefficient or activation energies can be attained, as well as the evolution of the degradation rate as a function of any of the studied variables.

The same procedure can be used to quantify the influence of some degradation accelerators, such as a small water content when polymers are generated at high anodic potentials in organic solvents.[90–92]

As in the case of commodity polymers, degradation and stabilization are opposed and related processes. The same described methodology can be used to quantify the actuation of different stabilizers included in the film during electrogeneration or present in the checking solution.

Many attempts in the literature have tried to follow the electrochemical degradation using spectroscopic methodologies. Degradation of the film electroactivity over 80% of the stored specific charge only produces very minor spectroscopic responses: the electrochemical control results are the most efficient way to follow the degradation processes.

The degradation gives the branching and cross-linking processes (Figure 3.9), with nucleophilic attack of the radical (generated during solvent, electrolyte or contaminant oxidation) on the radical cations of the oxidized chain, giving sp^3 carbons and loss of the chain conjugation.[93–103] Films showing flat voltammetric responses still have enough electronic conductivity to be used as electrodes for other analytical purposes, avoiding interferences from the redox activity of the film.

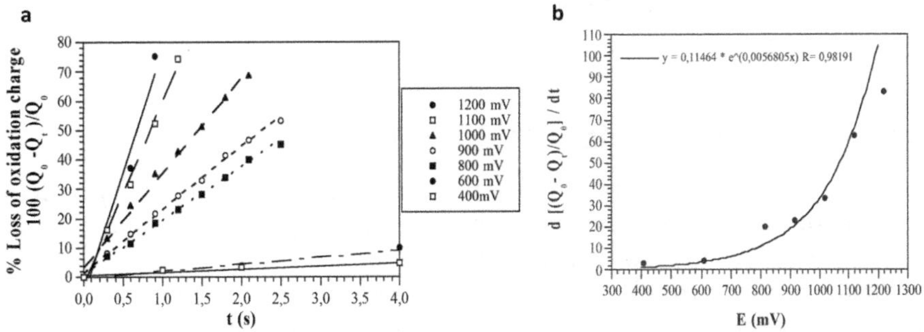

Figure 3.8 (a) Evolution of the charge storage degradation rate as a function of the polarization time for different potentials of degradation. (b) Evolution of the degradation rate, $d[(Q_0 - Q_t)/Q_0]/dt$, as a function of the degradation potential. Reprinted from *Port. Electrochim. Acta*, 1995, **13**, 403.

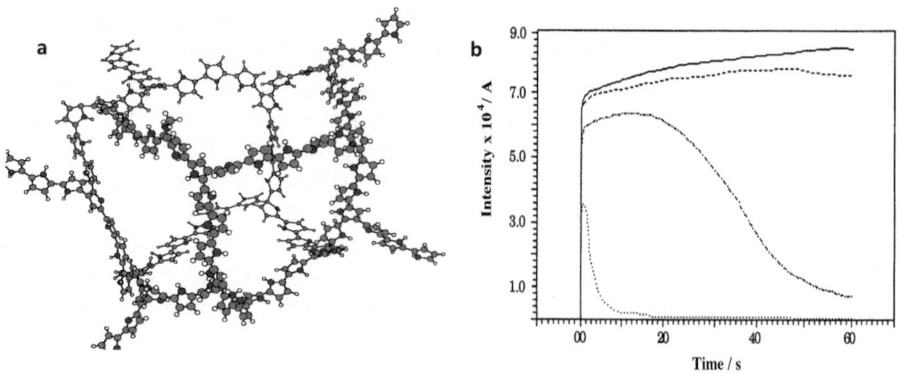

Figure 3.9 (a) Scheme of the cross-linked structure of a conducting polymer. (b) Chronoamperometric responses from a Pt electrode subjected to 1700 mV in 0.1 M thiophene and 0.1 M $LiClO_4$ acetonitrile solutions with different water contents: —, 0.04%; ---, 0.14%; -•-, 0.44% and •••, 1.04%. Reprinted with permission from ref. 92: *J. Electroanal. Chem.*, 1991, **310**, 219–237. Copyright (1991) Elsevier.

Some polymers require a high anodic potential to initiate the electropolymerization by monomeric oxidation, overlapping with oxygen evolution from water. Despite the use of organic solvents, the presence of residual water results in large variations of the polythiophene conductivity: concentrations higher than 0.4% promote the very fast degradation–passivation of polythiophene films (Figure 3.9b) generated at 1.7 V. At electropolymerization potentials lower than 1–2 V (*i.e.*, below the electrolysis of water), the effect of the water content on the film degradation is very low.

The same degradation processes overlapping here with the polymer electro-synthesis will influence the polymer life-time during its oxidation/reduction in electrochemical devices. In summary, a good exploration of polymer degradation during polymerization and of the degradation and stabilization kinetics under working conditions should become a key aspect to obtain life-time industrial products.

3.17 Parallel Chemical Polymerization

The use of organic solvents with low proton affinities results in very high productivity of the polymerization charge (w/Q_p). In parallel, the specific charge (Q_{ox}/w) stored in the resulting film is too low.

During the electropolymerization process, a brown cloud of polymer is observed dropping from around the electrode towards the bottom of the electrochemical cell.[90] By adding some protophilic compound, such as water (<1.5% w/w), to the pristine monomeric solution, the brown cloud doesn't appear around the electrode during the electropolymerization. The productivity of the consumed charge decreases and the specific charge stored in the electrogenerated film increases when compared to those from films electrogenerated in the absence of water.[104]

By adding some acid to the pristine monomeric solution, a homogeneous chemical polymerization process was followed with time by ultraviolet-visible (UV-vis) spectroscopy.[105] By changing one of the kinetic variables each time [*e.g.*, water content, reactant (monomer or electrolyte) concentration or temperature], the empirical kinetics for the acidic polymerization were attained.[105] The conclusion is that a chemical polymerization occurs around the electrode. In the absence of a more protophilic agent, the protons generated at the electrode surface by the radical cation polycondensation steps [eqn (3.3), (3.5) and (3.7)] diffuse towards the solution bulk. In this way, they meet the monomeric molecules diffusing in the opposite direction towards the electrode surface and a fraction of them become protonated. The protonated monomer initiates a chemical polymerization around the electrode, forming a brown cloud.[106] A fraction of the protonated polymer is adsorbed on and trapped by the film growing on the electrode by electropolymerization. The final result is that the same polymerization charge produces a higher polymer mass (electrochemical plus chemical), giving a higher productivity in the organic solvent than in the organic solvent with 1% water. Nevertheless, one mass unit of the polymer generated in the organic solution stores less charge due to the non-electroactive fraction (chemically generated), while a little water content avoids the formation of the non-electroactive fraction. The protonated carbons present sp^3 bonds, breaking the chain conjugation and its subsequent electroactivity.

When the electropolymerization potential is close to the water discharge potential, a water content greater than 2% gives films with decreasing specific charges due to the faster parallel degradation of the electrogenerated film. Electropolymerization processes at potentials higher than 1.2 V require more stable protophilic agents than water: water electrolysis above this potential promotes nucleophilic attack and polymer degradation, with a fast drop of the specific stored charge.

Whatever the selected initial conditions of synthesis, the electrogeneration of thick polymer films during long polymerization times in one compartment of the electrochemical cell promotes a change of the solution chemical conditions due to the reaction products (the WE and CE), such as pH variation, the chemical generation of oligomers and solution darkening.

3.18 Parallel Adsorption of Macroions

The electrochemical generation of a conducting polymer from a monomeric solution always requires the presence of a dissociated salt. The dissociated ions play two basic roles during the electrogeneration of conducting polymers: they drive the current required to oxidize the monomer and the generated polymer through the solution, and they incorporate some anions into the growing film in order to compensate for the polaronic charges on the polymer chains. The use of polyelectrolytes, salts of large anions or macroanions (organic or inorganic) during the electropolymerization results in polymeric blends (conducting polymer–polyelectrolyte, conducting polymer–amphiphilic anion) or hybrids (conducting polymer–inorganic macroions). The large macroions experience a strong van der Waals attraction with any surface, electrode or particle present in the solution: they adsorb at the electrode surface before and during the electropolymerization process.[107–111] The final result is that, in addition to the polyanion required for charge balance, the electrogenerated film has adsorbed and trapped an extra polyelectrolyte fraction (polyanion–cation couple).[110,112] The polyanion content decreases from decreasing electrolyte concentrations used during the electropolymerization, as deduced from the rising specific charges stored in the generated polymer blend. Both the composition and properties of the electrogenerated blend or hybrid material can be controlled by the synthesis conditions. The nature of the polyelectrolyte and its content in the final polymer blend can be used to control some of the properties (mechanical, electrical, optical) of the electrogenerated conducting polymer.

The strong adsorption of some polyelectrolytes coats the electrode, which prevents monomer oxidation up to such high overpotentials that it occurs together with the electrolyte discharge, promoting overoxidized or full degraded materials from the point of view of the electroactivity. Only at very low polyelectrolyte concentrations ($<10^{-2}$ M) does the adsorption equilibrium partially coat the electrode, allowing the electrogeneration of electroactive blends by electropolymerization of the monomer at low potentials below the electrolyte discharge.[112] In order to increase the electropolymerization rate (*i.e.*, the flow of high oxidation/polymerization currents at low overpotentials) and the conductivity of the electrogenerated material, a mixture of the low concentrated polyelectrolyte and a higher concentration of any other fully dissociated salt constituted by small ions in the solvent can be used.

3.19 Shift of the Molecular Interaction Forces: Electrodissolution

During the controlled polymer oxidation/reduction electrochemical reaction, the intermolecular forces (polymer–polymer, polymer–anion, polymer–cation, polymer–solvent and solvent–ion) inside the film change. Strong van der Waals attractive forces between the different film components can become weak interaction forces in the presence of different solvents. According to the empirical results, some stable polymer films can be solubilized after electrogeneration during electrochemical reduction in electrolytes constituted by different solvents.[113–118]

Thus, the thiophene–pyrrole–thiophene trimer gives, by electropolymerization, oxidized films constituted by oligomers that,[119,120] by reduction in electrolytes of different organic solvents, give soluble, reduced oligomers.[121]

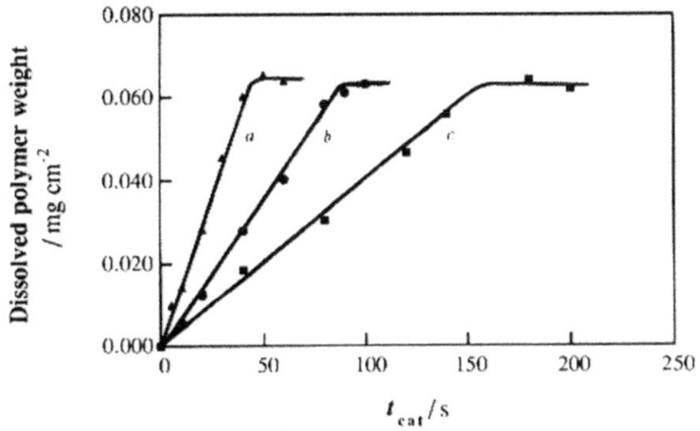

Figure 3.10 Evolution of the solved poly[2,5-di-(2-thienyl)pyrrole] [poly(SNS)] weight in a 0.1 M $LiClO_4$ acetonitrile solution under reduction by flow of different cathodic currents: (a) −0.2, (b) −0.1 or (c) −0.05 mA cm^{-2}. Every point was attained from a just-electrogenerated film on a Pt electrode by flow of a constant anodic current of 0.5 mA cm^{-2} for 60 s through a 5 mM SNS and 0.1 M $LiClO_4$ acetonitrile solution, generating 0.0685 ± 0.0005 mg poly(SNS). Reprinted with permission from ref. 123: *J. Electroanal. Chem.*, 1994, **370**, 231–239. Copyright (1994) Elsevier.

The oxidized film, constituted by oxidized chains and anions, is insoluble and, during reduction, the neutral oligomers solubilize in the electrolyte.[122] The electrodissolution is a faradaic process (Figure 3.10): the solved mass is a linear function of the involved charge.[123,124] From the attained solutions, the oligomers are faradically precipitated by oxidation, resulting in oxidized films.[125] These processes mimic metal electrodissolution and electrodeposition.[126] Looking at the large technological applications of the metal processes, it is surprising that the parallel polymeric processes are being ignored by today's technologies. One of the reasons for this could be that the literature uses alternative terms to identify electropolymerization (change of the polymer chain length and weight) and electrodeposition (no change) concepts.

3.20 Incorporation of Different Material Nanoparticles

Moreover, when the solution used for the polymerization process includes solvent, salt, monomer and particles of different materials or structures in suspension, these particles can also be adsorbed and incorporated into the electrogenerated polymer film, resulting in a polymeric composite. The procedure is used to incorporate carbon nanotubes, graphenes, metal nanoparticles, inorganic materials, organic materials and so on. The final aim is to provide a conducting polymer with some mechanical, optical, chemical, catalytic, electrical or electrochemical properties that hold the characteristics of the added nanoparticle.

3.21 Polymerization Mechanism

In summary, the flow of an anodic current through the monomeric solution (including a salt) initiates a fast monomeric oxidation–polymerization process to generate a mixed material of electroactive and non-electroactive fractions through a complex mechanism (Scheme 3.1), including several of the following steps:[90,127]

- Monomeric oxidation and polycondensation of the radical cations.
- Oxidation of the electrogenerated polymer with incorporation of balancing counterions.
- A partial polymeric degradation (branching, cross-linking, attack of radicals produced by solvent discharge on the polymeric radical ions).
- A parallel chemical polymerization in protophobic media initiated by the protonated monomer around the electrode with partial adsorption of the generated chains in the growing film.
- Parallel adsorption of large ions before and during the electropolymerization, giving blends or hybrid materials with a higher content than that required for charge balance.
- Parallel or consecutive electrodissolution of the electrogenerated oligomers.
- Parallel adsorption and incorporation of different nanoparticles suspended in the polymerization solution to generate conducting polymer composite films.

3.22 General Comments

The electropolymerization or electrogeneration of conducting polymers is a fast process that occurs *via* a complex mechanism (beyond the generally accepted polycondensation of radical cations) to form mixed materials of electroactive and non-electroactive (degraded, branched, cross-linked, adsorbed) fractions. The composition of the final material is the result of the relative reaction rates, followed by the different simultaneous processes (Scheme 3.1). The "best" conducting polymer is the one with the most suitable mechanical, optical and electrochemical properties required by the specific device of interest. Although it should be noted that the best conducting polymer doesn't exist, only the most appropriate (fulfilling the specific designed properties) for a defined application.

With this aim, the electro-synthesis conditions, the selected monomer (basic, substituted, self-doped, dimer or trimer from different monomers) and the selection of the background electrolyte (solvent, salt constituted by small ions, organic macroions, inorganic macroions, ionic liquid) in the presence, or not, of suspended particles or nanoparticles (carbon nanotubes, graphenes, fullerenes, metal particles, inorganic particles, plastics and so on) and additives (if required) to avoid parallel

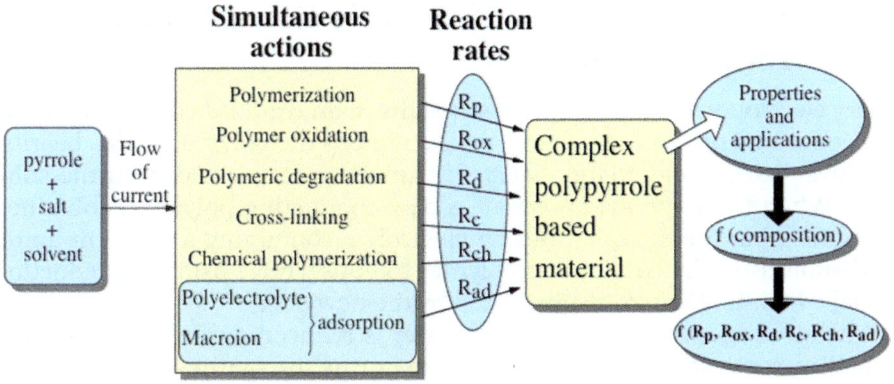

Scheme 3.1 The flow of an anodic current through a monomeric solution triggers several simultaneous reactions, the relative reaction rates of which define the final composition of the generated material, its properties and its applications.

chemical polymerization or to retard polymeric degradation must be carefully designed to produce the material with the best properties for the designed application. The design of the best conducting polymer required to produce a defined device starts by designing the best electro-synthesis conditions. A good polymer scientist, specialized in polymerization and/or electropolymerization kinetics, is an excellent investment for any laboratory interested in producing the best electrochemical devices, such as those described in Chapters 8 and 9. The composition, structure and properties of the electrogenerated material are a function, which is still not well quantified, of the synthesis conditions.

The water content in acetonitrile solvents changes between 0 and 1% (w/w) according to the laboratory's humidity level after the dry acetonitrile container is opened to the air for the first time. When used to electrogenerate polythiophene, the thiophene oxidation overlaps with the water oxidation potential, promoting partial degradation of the electrogenerated polymer.[128] This fact explains why the conductivity of polythiophene films generated during the 1980s in different laboratories spread across six orders of magnitude.[92]

In conclusion, even after 35 years of history, we still have a poor understanding of how the different chemical and physical variables of synthesis influence in a quantitative way the final composition and properties of the generated materials (*i.e.*, conducting polymers). One of the probable reasons for this is the low interest of both scientists and journal editors in empirical and theoretical chemical kinetics, including mechanistic studies: chemical kinetics is out of chemical fashion these days. Full empirical kinetic studies of complex reactions require laborious, lengthy and consuming work, and manuscripts in this area are often confined to low impact journals.

3.23 Synthesis of New Polymeric Compounds by Ionic Substitution

Every electropolymerization process results in an oxidized compound of the concomitant conducting polymer constituted by the polymer chains bearing positive charges and charge balancing anions incorporated from the solution. When these anions are small, a new conducting polymer is obtained by translating the material into an electrolyte containing a different anion and subjecting the conducting polymer to consecutive oxidation/reduction cycles: the initial anion is substituted by the new one.

The electrogenerated, oxidized polymer is reduced during the first cycle, expelling some counterions incorporated during the synthesis:

$$(\mathrm{Pol}^{n+})(\mathrm{A}^-)_n + ne^- \rightarrow \mathrm{Pol} + n\mathrm{A}^-. \tag{3.18}$$

Due to the great concentration of anion B^- in the new electrolyte, the concentration of newly incorporated A^- is very small. Thus, during the subsequent oxidation:

$$(\mathrm{Pol}) + (\mathrm{B}^-) \rightarrow (\mathrm{Pol}^{n+})(\mathrm{B}^-)_n + ne^-. \tag{3.19}$$

In fact, due to the reaction-driven structural process, a fraction of A^- remains trapped during the first reduction/oxidation cycle (see Chapter 6). Several cycles are required for the full removal of A^-. The final result is that, starting from conducting polymer $(\mathrm{Pol}^{n+})(\mathrm{A}^-)_n$, a new conducting polymer $(\mathrm{Pol}^{n+})(\mathrm{B}^-)_n$ is attained following a soft, clean and fast methodology.

A similar procedure can be followed to obtain different conducting polymers by exchange of cations [eqn (4.2) and (4.3) in Chapter 4].

3.24 Electropolymerization Initiated by Electrochemical Reduction

The electrochemical reduction of monomers with a high electroaffinity give radical anions that can initiate a polymerization process to generate reduced conducting polymers: reduced chains balanced by incorporated cations. Comparatively, the electropolymerization of conducting polymers initiated by electrochemical reduction of the monomer has been used quite scarcely.

Powders and films of conducting polymers have been obtained when a mercury pool or a conducting glass plate (indium tin oxide, ITO) and foil or a rod of lithium were used as the WE and a "sacrificial" CE, respectively. This technique is applicable for the preparation of other polymers, such as poly-pyridine,[129–131] poly(2,6-naphthylene),[132] poly(2,5-thienylene),[133] polyfuran,[134] polyfluorene, polycarbazole[135] or poly(*para*-phenylene vinylene).[136]

An unresolved point is that the generated reduced polymer can incorporate macro-cations inside the material during electropolymerization. In that case, a new general reaction with exchange of anions would be:

$$Pol^{n-}(MC)^{n+} + nA^{-} \rightleftarrows Pol(MC)^{n+}(A^{-})_n + ne^{-}, \quad (3.20)$$

which should be incorporated into those presented in Chapter 4.

References

1. T. F. Otero, in *Modern Aspects of Electrochemistry*, ed. R. E. White, J. O. Bockris and B. E. Conway, Springer US, New York, 1999, pp. 307–434.
2. *Electropolymerization: Concepts, Materials and Applications*, ed. S. Cosnier and A. Karyakin, Wiley Vch Verlag Gmbh, Weinheim, 1st edn, 2010.
3. B. Liu and G. C. Bazan, *Conjugated Polyelectrolytes: Fundamentals and Applications*, Wiley-VCH, Weinheim, 1st edn, 2013.
4. *Conjugated Polymers: A Practical Guide to Synthesis*, ed. K. Müllen, J. R. Reynolds and T. Masuda, Royal Society of Chemistry, Cambridge, 2013.
5. *Handbook of Conducting Polymers*, ed. T. A. Skotheim and J. Reynolds, CRC Press, New York, 3rd edn, 2006.
6. G. G. Wallace, G. M. Spinks and L. A. P. Kane-Maguire, *Conductive Electroactive Polymers: Intelligent Polymer Systems*, CRC Press, London, 3rd edn, 2008.
7. T. F. Otero, *Polym. Rev.*, 2013, **53**, 311–351.
8. T. F. Otero and J. M. Sansinena, *J. Electroanal. Chem.*, 1996, **412**, 109–116.
9. J. Baudoin, F. Chao and M. Costa, *J. Chim. Phys. Phys.-Chim. Biol.*, 1990, **87**, 1125–1141.
10. T. Otero and E. Delarretaazelain, *Polym. Commun.*, 1988, **29**, 21–24.
11. K. Kanazawa, A. Diaz, W. Gill, P. Grant, G. Street and G. Gardini, *Synth. Met.*, 1980, **1**, 329–336.
12. A. Diaz, J. Vallejo and A. Duran, *IBM J. Res. Dev.*, 1981, **25**, 42–50.
13. T. F. Otero, J. Arias-Pardilla, H. Herrera, J. Luis Segura and C. Seoane, *Phys. Chem. Chem. Phys.*, 2011, **13**, 16513–16515.
14. J. Tanguy, G. Deniau, G. Zalczer and G. Lecayon, *J. Electroanal. Chem.*, 1996, **417**, 175–184.
15. J. Arias-Pardilla, P. A. Gimenez-Gomez, A. de la Pena, J. L. Segura and T. F. Otero, *J. Mater. Chem.*, 2012, **22**, 4944–4952.
16. A. Diaz, J. Castillo, J. Logan and W. Lee, *J. Electroanal. Chem.*, 1981, **129**, 115–132.
17. F. Li and W. J. Albery, *Electrochim. Acta*, 1992, **37**, 393–401.
18. R. Córdova, M. A. del Valle, A. Arratia, H. Gómez and R. Schrebler, *J. Electroanal. Chem.*, 1994, **377**, 75–83.
19. K. Bade, V. Tsakova and J. W. Schultze, *Electrochim. Acta*, 1992, **37**, 2255–2261.
20. B. J. Hwang, R. Santhanam and Y.-L. Lin, *J. Electrochem. Soc.*, 2000, **147**, 2252.

21. H. Randriamahazaka, V. Noël and C. Chevrot, *J. Electroanal. Chem.*, 1999, **472**, 103–111.
22. J. P. Soto, F. R. Díaz, M. A. del Valle, J. H. Vélez and G. A. East, *Appl. Surf. Sci.*, 2008, **254**, 3489–3496.
23. T. F. Otero, *Polym. Rev.*, 2013, **53**, 311–351.
24. T. Otero, E. Delarreta and R. Tejada, *Makromol. Chem., Macromol. Symp.*, 1988, **20–21**, 615–622.
25. T. Otero and E. Delarretaazelain, *J. Chim. Phys. Phys.-Chim. Biol.*, 1989, **86**, 131–141.
26. T. Otero and E. Delarreta, *Synth. Met.*, 1988, **26**, 79–88.
27. T. Otero and J. Rodriguez, *Synth. Met.*, 1993, **55**, 1436–1440.
28. K. K. Kanazawa, A. F. Diaz, R. H. Geiss, W. D. Gill, J. F. Kwak, J. A. Logan, J. F. Rabolt and G. B. Street, *J. Chem. Soc., Chem. Commun.*, 1979, 854.
29. A. F. Diaz, K. K. Kanazawa and G. P. Gardini, *J. Chem. Soc., Chem. Commun.*, 1979, 635.
30. R. E. Noftle and D. Pletcher, *J. Electroanal. Chem. Interfacial Electrochem.*, 1987, **227**, 229–235.
31. G. G. Wallace, P. R. Teasdale, G. M. Spinks and L. A. P. Kane-Maguire, *Conductive Electroactive Polymers: Intelligent Materials Systems*, CRC Press, 2nd edn, 2002.
32. P. Chandrasekhar, *Conducting Polymers, Fundamentals and Applications: A Practical Approach*, Springer Science & Business Media, 2013.
33. T. A. Skotheim and J. Reynolds, *Conjugated Polymers: Processing and Applications*, CRC Press, 3rd edn, 2006.
34. *Handbook of Organic Conductive Molecules and Polymers, Four Volume Set*, ed. H. S. Nalwa, Wiley, Chichester, New York, 1st edn, 1997.
35. F. Carpi and E. Smela, *Biomedical Applications of Electroactive Polymer Actuators*, John Wiley & Sons, Chichester, 1st edn, 2009.
36. R. H. Baughman, L. W. Shacklette, R. L. Elsenbaumer, E. Plichta and C. Becht, in *Conjugated Polymeric Materials: Opportunities in Electronics, Optoelectronics, and Molecular Electronics*, ed. J. L. Brédas and R. R. Chance, Springer Netherlands, Dordrecht, 1990, pp. 559–582.
37. J. Jang, in *Emissive Materials: Nanomaterials*, Springer-Verlag Berlin, Berlin, 2006, vol. 199, pp. 189–259.
38. G. Inzelt, *Conducting Polymers*, Springer Berlin Heidelberg, Berlin, Heidelberg, 2012.
39. J. Przyluski, *Conducting Polymers: Electrochemistry*, Sci-Tech Publications, 1991.
40. R. Azooz, S. M. Sayyah and S. S. Abd-ElRehem, *Conducting Polymers: Electropolymerization, Kinetics, Characterizations, Corrosion Inhibitors*, LAP LAMBERT Academic Publishing, 2011.
41. *Conducting Polymers: Synthesis, Properties and Applications*, ed. L. C. P. Almeida, Nova Science Pub Inc., New York, 2013.
42. H. Kiess, *Conjugated Conducting Polymers*, Springer Science & Business Media, 2012.

43. *Conjugated Polymers: A Practical Guide to Synthesis*, ed. K. Müllen, J. R. Reynolds and T. Masuda, Royal Society of Chemistry, Cambridge, 2013.
44. K. Loganathan, *Electrochemical Studies of Conducting Polymers*, Memorial University of Newfoundland, Canada, 2005.
45. *Electropolymerization: Concepts, Materials and Applications*, ed. S. Cosnier and A. Karyakin, Wiley Vch Verlag Gmbh, Weinheim, 1st edn, 2010.
46. T. F. Otero and E. De Larreta, *J. Electroanal. Chem. Interfacial Electrochem.*, 1988, **244**, 311–318.
47. D. Fichou, G. Horowitz, B. Xu and F. Garnier, *Synth. Met.*, 1990, **39**, 243–259.
48. G. Zotti and G. Schiavon, *Synth. Met.*, 1990, **39**, 183–190.
49. R. J. Waltman, J. Bargon and A. F. Diaz, *J. Phys. Chem.*, 1983, **87**, 1459–1463.
50. M. Akimoto, Y. Furukawa, H. Takeuchi, I. Harada, Y. Soma and M. Soma, *Synth. Met.*, 1986, **15**, 353–360.
51. K. Imanishi, M. Satoh, Y. Yasuda, R. Tsushima and S. Aoki, *J. Electroanal. Chem. Interfacial Electrochem.*, 1989, **260**, 469–473.
52. Y. Furukawa, M. Akimoto and I. Harada, *Synth. Met.*, 1987, **18**, 151–156.
53. T. F. Otero, P. Herrasti, P. Ocon and C. R. Alves, *Electrochim. Acta*, 1998, **43**, 1089–1100.
54. T. Otero and C. Santamaria, *Electrochim. Acta*, 1992, **37**, 297–307.
55. T. Otero and C. Santamaria, *J. Electroanal. Chem.*, 1991, **312**, 285–291.
56. T. F. Otero, in *Modern Aspects of Electrochemistry*, ed. R. E. White, J. O. Bockris and B. E. Conway, Springer US, New York, 1999, pp. 307–434.
57. T. Otero and E. Angulo, *J. Appl. Electrochem.*, 1992, **22**, 369–375.
58. T. Otero and C. Santamaria, *Synth. Met.*, 1992, **51**, 313–319.
59. T. Otero and E. Delarretaazelain, *Polymer*, 1988, **29**, 1522–1527.
60. T. F. Otero and F. D'Eramo, *J. New Mater. Electrochem. Syst.*, 2001, **4**, 51–60.
61. S. A. Cheng and T. F. Otero, *J. Mater. Sci. Lett.*, 2001, **20**, 1055–1057.
62. C. R. Alves, P. Herrasti, P. Ocon, L. A. Avaca and T. F. Otero, *Polym. J.*, 2001, **33**, 255–262.
63. T. Otero and J. Rodriguez, *Synth. Met.*, 1992, **51**, 307–312.
64. S. A. Cheng and T. F. Otero, *Synth. Met.*, 2002, **129**, 53–59.
65. J. Carrasco, T. F. Otero, E. Brillas and M. Montilla, *J. Electroanal. Chem.*, 1996, **418**, 115–122.
66. T. Otero and M. Vazquez, *J. Electroanal. Chem.*, 1995, **397**, 171–176.
67. E. Brillas, G. Anton, T. F. Otero and J. Carrasco, *J. Electroanal. Chem.*, 1998, **445**, 125–133.
68. T. F. Otero, J. Carrasco, A. Figueras and E. Brillas, *Synth. Met.*, 1996, **83**, 193–196.
69. I. Villareal, E. Morales, T. F. Otero and J. L. Acosta, *Synth. Met.*, 2000, **108**, 57–65.
70. T. F. Otero and V. Olazabal, *Electrochim. Acta*, 1996, **41**, 213–220.

71. T. Otero, J. Rodriguez and E. Delarretaazelain, *Polymer*, 1990, **31**, 220–222.
72. I. Villarreal, E. Morales, T. F. Otero and J. L. Acosta, *Synth. Met.*, 2001, **123**, 487–492.
73. T. F. Otero, S. A. Cheng and F. Huerta, *J. Phys. Chem. B*, 2000, **104**, 10522–10527.
74. T. Otero, C. Santamaria and R. Bunting, *J. Electroanal. Chem.*, 1995, **380**, 291–294.
75. T. Otero and A. Arevalo, *J. Braz. Chem. Soc.*, 1994, **5**, 183–189.
76. T. F. Otero, I. Cantero and B. Azcano, *Int. J. Hydrogen Energy*, 2000, **25**, 221–233.
77. T. Otero and J. Rodriguez, *Electrochim. Acta*, 1994, **39**, 245–253.
78. T. Otero and J. Rodriguez, *Synth. Met.*, 1993, **55**, 1418–1423.
79. T. Otero and A. Arevalo, *Synth. Met.*, 1994, **66**, 25–32.
80. T. F. Otero and M. J. GonzalezTejera, *J. Electroanal. Chem.*, 1996, **410**, 69–77.
81. T. F. Otero and M. J. GonzalezTejera, *J. Electroanal. Chem.*, 1997, **429**, 19–25.
82. T. F. Otero and O. Roig, Polypyrrole-poly(styrene-alt-maleic acid) Blend: Electrogeneration and Kinetic Study for a Tailored Synthesis, in *Proceedings of the MRS*, 1999, vol. 98.
83. T. F. Otero, S. Villanueva, E. Brillas and J. Carrasco, in *Electroactive Polymers (EAP)*, ed. Q. M. Zhang, T. Furukawa, Y. BarCohen and Y. Scheinbeim, 2000, vol. 600, pp. 167–172.
84. F. D. D'eramo and T. F. Otero, *Bol. Soc. Esp. Ceram. Vidrio*, 2000, **39**, 377–381.
85. J. Arias-Pardilla, T. F. Otero, R. Blanco and J. L. Segura, *Electrochim. Acta*, 2010, **55**, 1535–1542.
86. T. F. Otero, C. Santamaria, E. Angulo and J. Rodriguez, *Synth. Met.*, 1991, **43**, 2947–2952.
87. T. F. Otero and M. Bengoechea, *Electrochim. Acta*, 1996, **41**, 1871–1876.
88. J. M. Sansinena, J. B. Gao and H. L. Wang, *Adv. Funct. Mater.*, 2003, **13**, 703–709.
89. T. F. Otero and I. Boyano, *Electrochim. Acta*, 2006, **51**, 6238–6242.
90. T. Otero and J. Rodriguez, *Synth. Met.*, 1992, **51**, 307–312.
91. T. Otero and J. Rodriguez, *Synth. Met.*, 1993, **55**, 1418–1423.
92. T. Otero and J. Rodriguez, *J. Electroanal. Chem.*, 1991, **310**, 219–237.
93. H. Tang, A. Kitani and M. Shiotani, *Electrochim. Acta*, 1996, **41**, 1561–1567.
94. D. E. Stilwell, *J. Electrochem. Soc.*, 1988, **135**, 2497.
95. G. Zotti and G. Schiavon, *Synth. Met.*, 1990, **39**, 183–190.
96. Y. Takenaka, T. Koike, T. Oka, M. Tanahashi, K.-I. Kaneto and K. Yoshino, *Synth. Met.*, 1987, **18**, 207–212.
97. X. Lu, H. Y. Ng, J. Xu and C. He, *Synth. Met.*, 2002, **128**, 167–178.
98. A. Kabumoto, K. Shinozaki, K. Watanabe and N. Nishikawa, *Synth. Met.*, 1988, **26**, 349–355.

99. D. E. Stilwell, *J. Electrochem. Soc.*, 1989, **136**, 688.
100. S. Ghosh, G. A. Bowmaker, R. P. Cooney and J. M. Seakins, *Synth. Met.*, 1998, **95**, 63–67.
101. L. A. Samuelson and M. A. Druy, *Macromolecules*, 1986, **19**, 824–828.
102. A. Punning, I. Must, I. Poldsalu, V. Vunder, R. Temmer, K. Kruusamae, F. Kaasik, J. Torop, P. Rinne, T. Lulla, U. Johanson, T. Tamm and A. Aabloo, *J. Intell. Mater. Syst. Struct.*, 2014, **25**, 2267–2275.
103. H. N. Dinh, J. Ding, S. J. Xia and V. I. Birss, *J. Electroanal. Chem.*, 1998, **459**, 45–56.
104. T. Otero, E. Angulo, J. Rodriguez and C. Santamaria, *J. Electroanal. Chem.*, 1992, **341**, 369–375.
105. T. Otero and J. Rodriguez, *J. Electroanal. Chem.*, 1994, **379**, 513–516.
106. G. A. Snook, A. I. Bhatt, M. E. Abdelhamid and A. S. Best, *Aust. J. Chem.*, 2012, **65**, 1513–1522.
107. T. F. Otero and V. Olazabal, *Electrochim. Acta*, 1996, **41**, 213–220.
108. T. F. Otero and J. M. Sansinena, *J. Electroanal. Chem.*, 1996, **412**, 109–116.
109. T. Otero and M. Vazquez, *J. Electroanal. Chem.*, 1995, **397**, 171–176.
110. T. F. Otero and M. J. GonzalezTejera, *J. Electroanal. Chem.*, 1996, **410**, 69–77.
111. T. F. Otero and M. J. GonzalezTejera, *J. Electroanal. Chem.*, 1997, **429**, 19–25.
112. T. F. Otero and F. D'Eramo, *J. New Mater. Electrochem. Syst.*, 2001, **4**, 51–60.
113. R. L. Elsenbaumer, K. Y. Jen and R. Oboodi, *Synth. Met.*, 1986, **15**, 169–174.
114. M. Sato, S. Tanaka and K. Kaeriyama, *J. Chem. Soc., Chem. Commun.*, 1986, 873.
115. P. Chandrasekhar and R. Gumbs, *J. Electrochem. Soc.*, 1991, **138**, 1337–1346.
116. P. Bross, U. Schoberl and J. Daub, *Adv. Mater.*, 1991, **3**, 198–200.
117. M. Reghu, Y. Cao, D. Moses and A. Heeger, *Phys. Rev. B*, 1993, **47**, 1758–1764.
118. M. Fukuda, K. Sawada and K. Yoshino, *J. Polym. Sci., Part A: Polym. Chem.*, 1993, **31**, 2465–2471.
119. G. G. McLeod, M. G. B. Mahboubian-Jones, R. A. Pethrick, S. D. Watson, N. D. Truong, J. C. Galin and J. François, *Polymer*, 1986, **27**, 455–458.
120. J. Ferraris and T. Hanlon, *Polymer*, 1989, **30**, 1319–1327.
121. S. Villanueva, T. F. Otero, E. Brillas and J. Carrasco, *Bol. Soc. Esp. Ceram. Vidrio*, 2000, **39**, 382–385.
122. T. F. Otero, S. Villanueva, E. Brillas and J. Carrasco, *Synth. Met.*, 1999, **102**, 1402–1403.
123. T. Otero, J. Carrasco, A. Figueras and E. Brillas, *J. Electroanal. Chem.*, 1994, **370**, 231–239.
124. J. Carrasco, A. Figueras, T. Otero and E. Brillas, *Synth. Met.*, 1993, **61**, 253–258.

125. E. Brillas, J. Carrasco, A. Figueras, F. Urpi and T. Otero, *J. Electroanal. Chem.*, 1995, **392**, 55–61.
126. T. F. Otero, S. Villanueva, E. Brillas and J. Carrasco, *Acta Polym.*, 1998, **49**, 433–438.
127. R. E. White, J. O. 'M. Bockris and B. E. Conway, *Modern Aspects of Electrochemistry*, Kluwer Academic/Plenum Publisher, NY, 1999, vol. 33.
128. T. Otero, J. Rodriguez, E. Angulo and C. Santamaria, *Synth. Met.*, 1991, **43**, 2831–2836.
129. G. Schiavon, G. Zotti and G. Bontempelli, *J. Electroanal. Chem. Interfacial Electrochem.*, 1985, **194**, 327–338.
130. G. Schiavon, G. Zotti, G. Bontempelli and F. Lo Coco, *J. Electroanal. Chem. Interfacial Electrochem.*, 1988, **242**, 131–142.
131. G. Schiavon, G. Zotti, G. Bontempelli and F. Lo Coco, *Synth. Met.*, 1988, **25**, 365–373.
132. G. Schiavon, G. Zotti and G. Bontempelli, *J. Electroanal. Chem. Interfacial Electrochem.*, 1984, **161**, 323–335.
133. G. Zotti and G. Schiavon, *J. Electroanal. Chem. Interfacial Electrochem.*, 1984, **163**, 385–388.
134. G. Zotti, G. Schiavon, N. Comisso, A. Berlin and G. Pagani, *Synth. Met.*, 1990, **36**, 337–351.
135. E. Ngbilo, D. Ades, C. Chevrot and A. Siove, *Polym. Bull.*, 1990, **24**, 17–22.
136. R. Tomat, S. Zecchin, G. Schiavon and G. Zotti, *J. Electroanal. Chem. Interfacial Electrochem.*, 1988, **252**, 215–219.

CHAPTER 4

Gel Membrane Electrodes: Electrochemical Reactions

4.1 Introduction

The different families of conducting polymers can be oxidized (p-doping) or reduced (n-doping) from the neutral state of the chains following different reactions, which will be described here. Each reaction results in radical anions or radical cations (polarons) in the chains, driving simultaneous ionic exchanges with the electrolyte required for charge balance inside the conducting polymer. The presence of ions inside the polymer forces solvent exchange for osmotic balance. The reversibility of every electrochemical reaction results in the reversible volume variation of the material in order to lodge/expel balancing counterions and solvent: the reaction drives reversible swelling and shrinking structural changes. Each electrochemical reaction is a faradaic process: the counterion composition in the polymer changes in a continuous and reversible way under control of the consumed charge, which means that it's under reaction control and conducting polymers can be described as giant non-stoichiometric compounds. The use of different electrochemical methodologies allows the identification of the different reactions followed by the different families of conducting polymers and, for every reaction, the identification of the basic reaction-driven structural processes, including conformational relaxation, swelling, shrinking and conformational compaction. Like biochemical reactions, these electrochemical reactions drive the conformational changes of the reacting chains and the structural changes of the reacting material.

In electrochemical processes, the selected electrodes can act in two different ways: as supports of electron transfer to/from molecules or ions present in solution or as part of the reaction in the form of polymer reactants.

RSC Smart Materials No. 19
Conducting Polymers: Bioinspired Intelligent Materials and Devices
By Toribio Fernández Otero

Published by the Royal Society of Chemistry, www.rsc.org

4.1.1 Inert and Reactive Electrodes

In electrochemical processes, working electrodes and counter-electrodes can be selected from:

- *Inert electrodes*: metals or semiconductors (noble metals, other metals, carbon-based materials or semiconducting inorganic or organic materials) that are inactive in the studied potential range of the used electrolyte. The electrode supports the electronic transfer at the electrode/electrolyte interface towards or from the reactants present in the electrolyte. For most electrochemical reactions, the reaction products are also solved in the electrolyte at the end of the reaction. When the reaction product is deposited on the electrode during the reaction, the process is an electrogeneration–deposition (metals, semiconductors, conducting polymers).
- *Electroactive electrodes*: the electrode material supports the current flow and participates (its atoms or molecules are one of the reactants) in the electrochemical reaction: metal corrosion, redox metal oxides or metal compounds from battery electrodes, or redox conducting polymers studied here.

Thus, for the study of conducting polymers, inert metals or semiconducting (indium tin oxide, ITO) electrodes must be selected for their electropolymerization–deposition from the monomeric solutions. After electrogeneration, the same inert electrode coated with the conducting polymer film can be used as the new working electrode for control of the conducting film electroactivity (the reversible film oxidation/reduction). The film oxidation/reduction must occur without any interference from the electrolyte (solvent or ions) discharge on the metal.

Thick electrogenerated films can also be peeled off from the inert metal or semiconducting electrode and used as self-supported electrodes for the characterization of their electroactivity (oxidation/reduction). The use of self-supported polymeric electrodes eliminates any electrochemical reaction from the electrolyte components on the metal occurring in parallel to the polymeric reactions.

4.2 Conducting Polymers as Electroactive Electrodes

After electropolymerization, a conducting polymer film coats the metal electrode. The electrode coated with a thin film of conducting polymer is rinsed after electropolymerization with the same solvent used as the checking electrolyte and transferred inside a fresh electrolyte for checking of the electrochemical (oxidation/reduction) behavior of the generated film. The checking solution can be the same background electrolyte (without monomer) used during the electropolymerization or different (*e.g.*, the same solvent and a different salt, a different solvent and the same salt, or a different solvent and a different salt). It can also be an ionic liquid (without solvent), a gel electrolyte or a solid electrolyte.

When the electrogenerated film is several microns thick and its adherence to the metal electrode is low enough, the polymer can be scraped from the metal electrode borders and two films, one coming from each metal side plate, can be peeled off from the metal. These films, or longitudinal strips cut from them, can be used as self-supported electrodes for electrochemical characterization.[1–18]

In order to obtain good freestanding films by electropolymerization, stainless steel electrodes give better results than gold or platinum electrodes. The low adherence behavior of steel is improved after treatment at high temperature or in an oxidizing media (such as concentrated hot nitric acid): a thick, compact, stable and conducting layer of metal oxide (the passive layer) is formed. This layer supports the electropolymerization process, yielding low adherence of the electrogenerated film. The use of Au, Pt or freshly polished stainless steel electrodes gives, by electropolymerization, quite adhesive films that break and fissure while being peeled off from the metal electrode.[19]

Whatever the conducting polymer electrode, whether it is coating a metal or prepared as a self-supported film, the checking solution and the polymer must be subjected to several oxidation/reduction cycles before steady state responses are attained (erasing any structural or chemical material memory; see Section 6.5).

4.3 Electrochemical Reactions

Conducting polymers are used as working electrodes in three electrode electrochemical cells containing an electrolyte. The working electrode is a conducting polymer film coating an inert metal or inert semiconducting electrode, or a self-supported film of any conducting polymer (Figure 4.1). The polymer can then be oxidized (p-doping) or reduced (n-doping) by extraction or injection of *n* electrons from or to, respectively, each of its constituent polymer chains using different electrochemical methodologies (Chapter 2). Each electron extracted/injected from/to a neutral film generates a radical cation or a radical anion, respectively, conjugated structure (polarons in physical terminology) involving several (2–4) monomeric units (Figure 1.1b) of a film chain. These positive or negative charges in the chains can only emerge with the simultaneous entrance in the film of counterions (anions or cations, respectively) from solution, which are required for charge balance. That means that the conducting polymer film only reacts if the extraction/injection of electrons and the exchange of charge balance ions with the electrolyte occur simultaneously. The presence of charge balance ions inside any polymeric film requires the subsequent entrance of solvent from the solution in order to balance the osmotic pressure

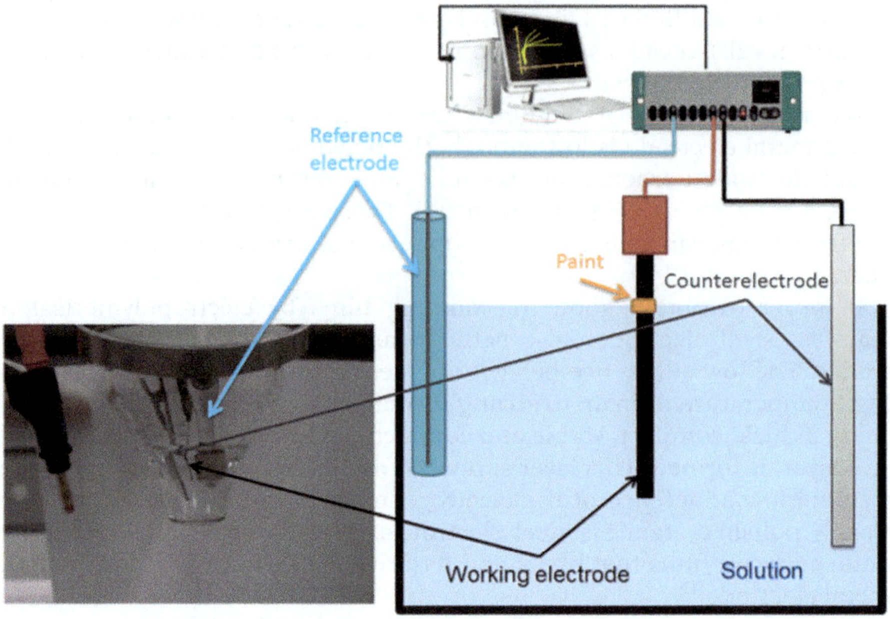

Figure 4.1 Scheme (right) and picture (left) of a three electrode electrochemical cell: the working electrode is a self-supported conducting polymer film and the transversal paint strip avoids the electrolyte ascension by capillarity until the metal clamp electrical contact.

(Section 8.2.9). The polymer film becomes a dense membrane (gel) electrode constituted by a polyelectrolyte, *i.e.*, polymer chains bearing charges, counterions and solvent.

Under these circumstances, different basic oxidation (also named p-doping) or reduction (n-doping) reactions with exchange of different ions have been identified from the different families of conducting polymers (Section 1.5.1).[20–23] Considering that the conducting polymer is the only electroactive component (hybrid conducting polymers can also include electroactive inorganic macroions) and using water as a solvent, the four basic characteristic reactions at the moment are:

- *Oxidation/reduction (p-doping/p-dedoping) with exchange of anions (A^-)*: for basic conducting polymers and copolymers synthesized in the presence of small anions and characterized in the presence of another small anion A^-:

$$\mathrm{Pol} + nA^- + mH_2O \rightleftarrows [\mathrm{Pol}^{n+}(A^-)_n(H_2O)_m]_{\mathrm{gel}} + ne^-, \tag{4.1}$$

  where Pol represents any polymeric chain in the film. The material swells by oxidation to lodge anions and solvent, and shrinks by reduction to expel anions and water towards the solution.

- *Oxidation/reduction (p-doping/p-dedoping) with exchange of cations (C^+)*: followed by polymer (Pol) blends or hybrids (conducting polymers synthesized in the presence of organic or inorganic macro-anions, MA^{n-}):

$$[(\mathrm{Pol})(\mathrm{MA}^{n-})(\mathrm{C}^+)_n(\mathrm{H_2O})_m]_{\mathrm{gel}} \rightleftarrows (\mathrm{Pol}^{n+})(\mathrm{MA}^{n-}) + n\mathrm{C}^+ + m\mathrm{H_2O} + ne^-. \quad (4.2)$$

The material shrinks by oxidation, expelling cations and water towards the solution, and swells by reduction to lodge cations and water.

- *Reduction/oxidation (n-doping/n-dedoping) with exchange of cations (C^+)*: some families of conducting polymers have electronic affinities high enough to accept electrons in the potential range of most organic electrolytes following the reaction in eqn (4.3):

$$\mathrm{Pol} + n\mathrm{C}^+ + m\mathrm{H_2O} + ne^- \rightleftarrows [(\mathrm{Pol}^{n-})(\mathrm{C}^+)_n(\mathrm{H_2O})_m]_{\mathrm{gel}}. \quad (4.3)$$

The material shrinks by oxidation and swells by reduction.

- *Oxidation/reduction with exchange of protons (H^+)*: polyaniline (PANI) and polyaniline derivatives before the exchange of anions experience a reversible exchange of protons:

$$\mathrm{PANIH}_n \rightleftharpoons \mathrm{PANI} + n\mathrm{H}^+ + ne^-. \quad (4.4)$$

4.4 Some Considerations Related to Conducting Polymer Reactions

One of these reactions usually prevails during the electrochemical reaction from each conducting polymer family. The polyaniline family follows the reactions shown in eqn (4.1) and (4.4). Low band-gap materials, as those from the polythiophene family, follow the reactions shown in eqn (4.1) and (4.3) in different potential ranges.

These reversible oxidation/reduction reactions, in particular from eqn (4.1) to (4.3), induce large and reversible structural volume variations,[19,24–31] constituting the first basic material and reaction models for the characterization of reaction-driven structural changes and for the study of reaction-driven structural chemical kinetics.[22,32,33]

Some basic methodologies are required in order to determine which reaction (or reactions) and which reaction-driven structural changes are present and in which potential range during polymeric reactions.

Hereinafter, unless otherwise stated, eqn (4.1) will be selected for the presentation and discussion of the basic reaction-driven structural changes, the reaction-driven variation of the polymer composition, the reaction chemical kinetics under structural influence, the biomimetic composition-dependent material properties and the reaction-driven actuation of the new electro-chemo-biomimetic devices based on each of the composition-dependent properties. Concepts, properties and devices here presented can be easily translated to other conducting polymers and materials following the reactions shown in eqn (4.2), (4.3) or (4.4).

In those reactions, water cannot be considered as a reactant: its chemical structure is the same whatever side of the reaction it is on. Water exchange is driven by osmotic pressure under a physical law, with the osmotic pressure proportional to the ionic concentration inside the film (Section 8.2.9). Given that this ionic concentration is controlled by each of the above-presented faradaic reactions, the entrance or expulsion of water must follow the faradaic entrance or exit, respectively, of the balancing counterions. Consequently, the membrane osmotic pressure and the concomitant water exchange are expected to occur under quantitative reaction control.

4.5 Giant Non-Stoichiometry: Transfer of Consecutive Electrons and Continuous Polymer/Ion Composition Evolution

The n electrons exchanged per polymeric chain during any of the reactions shown in eqn (4.1)–(4.4) are not equivalent. Each polymeric chain has a first, second, third...nth ionization potential. By assuming a mono-disperse polymer film (the m constituent chains have the same length constituted by the same number of monomeric units), only after extraction of the first electron from the last (mth) chain can one of the film chains lose its second electron. The chain oxidation (not considering, for simplicity, solvent exchange) can be decomposed into n consecutive equilibrium states:

$$\mathrm{Pol} + \mathrm{A}^- \rightleftharpoons \mathrm{Pol}^{\bullet+}\mathrm{A}^- + \mathrm{e}^- \qquad \Rightarrow E_1, E_{01} \tag{4.5}$$

$$\mathrm{Pol}^{\bullet+}\mathrm{A}^- + \mathrm{A}^- \rightleftharpoons \mathrm{Pol}^{2(\bullet+)}(\mathrm{A}^-)_2 + \mathrm{e}^- \qquad \Rightarrow E_2, E_{02} \tag{4.6}$$

$$\mathrm{Pol}^{2(\bullet+)}(\mathrm{A}^-)_2 + \mathrm{A}^- \rightleftharpoons \mathrm{Pol}^{3(\bullet+)}(\mathrm{A}^-)_3 + \mathrm{e}^- \qquad \Rightarrow E_3, E_{03} \tag{4.7}$$

and so on, following the rule:

$$\mathrm{Pol}^{(n-1)(\bullet+)}(\mathrm{A}^-)_{(n-1)} + \mathrm{A}^- \rightleftharpoons \mathrm{Pol}^{n(\bullet+)}(\mathrm{A}^-)_n + \mathrm{e}^- \quad \Rightarrow E_n, E_{0n} \tag{4.8}$$

To achieve completion of the material oxidation, every chain must lose n electrons through n consecutive intermediate states. Each of those intermediate states constitutes an equilibrium state at increasing equilibrium potentials: $E_1 < E_2 < E_3 <...< E_{n-1} < E_n$, where E_{01}, $E_{02}...E_{0n}$ are the standard potentials for the concomitant equilibriums. It can be assumed that the potential step between the potential required for the extraction of the last ith electron from each of the m constituent chains and the potential required to extract the first $i + 1$ electron from one of the m film chains is constant ($\Delta E_i = \Delta E$, for $1 \leq i \leq n$).

In conclusion, a mono-disperse polymeric film should pass through n consecutive oxidation steps to move from a theoretical neutral state of every constitutive chain (without any positive charge) to an oxidized state, where each chain stores n positive charges balanced by n monovalent anions. This means that the material passes through $n + 1$ stoichiometric states, for which the chain/ion (Pol/A^-) relative content moves step-by-step from 1/0 (Pol) to $1/i$ [$Pol^{i(\bullet+)}(A^-)_i$] for $0 \leq i \leq n$.

Nevertheless, actual films of conducting polymers are poly-disperse: the chain-conjugated lengths must follow a bell distribution. According to Hückel's principle, the first ionization potential (IP), the transition from the highest occupied molecular orbital (HOMO) to the lowest unoccupied molecular orbital (LUMO; Figure 4.2) and the oxidation potential decrease for chains of increasing length.[34]

In accordance with this evolution, when the oxidation of the shorter chains from a polymer film begins, the longer chains present a high oxidation state. Chains of different length present at any oxidation potential and at any oxidation time different oxidation states. In summary, the polymer/ion relative content shifts continuously through infinite non-stoichiometric intermediate stable states with the reaction time.

After consideration of the reaction-driven structural processes in Chapter 6, electrochromic films could be used to visualize this non-stoichiometric nature: any final film's composition-dependent color could be attained by

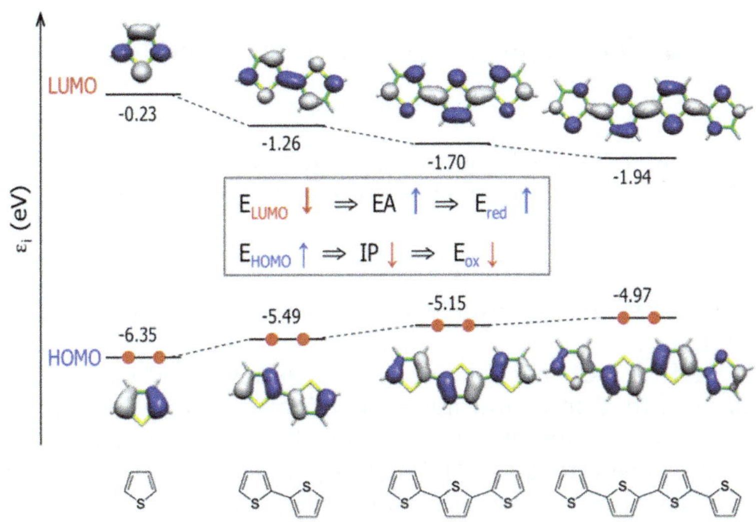

Figure 4.2 Evolution of the bonding (single bonds) and antibonding (double bonds) electron densities (ε_i), HOMO and LUMO energy levels, electron affinity (EA), ionization potential (IP), oxidation potential (E_{ox}) and reduction potential (E_{red}) with the number of monomeric units.

stopping the oxidation–nucleation at any intermediate time before nuclei coalescence.

4.6 Ionic Composition Variation with Stable Physical Integrity

The polymeric content of the film remains constant during reaction, being defined by the experimental mass of a reduced and dry film (w, in g). After flow of an oxidation charge Q (in C), the number of ions (n') that have penetrated the film from the solution driven by eqn (4.1), according to Faraday's law, is:

$$n' = Q/ze^{-}, \tag{4.9}$$

where z is the ion charge and e^{-} is the electron charge. The number of ionic equivalents exchanged by the film (n'') driven by the reaction is:

$$n'' = Q/F, \tag{4.10}$$

where F is Faraday's constant. The number of exchanged ionic moles is, according to eqn (2.2) in Chapter 2, $n = Q/zF$.

The specific (per unit of polymer mass) ionic concentration variation in the polymer film after consuming the oxidation charge Q (in M g^{-1}) can be attained from eqn 2.3 by:

$$c = Q/wzF, \tag{4.11}$$

promoting the film mass increase during oxidation ($\Delta w'$, in g) by:

$$\Delta w' = (Q/zF)(A^{-}), \tag{4.12}$$

where A^{-} is the molar weight of the exchanged ion.

In order to lodge the ions, the film volume (V) must increase by ΔV:

$$\Delta V = (Q/zF)V_{\mathrm{A}}, \tag{4.13}$$

where V_{A} gives the molar crystallographic volume of the anion incorporated by the film.

The actual ion/polymer relative weight content, $\Delta w'/w$, in the reacting film can be shifted in a continuous way from 10^{-4} to 0.4–0.6 (depending on the ionic weight) under control of the charge (eqn 4.11), consumed by the reaction shown in eqn (4.1). The ion/polymer relative weight variation during the reaction is controlled at any time, t (in s), by the driving current, I (in A):

$$\frac{\Delta w'}{w} = \frac{Q}{zF}A^{-} = \frac{It}{zF}A^{-}. \tag{4.14}$$

Such large concentration variations only can be found inside living cells during biochemical reactions (rest-to-function transition). Relative concentrations lower than 10^{-3} M are difficult to attain due to structural limitations (see Chapter 6, Section 6.4.1). Inside this composition range, the flow of infinitesimal amounts of charge (anodic or cathodic) allows infinitesimal control of the ionic film composition variation (an increase or decrease).

Scientists have never before had at their disposal a material with a composition that could be changed under perfect electrochemical control and inside such a large range whilst maintaining the material integrity. Chemical compounds are characterized by their constant composition defined by the chemical formula according to the valence (*i.e.*, Fe, $FeCl_2$, $FeCl_3$). The result is that every material has a constant composition, providing constant physical and chemical properties. These properties change dramatically when the material composition changes, *e.g.*, a shift of the mechanical properties from Fe to $Fe(OH)_2$ promotes building collapse. However, in giant non-stoichiometric conducting polymers the material composition can be shifted whilst maintaining the material's integrity in a continuous and reversible way through several orders of magnitude: a new way is open for the reversible tuning of the physical and chemical composition-dependent properties within a large range.

The past century brought an explosion of non-stoichiometric extrinsic semiconductors. These inorganic semiconductors have a crystallographic structure. The extrinsic semiconductivity is attained by doping: a new atom is lodged at the crystal interstices or one of the crystal constitutive ions is substituted, lodging in its place another one with one electron more (n-doping) or less (p-doping) than the original one. But both the crystallographic diameter and the chemical interactions of the doping agent are different from those of the original one, inducing mechanical and chemical stress in the crystal network. The crystallographic structure can only support a low concentration variation of the dopant agent (up to 0.1–0.3% w/w) before the crystal collapses. Variation of the dopant concentration promotes a shift of some material properties, such as conductivity. A relatively low concentration (and related electronic conductivity) variation has supported the parallel development and explosion of electronic, microelectronic and optoelectronic technologies.

By contraposition, conducting polymers are amorphous materials. The entangled chain structure supports the entrance and expulsion of massive amounts of doping counterions (up to 40–60% w/w), while maintaining a similar physical consistency and the polymeric material integrity in the full composition range.[23,35–40] With variation of the material's composition, most of the physical properties will also shift. In inorganic semiconductors the control of the dopant concentration was initially obtained through complex and expensive physical processes using very high electric fields and high vacuum conditions, such as ionic bombardment. Now, very low electric field (in mV) gradients in liquid solutions and under atmospheric pressure are used for reversible control of the dopant composition variation driven by reversible electrochemical reactions in conducting polymers. In parallel, the value of each composition-dependent property of the material is expected to change in a reversible way under reaction control [eqn (4.1)] in a similarly large range. New technological developments are expected based on the reversible change under reaction control of these composition-dependent properties: the new devices could work under reaction control of the constitutive materials, as occurs in biological cells and biological organs.[23,35]

Akin to organic semiconductors, conducting polymers are attaining massive use in oligomeric solutions for the development of a new generation of soft, flexible and transparent electronic and microelectronic[41–51] or optoelectronic[50,52–54] devices. The electronic, microelectronic or optoelectronic components work, as inorganic components do, under constant composition of the conducting polymer. In contrast, the astonishing giant non-stoichiometric nature of the materials and the concomitant electrochemical control of both composition and composition-dependent properties are still waiting for a parallel explosion of new chemical and electrochemical biomimetic devices and products.

Due to film volume variations during the driving reaction, the use of specific concentrations (related to the constant mass of a reference in the reduced polymeric state) is more reliable than the volumetric concentrations.

For every film and material, control of the concentration variation requires preliminary identification of the basic reactions [eqn (4.1)–(4.4)] that are involved, characterization of the related potential domains and quantification of the charge consumed to complete each reaction-driven process in the film. Considering the electrochemical nature of the involved reactions, the use of electrochemical methodologies seems the most suitable initial approach to obtain the required information from the concomitant electrochemical responses.

4.7 Electrochemical Responses from Different Methodologies

The same electrochemical methodologies used above for the film electrogeneration are used to check the electrochemical (oxidation/reduction) behavior of the electrogenerated materials. The conducting polymer electrode can be subjected to linear (anodic or cathodic) or cyclic potential sweeps, potential steps (anodic or cathodic) and consecutive square potential waves, and the flow of a constant current (anodic or cathodic) or the flow of square current waves. Sweeping frequencies of sinusoidal potential waves (impedance measurements) are also useful methodologies but are less well used for basic studies: impedance responses include the overall system information from the fastest physical processes, such as electronic orbital polarizations, to the slower chemical or physical ones, such as reaction and diffusion processes. But an equivalent circuit is required to obtain the required information and different equivalent circuits with very different components (*e.g.*, different physical and chemical meanings) give the same impedance response. In order to choose one of them, some previous knowledge of the reaction components is required: it does not appear to be the best methodology to initiate the clarification of such a complex reacting system as conducting polymers.

The concomitant voltammetric, coulovoltammetric, chronoamperometric and chronopotentiometric responses from the conducting polymer electrode will give qualitative and quantitative information about the different reactions and structural processes taking place during the film oxidation/reduction.

4.7.1 Voltammetric Responses

The selected electrolyte must have a potential window large enough to allow the exhibition of the different possible p-doping/p-dedoping and n-doping/n-dedoping processes from the studied conducting polymer, avoiding electrolyte discharge overlapping with the film reactions (Figure 4.3).[55]

When the anodic potential sweep pushes the electrode Fermi level below the polymer HOMO level, the first electron is extracted from the polymer chain, generating a positive charge on the chain. The oxidation potential of the largest polymer chain is attained and polymer oxidation (p-doping) starts [eqn (4.1), forwards]. The increase of the anodic current results in a very large maximum. For some conducting polymers, the amplitude of the p-doping maximum is around 1 V (0.75 V in Figure 4.3), corresponding to the extraction of n consecutive electrons [eqn (4.5)–(4.8)] from chains of different lengths.

If the anodic potential sweep pushes the electrode Fermi level below the highest HOMO level of one of the electrolyte components (solvent, anion or cation), the electrolyte discharge starts. Thus, before attaining the anodic

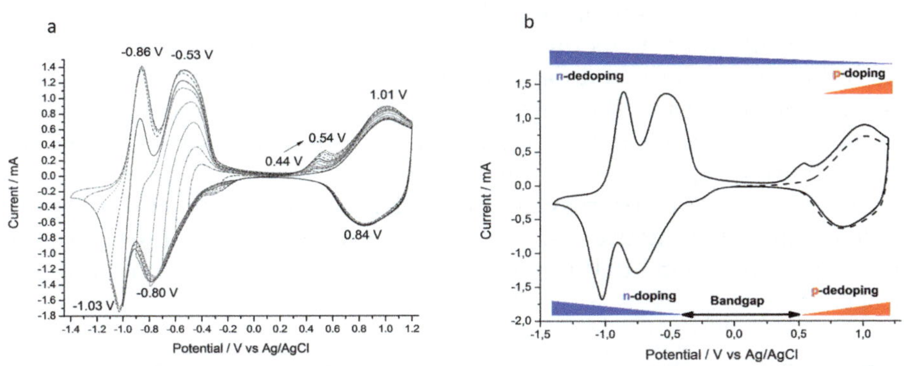

Figure 4.3 (a) Voltammetric control obtained from a poly(naphthalenediimide-terthiophene) [poly(NDI-3T)] film in 0.1 M tetrabutylammonium hexafluorophosphate ($TBAPF_6$) in an acetonitrile solution at 50 mV s^{-1} and 20 °C for different cathodic potential limits. (b) p-Doping, p-dedoping, band-gap, n-doping and n-dedoping potential ranges. Reproduced from ref. 55 with permission from the *Phys. Chem. Chem. Phys.* owner societies.

limit of the electrolyte potential window, the potential sweep is reversed in order to avoid degradation processes as a result of electrolyte discharge. At the beginning of the cathodic potential sweep, the flow of decreasing anodic currents continues (the polymer oxidation goes on), indicating the great electrochemical inertia of the material. After crossing the zero current axis, a large reduction maximum extends in a potential range close to 1 V due to polymer reduction by the consecutive injection of electrons to the polymer chains [eqn (4.1), backwards]. After the reduction maximum, the current never drops to zero and remains a low and constant cathodic current at rising cathodic potentials.

When the cathodic potential sweep shifts the Fermi level above the polymer LUMO, the first electron is transferred from the metal to the neutral polymer chains, initiating the polymer reduction [eqn (4.3), backwards], storing negative charges (n-doping) on the chains. A large reduction process involving two maxima and extending over 1 V indicates the progressive injection of consecutive electrons to the polymer chains with different conjugation lengths (polarons and polaron pairs or bipolarons).

The potential sweep is reversed before the cathodic limit of the electrolyte potential window is attained. The polymer reduction continues (the flow of decreasing cathodic currents) at the beginning of the anodic potential sweep, indicating the great electrochemical inertia of the material. Two n-dedoping anodic peaks, apparently due to the oxidation of polaronic species and polaronic pairs (or bipolaronic), evolve. After the second peak, the current never drops to zero.

Potential sweeps until different cathodic potential limits provide a good correlation between each reduction process (reduction maximum) and its oxidation maximum (Figure 4.3a).

After going into the n-doping/n-dedoping region, a new anodic peak is formed at the beginning of the p-doping process and the original p-doping peak presents higher anodic currents than when the cathodic potential limit did not reach the n-doping region. When the cathodic potential limit is less cathodic than the beginning of the n-doping process, the large p-doping peak recovers the original one-maximum structure and low current (Figure 4.3b, dashed voltammograms).

Most polypyrrole or polyaniline derivatives only present p-doping/p-dedoping voltammetric responses. Their low electronic affinities should require cathodic potentials beyond the cathodic potential limits of most typical electrolytes.

4.7.2 Coulovoltammetric Responses: Full Electrochemical Reversibility

Eqn (4.1)–(4.4) are reversible: the anodic voltammetric charges must equal the cathodic voltammetric charges. In that case, by integration of the voltammetric response, a coulovoltammetric (charge/potential, Q/E) response is attained. Most of the present potentio-galvanostats directly give both voltammetric and coulovoltammetric responses. The Q/E response is a closed loop if inside the studied potential range the conducting polymer film only experiences reversible oxidation/reduction reactions (Figure 4.4).[33] Normalized Q/E responses can be presented by taking the Q/E minimum as the charge origin. Positive charge increments refer to charges spent by the film oxidation and negative charge increments refer to the film reduction reactions. Closed Q/E loops under different chemical or electrochemical conditions unequivocally guarantee the full electrochemical reversibility of the polymer film: the film oxidation charge equals the film reduction charge. This reversibility contradicts the general belief established at the beginning of the study of conducting polymer electrochemistry[56] that the electrochemistry of conducting polymers is an irreversible process. The contradiction may be due to not taking into account the electrochemical inertial oxidation and reduction charges (see Sections 6.4.1 and 6.4.2 in Chapter 6) or due to the presence of irreversible reactions overlapping the reversible film oxidation/reduction. Coulovoltammetric responses will be used in Section 4.8 for the detection of parallel irreversible reactions.

In this way, Q/E responses can be used to clarify some difficult questions, like how the electrolyte potential window changes by using metal,

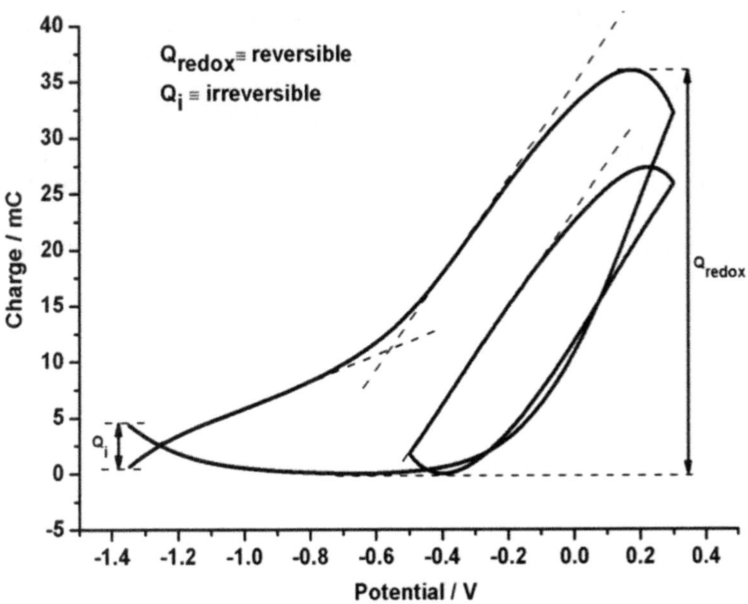

Figure 4.4 Coulovoltammetric responses from a polypyrrole film coating a Pt electrode in a 0.1 M NaCl aqueous solution at 50 mV s^{-1} from −0.5 V showing a closed loop, and from −1.35 V showing two reduction kinetics (two Q/E slopes) inside a closed loop, giving the film reduction/oxidation reversible charge (Q_{redox}); moreover, the open part gives the irreversible charge (Q_i). Reprinted with permission from ref. 33: *Adv. Funct. Mater.*, 2013, **23**, 3929–3940. Copyright (2013) John Wiley and Sons.

polymer-on-metal or self-supported polymeric electrodes, which families of conducting polymers give overlapping irreversible processes (in addition to the reversible self-supported film oxidation/reduction reactions), and in which potential range these parallel processes occur.

4.7.3 Chronoamperometric Responses

Voltammetric exploration (Figure 4.3) gives the potential range for each of the p-doping, p-dedoping, n-doping and n-dedoping peaks. The *Q*/*E* responses reveal whether, in those potential ranges, only reversible film reactions occur. Now each peak reaction rate can be explored by consecutive square potential steps and the kinetic magnitudes can be quantified from the chronoamperometric responses.

As indicated in Section 2.5, the chronoamperometric responses (Figure 4.5) give the oxidation or reduction kinetics, allowing the determination of the empirical kinetic equation for the concomitant film oxidation or reduction process. A basic principle can be stated: any physical or chemical variable from the film or from the ambient acting on the studied reaction rate will be detected by a change in the concomitant chronoamperometric response.

This principle will be used later for the study of structural reaction-induced transformations. When both the reduction and oxidation potentials are selected from inside the voltammetric p-doping maximum (Figure 4.3), the chronoamperometric response shows the usual shape described by electrochemical textbooks (Figure 4.5c, dashed line). When the reduction potential is more cathodic than the reduction maximum, the oxidation chronoamperometric response shows a maximum (Figure 4.5c, solid line) for long times and is considered an anomalous relaxation, memory or insulator/conducting transition response.[57–61]

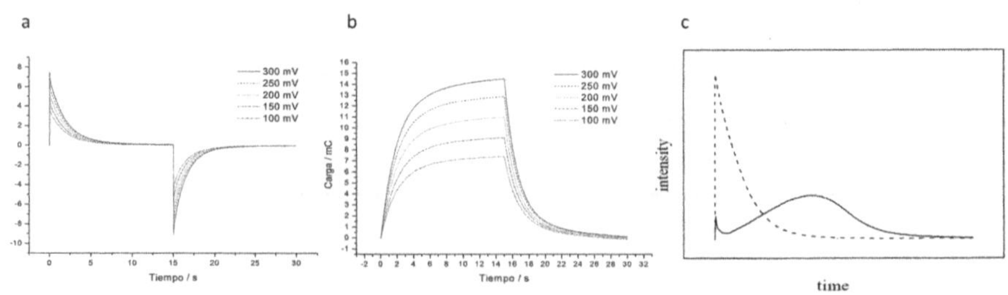

Figure 4.5 (a) Chronoamperometric and (b) chronocoulometric responses from a polypyrrole film (0.18 mg) coating a Pt electrode in a 0.1 M $NaPF_6$ acetonitrile solution subjected to potential steps from −100 mV (kept for 15 s) to different anodic potentials, as indicated on the figure (also kept for 15 s). (c) Anodic chronoamperometric responses after reduction at −100 mV (dashed line) or at −1600 mV and subsequent oxidation by potential steps to 300 mV.

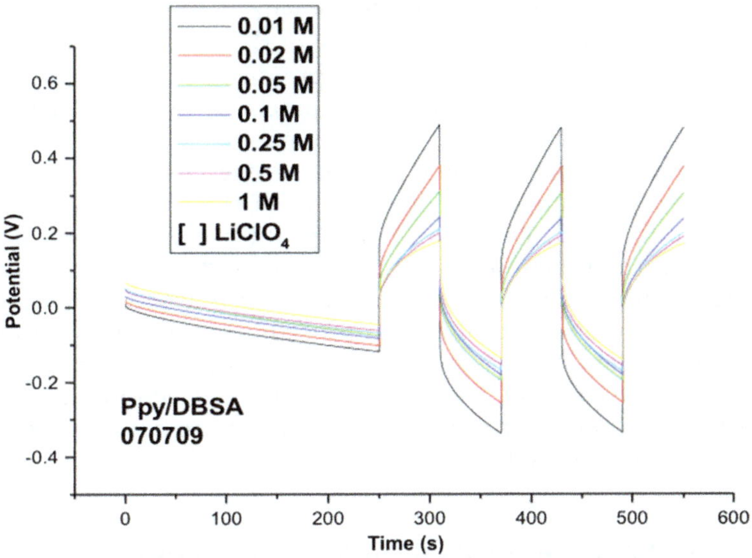

Figure 4.6 Chronopotentiometric responses obtained from a polypyrrole/dodecylbenzenesulfonic acid (Ppy/DBS) self-supported electrode in aqueous solutions with different concentrations of $LiClO_4$ when subjected to square current waves of ±0.8 mA.

4.7.4 Chronopotentiometric Responses

Any reduced conducting polymer film can be oxidized by the flow of a constant anodic current. The membrane electrode responds to the constant current flow by anodically jumping the membrane potential to meet the polymer oxidation potential (*i.e.*, a chronopotentiometric response). From there, a slow potential increase is initiated until completion of polymer oxidation (Figure 4.6).[62] Then polymer oxidation–degradation (overoxidation) begins, resulting in a faster potential increase. When the electrolyte oxidation potential is attained, a new constant potential evolution is observed.

Oxidized conducting polymer films can be reduced by the flow of a constant cathodic current. The membrane electrode responds by cathodically jumping the membrane potential to meet the polymer reduction potential (Figure 4.6). From there, a slow potential increase is initiated until complete reduction. Then the potential jumps again to meet the electrolyte reduction potential.

Consecutive oxidation/reduction processes of the membrane electrode, avoiding those regions related to polymer degradation and electrolyte discharges, are known in batteries as charge/discharge processes.

4.8 Detecting Parallel Irreversible Reactions

Long-term stability under oxidation/reduction cycling of both conducting polymers and their electrochemical devices is related to the absence of parallel irreversible reactions generating products that can result in chemical attacks on the polymer chains with subsequent material degradation. Despite the general acceptance of this statement, which is corroborated by results (for example, those presented in Figure 3.3), very few efforts have been focused on the development of methodologies able to detect the presence of irreversible reactions, the characterization of the potential domains and whether they occur overlapping with reversible material reactions, and the quantification of the fraction of consumed charge there involved. These seem significant points in order to investigate two important technological aspects: the induced material degradation kinetics and the mechanism, and these two aspects are required to develop the subsequent stabilization procedures. Keeping in mind these points, coulovoltammetric responses from conducting polymer films coating metal electrodes or self-supported membrane electrodes may be used to detect the presence of irreversible reactions and to quantify the irreversible charge there involved.

4.8.1 Parallel Irreversible Reactions from Films Coating Metal Electrodes

It is generally accepted that water electrolysis results in most of the irreversible reactions and degradation processes. In order to illustrate the identification of any irreversible reaction here, aqueous solutions will be used. When the polymer oxidation/reduction is checked by potential sweeps in aqueous solutions up to different cathodic potential limits (Figure 4.7a), closed coulovoltammetric responses (Figure 4.7b) are attained until a potential limit of −1.2 V, indicating the presence of reversible film oxidation/reduction processes only. Beyond this limit, the Q/E responses starting from the cathodic potential limit and finishing at the same potential show a closed loop and an open part (Figure 4.7a).[33] The charge gradient from the minimum to the maximum of the closed loop was consumed by the reversible redox reactions of the film. The charge gradient between the initial and the final Q/E points of the open part was consumed by the irreversible reduction reactions. As expected, this charge increases for increasing cathodic potential limits.

By selecting a cathodic potential limit avoiding these irreversible reduction reactions, the membrane redox behavior can be checked by shifting the anodic potential limit (Figure 4.7c). The potential corresponding to the beginning of the irreversible oxidation reactions is detected by the first open Q/E response (Figure 4.7d). Until that potential, only closed Q/E responses are attained. From that potential, the charge gradient at the cathodic potential open limits, corresponding to the irreversible oxidation reactions, increases for increasing potential limits. The charge involved in the reversible redox processes is the full Q/E charge gradient minus the irreversible charge.

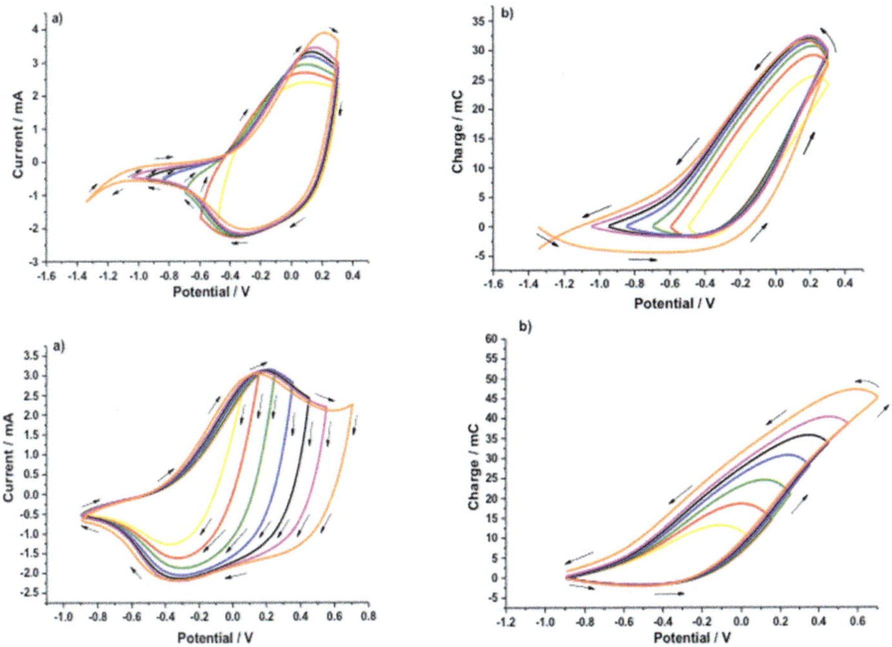

Figure 4.7 (a) Voltammograms obtained from a polypyrrole film coating a Pt electrode in a 0.1 M NaCl aqueous solution at 50 mV s^{-1} from different cathodic potential limits to 0.3 V. (b) Coulovoltammetric responses obtained by integration of the voltammograms shown in panel (a). (c) Voltammetric responses from −0.8 V up to different anodic potential limits and (d) concomitant coulovoltammetric responses. Adapted with permission from ref. 33: *Adv. Funct. Mater.*, 2013, **23**, 3929–3940. Copyright (2013) John Wiley and Sons.

4.8.2 Parallel Irreversible Reactions from Self-Supported Polymeric Electrodes

The same procedure can be used to detect the presence of irreversible reactions from self-supported film electrodes. In this case, a transversal paint strip below the metal clamp used for the electrical contact guarantees the isolation of the metal from the electrolyte contact by capillarity through the polymer.

4.8.2.1 Basic Conducting Polymers

By repeating the same experiments from Section 4.8.1 using self-supported electrodes of basic conducting polymers, such as polypyrrole, with a transversal strip painted on the dry film below the metal clamp contact to close the membrane pores, thus avoiding electrolyte contact with the clamp by

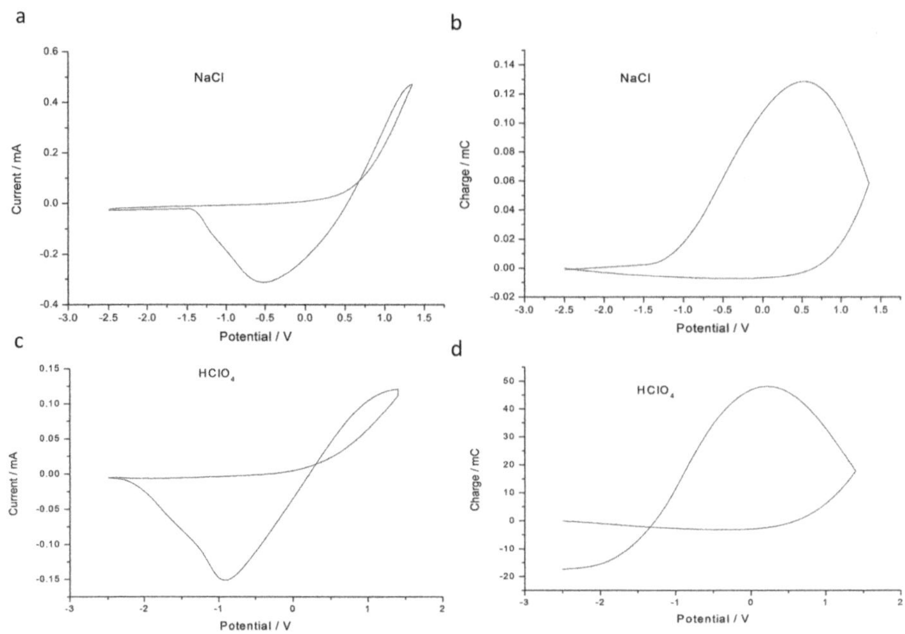

Figure 4.8 Voltammetric (a) and (b), and coulovoltammetric (c) and (d) responses of self-supported polypyrrole films in 0.1 M aqueous solutions of NaCl (a) and (c), and $HClO_4$ (b) and (d) at a constant sweep rate of 3 mV s^{-1}. Adapted with permission from ref. 18: *Adv. Funct. Mater.*, 2014, **24**, 1265–1274. Copyright (2014) John Wiley and Sons.

capillarity, the irreversible reduction reaction doesn't occur up to very high cathodic potential limits. No matter what the cathodic potential limit in aqueous solutions of NaCl, stationary voltammetric (Figure 4.8a) and closed coulovoltammetric (Figure 4.8b) responses are attained.[33,63]

The use of acid solutions, such as $HClO_4$, recovers some slow hydrogen evolution at high cathodic potentials, as shown by the open coulovoltammetric response (Figure 4.8d).[33] The original voltammogram (Figure 4.8c) does not present any evidence of the hydrogen evolution.

These results indicate that the irreversible reduction reaction starting at −1.2 V in the metal-coated electrode is due to hydrogen evolution from water taking place at the metal/polymer interface: the irreversible reaction does not exist when the back metal was eliminated. Cyclic voltammetric responses using a clean metal electrode in the same aqueous solution corroborate the flow of higher reduction currents through the uncoated electrode in the same potential range. Hydrogen evolution at the metal/electrode interface and the concomitant effects on the responses from quartz crystal microbalance requires some reconsideration: the literature used to attribute this charge to ionic exchanges with the conducting polymer.

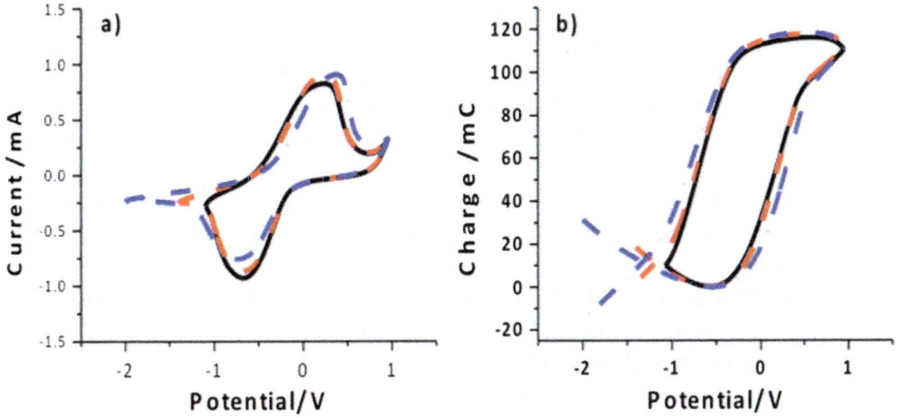

Figure 4.9 (a) Voltammetric and (b) coulovoltammetric responses of a Ppy/DBS self-supported film from different cathodic potential limits (−1, black solid line, −1.5, red dashed line and −2, blue-dashed line) in 0.1 M aqueous solutions of $LiClO_4$. Sweep rate, 6 mV s^{-1}. Reprinted with permission from ref. 18: *Adv. Funct. Mater.*, 2014, **24**, 1265–1274. Copyright (2014) John Wiley and Sons.

4.8.2.2 Polymer Blends with Organic Macroions

When the same control is repeated using a polypyrrole–organic macroion blend, such as polypyrrole/dodecylbenzenesulfonic acid (DBS) (Ppy/DBS), different electrochemical responses are attained (Figure 4.9). The observed irreversible charge was attributed to the hydrogen release from DBS at cathodic potentials and the irreversible release of protons (visualized by indicators in solution) from DBS at anodic potential limits.[33]

References

1. T. F. Otero and M. J. Ariza, *J. Phys. Chem. B*, 2003, **107**, 13954–13961.
2. H. Zhu, J. Hu, L. Zhong and F. Gana, *Russ. J. Electrochem.*, 2008, **44**, 1120–1126.
3. L. Wen and N. Kocherginsky, *Synth. Met.*, 1999, **106**, 19–27.
4. L. Zhang and Z. Zhang, *Synth. Met.*, 2011, **161**, 724–730.
5. I. Sultana, M. M. Rahman, S. Li, J. Wang, C. Wang, G. G. Wallace and H.-K. Liu, *Electrochim. Acta*, 2012, **60**, 201–205.
6. J. Xu, W. Zhou, J. Hou, S. Pu, Z. Wei and J. Wang, *Mater. Lett.*, 2006, **60**, 2569–2572.
7. X. Yan, J. Chen, J. Yang, Q. Xue and P. Miele, *ACS Appl. Mater. Interfaces*, 2010, **2**, 2521–2529.
8. H.-P. Cong, X.-C. Ren, P. Wang and S.-H. Yu, *Energy Environ. Sci.*, 2013, **6**, 1185.

9. L. Jiang, X. Lu, J. Xu, Y. Chen, G. Wan and Y. Ding, *J. Mater. Sci.: Mater. Electron.*, 2015, **26**, 747–754.
10. Z. Wang, P. Tammela, P. Zhang, J. Huo, F. Ericson, M. Strømme and L. Nyholm, *Nanoscale*, 2014, **6**, 13068–13075.
11. S. Hara, T. Zama, W. Takashima and K. Kaneto, *Synth. Met.*, 2005, **149**, 199–201.
12. P. Veerender, V. Saxena, P. Jha, S. P. Koiry, A. Gusain, S. Samanta, A. K. Chauhan, D. K. Aswal and S. K. Gupta, *Org. Electron.*, 2012, **13**, 3032–3039.
13. C. Ehrenbeck and K. Jüttner, *Electrochim. Acta*, 1996, **41**, 1815–1823.
14. I. Sultana, M. M. Rahman, J. Wang, C. Wang, G. G. Wallace and H.-K. Liu, *Solid State Ionics*, 2012, **215**, 29–35.
15. H. Huang, X. Zeng, W. Li, H. Wang, Q. Wang and Y. Yang, *J. Mater. Chem. A*, 2014, **2**, 16516–16522.
16. J. Arias-Pardilla, C. Plesse, A. Khaldi, F. Vidal, C. Chevrot and T. F. Otero, *J. Electroanal. Chem.*, 2011, **652**, 37–43.
17. J. G. Martinez and T. F. Otero, *RSC Adv.*, 2014, **4**, 29139.
18. T. F. Otero, J. G. Martinez, M. Fuchiwaki and L. Valero, *Adv. Funct. Mater.*, 2014, **24**, 1265–1274.
19. T. Otero, E. Angulo, J. Rodriguez and C. Santamaria, *J. Electroanal. Chem.*, 1992, **341**, 369–375.
20. T. F. Otero, *Polym. Rev.*, 2013, **53**, 311–351.
21. T. F. Otero, J. G. Martinez and J. Arias-Pardilla, *Electrochim. Acta*, 2012, **84**, 112–128.
22. T. F. Otero, *J. Mater. Chem. B*, 2013, **1**, 3754–3767.
23. T. F. Otero, in *Modern Aspects of Electrochemistry*, ed. R. E. White, J. O. Bockris and B. E. Conway, Springer US, New York, 1999, pp. 307–434.
24. E. Smela and N. Gadegaard, *Adv. Mater.*, 1999, **11**, 953–957.
25. E. Smela and N. Gadegaard, *J. Phys. Chem. B*, 2001, **105**, 9395–9405.
26. L. Lizarraga, E. M. Andrade and F. V. Molina, *J. Electroanal. Chem.*, 2004, **561**, 127–135.
27. D. S. H. Charrier, R. A. J. Janssen and M. Kemerink, *Chem. Mater.*, 2010, **22**, 3670–3677.
28. M. F. Suarez and R. G. Compton, *J. Electroanal. Chem.*, 1999, **462**, 211–221.
29. P. R. Singh, S. Mahajan, S. Raiwadec and A. Q. Contractor, *J. Electroanal. Chem.*, 2009, **625**, 16–26.
30. X. Chen and O. Inganas, *Synth. Met.*, 1995, **74**, 159–164.
31. P. Chiarelli, D. Derossi, A. Dellasanta and A. Mazzoldi, *Polym. Gels Networks*, 1994, **2**, 289–297.
32. T. F. Otero and J. G. Martinez, *Adv. Funct. Mater.*, 2013, **23**, 404–416.
33. T. F. Otero, M. Alfaro, V. Martinez, M. A. Perez and J. G. Martinez, *Adv. Funct. Mater.*, 2013, **23**, 3929–3940.
34. C. Quattrocchi, R. Lazzaroni, J. Bredas, R. Zamboni and C. Taliani, *Macromolecules*, 1993, **26**, 1260–1264.

35. T. F. Otero, in *Structural Biological Materials. Design and Structure-Properties Relationships*, ed. M. Elices and R. W. Cahn, Pergamon, Amsterdam, 2000.
36. R. H. Baughman and L. W. Shacklette, in *Science and Application of Conducting Polymers*, ed. W. R. Salaneck, D. T. Clark and E. J. Samuelsen, IOP Publishing, Bristol, 1991, pp. 47–61.
37. T. F. Otero and J. Rodriguez, in *Intrinsically Conducting Polymers: An Emerging Technology*, ed. M. Aldissi, Kluwer Academic Publishers, Dordrecht, 1993, vol. 246, pp. 179–190.
38. *Electropolymerization: Concepts, Materials and Applications*, ed. S. Cosnier and A. Karyakin, Wiley Vch Verlag Gmbh, Weinheim, 1st edn, 2010.
39. *Handbook of Conducting Polymers*, ed. T. A. Skotheim and J. Reynolds, CRC Press, New York, 3rd edn, 2006.
40. *Structural Biological Materials: Design and Structure-Property Relationships*, ed. M. Elices, Pergamon, Oxford, 2000.
41. A. Pron, P. Gawrys, M. Zagorska, D. Djurado and R. Demadrille, *Chem. Soc. Rev.*, 2010, **39**, 2577.
42. G. Horowitz, *J. Mater. Res.*, 2004, **19**, 1946–1962.
43. S. Kola, J. Sinha and H. E. Katz, *J. Polym. Sci., Part B: Polym. Phys.*, 2012, **50**, 1090–1120.
44. Z. Chen, V. Stepanenko, V. Dehm, P. Prins, L. D. A. Siebbeles, J. Seibt, P. Marquetand, V. Engel and F. Würthner, *Chem.–Eur. J.*, 2007, **13**, 436–449.
45. A. G. MacDiarmid, *Angew. Chem., Int. Ed.*, 2001, **40**, 2581–2590.
46. S. R. Forrest, *Nature*, 2004, **428**, 911–918.
47. A. Coskun, J. M. Spruell, G. Barin, W. R. Dichtel, A. H. Flood, Y. Y. Botros and J. F. Stoddart, *Chem. Soc. Rev.*, 2012, **41**, 4827.
48. M. Irimia-Vladu, *Chem. Soc. Rev.*, 2014, **43**, 588–610.
49. D. D. Chung, *Carbon*, 2001, **39**, 279–285.
50. K. Gurunathan, A. V. Murugan, R. Marimuthu, U. Mulik and D. Amalnerkar, *Mater. Chem. Phys.*, 1999, **61**, 173–191.
51. M. Gerard, *Biosens. Bioelectron.*, 2002, **17**, 345–359.
52. A. Moliton and R. C. Hiorns, *Polym. Int.*, 2004, **53**, 1397–1412.
53. D. H. Park, M. S. Kim and J. Joo, *Chem. Soc. Rev.*, 2010, **39**, 2439.
54. Y. K. Hong, D. H. Park, S. H. Lee and J. Joo, in *Controlled Polymerization and Polymeric Structures*, ed. A. Abe, K.-S. Lee, L. Leibler and S. Kobayashi, Springer International Publishing, Cham, 2013, vol. 259, pp. 201–244.
55. T. F. Otero, J. Arias-Pardilla, H. Herrera, J. Luis Segura and C. Seoane, *Phys. Chem. Chem. Phys.*, 2011, **13**, 16513–16515.
56. A. Diaz, J. Castillo, J. Logan and W. Lee, *J. Electroanal. Chem.*, 1981, **129**, 115–132.
57. C. Odin, M. Nechtschein and P. Hapiot, *Synth. Met.*, 1992, **47**, 329–350.
58. C. Odin and M. Nechtschein, *Phys. Rev. Lett.*, 1991, **67**, 1114–1117.
59. C. Odin and M. Nechtschein, *Synth. Met.*, 1993, **55**, 1287–1292.

60. K. Aoki, *J. Electroanal. Chem.*, 1991, **300**, 13–22.
61. K. Aoki, J. Cao and Y. Hoshino, *Electrochim. Acta*, 1994, **39**, 2291–2297.
62. J. G. Martinez and T. F. Otero, *RSC Adv.*, 2014, **4**, 29139–29145.
63. T. F. Otero, J. G. Martinez, M. Fuchiwaki and L. Valero, *Adv. Funct. Mater.*, 2014, **24**, 1265–1274.

CHAPTER 5

Membrane Composition-Dependent Electrochemical Properties

5.1 Introduction

Any one of the oxidation or reduction reactions (see Chapter 4, Section 4.3) may drive the reversible variation of the film's relative polymer/counterion composition through several orders of magnitude (Section 4.5). The chemical composition of a material determines its physical and chemical properties. The electrochemical reaction pushes a reversible change of the value of each composition-dependent property, which is defined as a physical or chemical property of the material that can be changed in a reversible way within a range with the material composition under reaction control.

Some of these composition-dependent properties are shown in Table 5.1. Each of the composition-dependent properties mimics a biological function that results from biochemical reactions in the cells of the related biological organ. Table 5.1 also presents the mimicked functions. Consequently, based on the variation of each of the chemical-driven properties, artificial devices mimicking the concomitant biological organs are being developed.

Each of these properties and the dependence of its value with the material composition will now be presented.

5.2 Electronic Conductivity

Given that our technological world is mainly based on the physical properties of materials, the attention of most researchers has focused on the reversible change of electronic conductivity for a conducting polymer by several

RSC Smart Materials No. 19
Conducting Polymers: Bioinspired Intelligent Materials and Devices
By Toribio Fernández Otero

Published by the Royal Society of Chemistry, www.rsc.org

Table 5.1 Electrochemical properties of the material (the properties of a conducting polymer for which the values shift in a reversible way during oxidation/reduction) with the biological organ experiencing a similar property shift during actuation, and biomimetic artificial devices. Actuation is based on variation of the concomitant composition-dependent property of the constituent conducting polymer under reaction control.

Electrochemical property	Shifting material property	Biological organ	Biomimetic device
	Electronic conductivity		
Electrochemomechanical	Volume	Muscle	Artificial muscle, actuator
Electrochromic	Color	Mimetic skin	Smart windows, mirrors, *etc.*
Electroactivity	Charge	Electric organ	Batteries, supercapacitors
Sensibility	Material potential	Sensing organs	Sensors, biosensors, sensing muscles
Electroporosity	Dialysis	Membrane	Smart membranes
Conformational packing	Electrochemo-conformational energetic states	Brain memory	EChemConf memories
Drug delivery	Ionic storage	Glands	Artificial glands
Wettability	Molecular interactions	Dry and clean leaves	Dry and clean surfaces
Ion delivery	Neurotransmitter storage	Synapse	Artificial chemical synapse

orders of magnitude during its reversible oxidation/reduction.[1–3] During polymeric oxidation [eqn (4.1) or (4.2)] or reduction [eqn (4.3)], positive or negative, respectively, conjugated polaronic species are generated along the chain. These generated conjugated sites extend along several (3–4) monomeric units (Figure 1.1b), acting as charge carriers: electrons jump between two neighbor charge carriers. In parallel with semiconductors or electrolytes, the resulting material conductivity (σ) is proportional to the concentration of charge carriers (c, number of carriers cm^{-3}) in the film and to the carrier mobility (v, $cm^2\ V^{-1}\ s^{-1}$):

$$\sigma = cev, \tag{5.1}$$

where e represents the electron charge.

But the carrier's concentration can be controlled [eqn (4.11)] in a reversible way by the imposed oxidation or reduction charge. The electronic conductivity of the material can be tuned through faradaic control (*i.e.*, an increase or decrease) of the charge carrier concentration through the oxidation/reduction charge:

$$\Delta\sigma = \Delta cev = (\Delta Q/wzF)ev, \tag{5.2}$$

where Q is the charge, w is the mass, z is the ion charge and F is Faraday's constant.

Thus, the carrier's concentration variation is managed by the charge consumed to change the oxidation state of the polymer. Taking into account that, whatever the oxidation state of the conducting polymer film, the material must be charge balanced, the polaronic concentration is related to the concentration of the counterions [eqn (2.2)] through the valence (z) of the counterion. The control of the electronic conductivity [eqn (5.2)] requires the use of the material as a working electrode in the presence of an electrolyte to attain, by film oxidation or reduction, the required concentration of electronic carriers. Once the expected value of the material composition and conductivity has been attained, the material may be extracted from the electrolyte, keeping (when the material is dry) the attained chemical composition and electronic conductivity.

In conclusion, the electronic conductivity of a conducting polymer is a composition-dependent property: the oxidized (doped) state determines its electronic conductivity [eqn (5.1)].

5.3 Volume Variation

The flow of anodic charge Q forces the entrance of n moles of anions according to $n = Q/zF$ [eqn (2.2)] and m moles of water [eqn (4.1)] from the electrolyte: the polymer swells (Figure 1.3). Thus, the specific film volume (ΔV, volume per unit of conducting polymer mass w reacting inside the electrolyte) increases by:[4]

$$\frac{\Delta V}{w} = \frac{nV_a + mV_w}{w} = \Delta c_a V_a + \Delta c_w V_w, \tag{5.3}$$

where V_a is the molar volume of the anion, V_w is the water molar volume, Δc_a is the ionic concentration variation (in M g^{-1}) per unit of polymer weight (the specific concentration) reacting inside the electrolyte and Δc_w is the variation of the specific concentration of water molecules in the reacting (dense gel) film. The entrance of water is driven through the osmotic pressure (Section 8.2.8) by the ionic concentration variation. Thus, the volume variation is a linear function of the specific ionic concentration variation (Δc_a) in the film [eqn (5.3)]: the reaction-driven volume variation of the conducting polymer film is a composition-dependent property.

5.4 Color Shift

The energy gap between the highest occupied molecular orbital (HOMO) and lowest unoccupied molecular orbital (LUMO) of most neutral conducting polymers matches that of ultraviolet (UV) photons, *i.e.*, it is transparent in the visible-infrared (vis-IR) regions (Figure 5.1).[5] Accepting the metal band model for conducting polymers, the HOMO levels from each of the film's chains constitute the occupied valence band and the LUMO levels should constitute the unoccupied valence band.

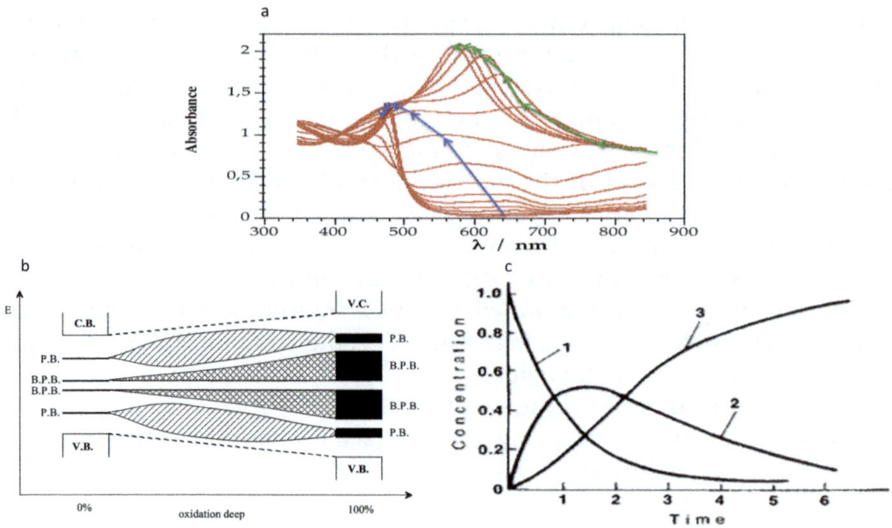

Figure 5.1 (a) Evolution of the absorption spectra of an electrochromic polypyrrole film as a function on the oxidation potential (one spectrum every 80 mV is presented) under potential cycling between −900 and 400 mV in a 2.5 M $LiClO_4$ aqueous solution at a scan rate of 20 mV s^{-1}. Blue arrows indicate the evolution of the polaronic band maximum and green arrows show the bipolarons. (b) Scheme of the polaronic band (PB) and bipolaronic band (BPB; number of molecular orbitals) variation with the oxidation depth of the conducting polymer film. (c) Theoretical variation of the polaronic (2) [eqn (5.9)] and bipolaronic (3) [eqn (5.11)] absorbance as a function of the oxidation time. Adapted with permission from ref. 10: *Langmuir*, 1999, **15**, 1323–1327. Copyright (1999) American Chemical Society.

During film oxidation by the flow of anodic charge Q, the extraction of one electron from the film generates a radical cation (polaronic species) in a film chain:

$$\mathrm{Pol} \rightarrow \mathrm{Pol}^{\bullet +} + e^{-} \tag{5.4}$$

The two unoccupied bonding and antibonding polaronic energetic levels are energetically located between the valence and conduction bands of the neutral chains (Figure 5.1a). The extraction of $n = Q/F$ moles of electrons generates the same number of polaronic ($\mathrm{Pol}^{\bullet +}$) species and polaronic energetic levels. The new electronic transitions from the occupied valence band (VB) to the unoccupied polaronic levels require less energy (visible region) than that required for exciting electrons from the VB to the conducting band. The oxidation progress results in new unoccupied polaronic levels constituting two rising populated unoccupied polaronic bands (Figure 5.1a). According to the Beer–Lambert law, light absorbance (A) is proportional to the chromophore (polaron) concentration (c) *via* $A = \alpha lc$, where α is the molar extinction coefficient of the conducting polymer and l is the length of the light path inside the polymer. Thus, the light absorption can be reversibly changed under electrochemical control of the polaron concentration variation:

$$\Delta A = \alpha l \Delta c. \tag{5.5}$$

This concentration increases with the oxidation charge, as described by eqn (4.10), with $z = 1$ (each polaron bears one positive charge). In conclusion, the light absorption is proportional to the polymer composition [eqn (5.5)], making it a composition-dependent property.

For increasing polaronic concentrations, two polaronic units recombine to form a polaron pair (or a bipolaron):[6–8]

$$\mathrm{Pol}^{2(\bullet+)} \rightarrow \mathrm{Pol}^{2+}. \tag{5.6}$$

The two new unoccupied molecular orbitals (bonding and antibonding) are located in between the two polaronic bands. New possibilities for electronic transitions according to the selection rules and requiring still lower energy give new absorption bands inside the IR region. When the oxidation advances, the concentration of the polaronic (radical) species in the film decreases by its recombination to form polaron pairs (the radical concentration decreases, as proved by electron paramagnetic resonance, EPR).[9]

Eqn (5.4) and (5.6) are consecutive reactions. According to the kinetics of consecutive reactions, starting from a theoretically fully reduced film (*i.e.*, the maximum [Pol]) the concentration of the chain electroactive centers follows (Figure 5.1b, curve 1) an exponential decay with time t {$[\mathrm{Pol}] = [\mathrm{Pol}]_0 e^{-kt}$, where k is the kinetic coefficient for the forward reaction in eqn (5.4)} during the film oxidation (by potential step, flow of a constant anodic current or under potential cycle). And the absorbance of the active centers will follow the concomitant decay during the film oxidation:

$$A(t) = \alpha l [\mathrm{Pol}]_0 \, e^{-k_1 t}, \tag{5.7}$$

where $[\mathrm{Pol}]_0$ is the specific concentration of polymeric active sites in the conducting polymer film at the beginning of the current flow, $[\mathrm{Pol}]_0 e^{-kt}$ describes the concentration evolution (decrease) with time t and α is the molar extinction coefficient of the polarons. In summary, the absorbance variation of the polymeric active sites during the electrochemical reaction follows the film concentration variation driven by the reaction, making it a concentration-dependent property.

The concentration of the polaronic species in the film evolves during the reaction according to:

$$[\mathrm{Pol}^{\bullet+}] = [\mathrm{Pol}]_0 \frac{k_1}{k_2 - k_1} \left(e^{-k_1 t - e^{-k_2 t}} \right) \tag{5.8}$$

where k_2 is the reaction coefficient of the backward reaction. Curve 2 in Figure 5.1b represents eqn (5.8), indicating the initial increase of the polaronic concentration, passing through a maximum and then decreasing for increasingly oxidized states of the film. Thus, the absorbance evolution of the polaronic band during the oxidation time, $A_P(t)$, is quantitatively described by combining eqn (5.5) and (5.8):

$$[A_P(t)] = \alpha' l[\text{Pol}^{\bullet+}] = [\text{Pol}]_0 \frac{k_1}{k_2 - k_1}\left(e^{-k_1 t - e^{-k_2 t}}\right) \tag{5.9}$$

where α' is the molar extinction coefficient of bipolarons. The absorbance of the polaronic band evolves with the oxidation time as the concentration of polarons passes through a maximum according to eqn (5.9). The polaronic absorbance is a concentration-dependent property.

The concentration of bipolarons (or polaron pairs) follows an initial exponential rise, passes through an inflection point with decreasing increments of the rising rate beyond this point and then attains a constant concentration at the end of oxidation (Figure 5.1b, curve 3), as described by:

$$[\text{Pol}^{2+}] = [\text{Pol}]_0\left[1 + \frac{1}{k_2 - k_1}(-k_2\, e^{-k_1 t} + k_1\, e^{-k_2 t})\right]. \tag{5.10}$$

Thus, the absorbance evolution of the polaron pairs during the oxidation time, $A_{PP}(t)$, is quantitatively described by combining eqn (5.5) and (5.9):

$$A_{pp}(t) = \alpha'' l[\text{Pol}^{2+}] = \alpha'' l[\text{Pol}]_0\left[1 + \frac{1}{k_2 - k_1}\left(-k_2\, e^{-k_1 t} + k_1\, e^{-k_2 t}\right)\right], \tag{5.11}$$

which is also a function of the film's concentration of bipolarons and its balancing counterions, so it's a composition-dependent property.

Experimental absorbance evolutions for polaronic and bipolaronic bands follow, during oxidation by potential step (Figure 5.1a), a fully parallel evolution (Figure 5.1c)[10] to those deduced from Figure 5.1b and eqn (5.9) and (5.11).

5.5 Charge Storage

Each electron extracted from the polymer film during the forward reactions in eqn (4.1) or (4.2) generates and stores a positive charge on one of the film's polymeric chains. These forwards and backwards reactions are also named charging/discharging reactions.

Each electron injected during the reaction shown in eqn (4.3) in the polymer film generates (stores) a negative charge in one of the film's polymer chains.

Thus, eqn (4.1)–(4.3) define the number of stored charges, Q_{ox} for eqn (4.1) and (4.2), and Q_{red} for eqn (4.3), and eqn (2.3) quantifies the specific concentration of positive or negative charge equivalents (Faraday's law) stored in the film. The stored charge results, whatever the storing reaction, in a function of the specific concentration of ions (c, in mol g^{-1}):

$$Q_{stored} = zcFw, \tag{5.12}$$

where z is the charge of the incorporated ion, w is the mass of the reacting conducting polymer film and F is the Faraday constant. The stored charge is also a composition-dependent property.

5.6 Ionic Storage

The reactions shown in eqn (4.1)–(4.3) only take place if the generated free volume in the film fits or overcomes the dimensions (Section 6.15 in Chapter 6) of the counterion unit (solvated counterion) penetrating the film from the solution for charge balance. The amount of ions (N) stored in the film during the reaction is, according to eqn (2.3), given by:

$$N = c'w, \tag{5.13}$$

where c' in mol g^{-1} is the specific ionic concentration attained with the consumed charge and w in g is the mass of the reacting conducting polymer. The ionic storage is also a composition-dependent property.

5.7 Transversal Ionic Conductivity and Diffusivity Tuning

Oxidation [eqn (4.1), forward reaction] and reduction [eqn (4.2) and (4.3), backwards reactions] promote film swelling, *i.e.*, the average distance between the polymeric chains increases. When the electrode is used as a membrane separating two electrolytes with different anions and cations and a potential gradient is applied across the membrane, the ionic flow across the membrane increases for the same transmembrane potential gradient for increasingly swollen films and decreases for increasingly shrunk-compacted films. But the swollen/shrunken states are related to the volume occupied between the chains by the counterions and solvent that, according to eqn (5.3), is under the control of the film's ionic concentration. As a result, both ionic conductivity and ionic diffusivity across the membrane will be a function of the membrane concentration: they are composition-dependent properties (see Section 8.3).[11]

5.8 Material Potential Shift

Each chain oxidation/reduction reaction occurs through n consecutive equilibrium states in eqn (4.5)–(4.8), which are related to the first, second, third *etc.* ionization (or electronic affinity) states. Each of these equilibrium states defines a Nernst potential:

$$E_n = E_{n0} + \frac{RT}{nF}\ln\frac{\left[\mathrm{Pol}^{n+}(\mathrm{A}^-)_n\right]}{[\mathrm{A}^-]\left[\mathrm{Pol}^{(n-1)+}(\mathrm{A}^-)_{n-1}\right]} \tag{5.14}$$

in such a way that $E_1 < E_2 < E_3 <...< E_n$. Any mono-disperse film should have n consecutive equilibrium potentials, one for each of the $0 \leq i \leq n$ different $[Pol^{i+}(A^-)_i]$ stoichiometric material compositions: the film potential is a function of the film composition.

Actual films are poly-disperse, giving different chain length intercalated equilibrium potentials: any intermediate potential between E_0 and E_n corresponds (in an initial approach) to an equilibrium potential related to a specific polymer/ion (w/w) composition. Whatever the mono-disperse or poly-disperse nature of the film, the electrode equilibrium potential results in a logarithmic function (Section 9.5.3) of the concentration of oxidized centers in the film: the electrode potential is a composition-dependent property.[12]

5.9 Surface Property Control

The most reduced state of any basic conducting polymer membrane is close to a fully organic material. If the dipolar component of the monomeric units is low, the polymer–water interaction will be very low, *i.e.*, it's a non-wetting surface. By oxidation, the surface concentration of charges on the chains and counterions increases: the water–dipole/ion and water–dipole/charge interactions increase, resulting in increased wetting of both the material surface and the material volume. Thus, the wetting/drying intermediate degrees can be tuned by the film's composition.

5.10 Ion Delivery

The ions stored inside the film [eqn (5.13)] can have some pharmacological, agricultural or chemical interest. In these cases, they can be liberated following the reactions in eqn (4.1) (backwards), and (4.2) or (4.3) (forwards). Eqn (5.13) also defines the number of moles delivered by the imposed charge, ΔQ_{del}, which now flows in the opposite direction:

$$\Delta N = \Delta c' w = c' \frac{\Delta Q_{\text{del}}}{zwF}. \tag{5.15}$$

As in previous cases, ion delivery is also a film composition-dependent property.

5.11 Packed Ionic-Conformational Energetic States

Chapter 6 is devoted to the conformational and macroscopic structural changes driven by electrochemical reactions in conducting polymers (conformational relaxation, swelling, shrinking and conformational compaction). Increasingly packed ionic-conformational energetic states will be attained under linear control of the reduction–compaction charge, which means under the control of a linear decrease of the ionic concentration.

Thus, it becomes a new composition-dependent property of the conducting polymers.

5.12 Chemo-Biomimetic Functions

In fact, the above-described composition-dependent (electrochemical) properties mimic similar drastic composition changes and properties occurring in cells from different biological organs to start or stop biological functions. Volume variations happen in muscles, color changes occur in skin and electrical energy storage happens in the electric organs of different fishes and animals. In addition, the opening and closing of ionic channels is the usual way of ion channel function in cell membranes, while electrical potentials occur in sensors. There is also chemical liberation in glands or during synaptic events, and packing conformational energetic levels could be considered as a new hypothesis to explain and quantify brain memory storage. This simultaneous chemical-driven multifunctionality makes conducting polymers some of the most intelligent constructs among artificial materials.

Thus, abrupt composition changes during reactions in conducting polymers mimic, in its simplest expression, the contents of the intracellular matrix (ICM) of living cells and some of the biological functions driven by chemical reactions and chemical composition changes in biological organs. Each of these reaction-driven and composition-dependent properties can be isolated for its individual study and possible application. This is typical nowadays in technological developments in order to develop different electro-chemo-biomimetic polymeric devices with a specific variable property (each device mimicking a biological organ), for example, artificial muscles; polymeric actuators or polymeric motors; smart windows, mirrors or color filters; full organic batteries and supercapacitors; smart membranes; sensors and biosensors; smart drug delivery devices; artificial chemical synapses; ionic pollutant scavengers; chemo-ionic-conformational memories; or self-cleaning and water-repellent surfaces, among others.

These devices constitute the forefront of a new technological world. Among so many millions of technological devices and products existing in our present technological world, only a few of them are based on the chemical reactions of the device's constituent materials: primary and secondary batteries and supercapacitors, smart mirrors (in cars), some electrochemical and chemical sensors or biosensors, and some photoelectrochemical cells.

Despite the fact that most of the above-described reaction-driven and composition-dependent properties were discovered during the 1980s, the concomitant biomimetic devices and products are still being developed quite slowly. This is particularly true when taking the fast parallel development of fully polymeric electronic and microelectronic dry, flexible and transparent devices and products based on conducting polymers as a reference.[13–23]

The main reason for the different development rates seems founded in the existence of a solid electronic, microelectronic and optoelectronic theoretical background developed during the past century. For the development of biomimetic devices, solid self-consistent physicochemical models do not exist that provide a full theoretical description and quantification of the new reactive materials, their reversible reactions, the variation of the composition-dependent properties of the material and the actuation of the new devices based on the reaction-driven properties of the device's constitutive materials. This is consequently the aim of this book: to establish solid foundations for the subsequent construction of new theoretical and technological electro-chemo-biomimetic and chemo-biomimetic structures. The final hope is that, after some adaptation, such models could be translated for the description of biochemical reactions and biological function, and for the correction of biological malfunctions.

5.13 ICM Electro-Chemo-Biomimicry

Most of the biomimetic properties, functions and devices described by today's literature are related to the physical or chemical aspects of the extracellular matrix (ECM) of living cells.[24–31] Currently, a new scientific and technological field is arising based on the aforementioned composition-dependent properties. The mimesis of either the contents of the ICM or reactions in living cells and that of their reaction-driven biological functions will allow the construction of artificial biomimetic devices. The new reactive dense gels from conducting polymers mimic the ICM contents in its simplest expression *via* reactive polymers and macromolecules, reaction-driven conformational movements, and ions and water exchange. The most important difference between reactive gels and the reactive states used for the development of the present chemical kinetic models (gaseous or in dilute solutions) is the presence and control of reaction-driven conformational movements. Now, these reaction-driven conformational movements and their macroscopic consequences require a deeper presentation, description and quantification, which will be given in the next chapter in order to explore in Chapter 7 the limits of the present chemical kinetic models when applied to biomimetic reactive gels.

References

1. S. Kivelson and A. Heeger, *Synth. Met.*, 1988, **22**, 371–384.
2. A. Diaz and J. Castillo, *J. Chem. Soc., Chem. Commun.*, 1980, 397–398.
3. G. Zotti and G. Schiavon, *Synth. Met.*, 1990, **39**, 183–190.
4. T. F. Otero and J. G. Martinez, *Chem. Mater.*, 2012, **24**, 4093–4099.
5. T. F. Otero, in *Modern Aspects of Electrochemistry*, ed. R. E. White, J. O. Bockris and B. E. Conway, Springer, US, New York, 1999, pp. 307–434.
6. S. S. Zade, N. Zamoshchik and M. Bendikov, *Acc. Chem. Res.*, 2011, **44**, 14–24.
7. C. Quattrocchi, R. Lazzaroni, J. Bredas, R. Zamboni and C. Taliani, *Macromolecules*, 1993, **26**, 1260–1264.

8. J. L. Bredas, D. Beljonne, V. Coropceanu and J. Cornil, *Chem. Rev.*, 2004, **104**, 4971–5003.
9. A. Zykwinska, W. Domagala, A. Czardybon, B. Pilawa and M. Lapkowski, *Chem. Phys.*, 2003, **292**, 31–45.
10. T. F. Otero and M. Bengoechea, *Langmuir*, 1999, **15**, 1323–1327.
11. M. J. Ariza and T. F. Otero, *J. Membr. Sci.*, 2007, **290**, 241–249.
12. T. F. Otero and M. T. Cortes, *Sens. Actuators, B*, 2003, **96**, 152–156.
13. M. Gerard, *Biosens. Bioelectron.*, 2002, **17**, 345–359.
14. A. Moliton and R. C. Hiorns, *Polym. Int.*, 2004, **53**, 1397–1412.
15. A. Pron, P. Gawrys, M. Zagorska, D. Djurado and R. Demadrille, *Chem. Soc. Rev.*, 2010, **39**, 2577.
16. Z. Chen, V. Stepanenko, V. Dehm, P. Prins, L. D. A. Siebbeles, J. Seibt, P. Marquetand, V. Engel and F. Würthner, *Chem.–Eur. J.*, 2007, **13**, 436–449.
17. K. Gurunathan, A. V. Murugan, R. Marimuthu, U. Mulik and D. Amalnerkar, *Mater. Chem. Phys.*, 1999, **61**, 173–191.
18. G. Horowitz, *J. Mater. Res.*, 2004, **19**, 1946–1962.
19. D. D. Chung, *Carbon*, 2001, **39**, 279–285.
20. P. Lin and F. Yan, *Adv. Mater.*, 2012, **24**, 34–51.
21. M. Irimia-Vladu, *Chem. Soc. Rev.*, 2014, **43**, 588–610.
22. D. H. Park, M. S. Kim and J. Joo, *Chem. Soc. Rev.*, 2010, **39**, 2439.
23. A. Coskun, J. M. Spruell, G. Barin, W. R. Dichtel, A. H. Flood, Y. Y. Botros and J. F. Stoddart, *Chem. Soc. Rev.*, 2012, **41**, 4827.
24. Y. Oaki, A. Kotachi, T. Miura and H. Imai, *Adv. Funct. Mater.*, 2006, **16**, 1633–1639.
25. J. R. Porter, T. T. Ruckh and K. C. Popat, *Biotechnol. Prog.*, 2009, **25**, 1539–1560.
26. B. Bhushan, *Philos. Trans. R. Soc., A*, 2009, **367**, 1445–1486.
27. F. Barthelat, *Philos. Trans. R. Soc., A*, 2007, **365**, 2907–2919.
28. Z. Tang, Y. Wang, P. Podsiadlo and N. A. Kotov, *Adv. Mater.*, 2006, **18**, 3203–3224.
29. S. Hecht and J. M. J. Fréchet, *Angew. Chem., Int. Ed.*, 2001, **40**, 74–91.
30. M. Sarikaya, C. Tamerler, A. K.-Y. Jen, K. Schulten and F. Baneyx, *Nat. Mater.*, 2003, **2**, 577–585.
31. N. F. Lepora, P. Verschure and T. J. Prescott, *Bioinspiration Biomimetics*, 2013, **8**, 013001.

CHAPTER 6

Reaction-Driven Conformational, Allosteric and Structural Changes

6.1 Introduction

The reactions in eqn (4.1)–(4.4) drive the entrance/expulsion of a huge amount of ions and solvent in the conducting polymeric membrane electrode quantified by eqn (2.2) and (2.3). The polymer film oxidation/reduction must induce the parallel generation/destruction of the required free volume [eqn (5.3)] to lodge/expel, respectively, balancing ions and solvent.[1,2] In fact, the reaction only occurs if the induced inter-chain volume variation is large enough to lodge the counterion coming from the electrolyte (Section 6.15). Thus, chain reactions must drive those required structural changes, from chain conformational movements to macroscopic structural changes (*e.g.*, swelling/shrinking), which allow polymer film reactions to occur. This chapter takes a comprehensive approach to look at the unexplored field of reaction-driven structural processes in an effort to identify some of its basic components.

6.2 Reversible Chain Molecular Motors

Taken as a reference, a polypyrrole chain with *m* consecutive pyrrole monomeric units (Figure 1.1a) constitutes a neutral and independent chain. Two consecutive monomeric units are linked by a C–C single bond, which allows free rotation in solution at ambient temperature. Any solved chain can pass

RSC Smart Materials No. 19
Conducting Polymers: Bioinspired Intelligent Materials and Devices
By Toribio Fernández Otero

Published by the Royal Society of Chemistry, www.rsc.org

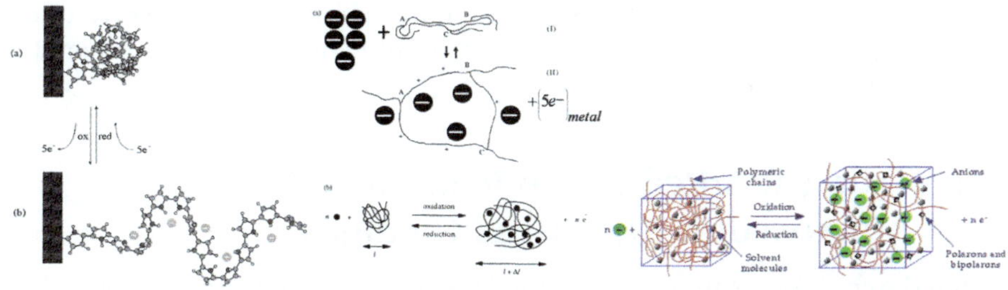

Figure 6.1 (a) Neutral chain of a conducting polymer inside an electrolyte and grafted to an electrode (black strip) showing a coil structure due to the high polymer–polymer interactions: reversible oxidation/reduction drive reversible coil-like/rod-like molecular motor actuation. (b) Bi-dimensional structure: reaction-driven conformational movements result in the required free volume to lodge/expel balancing counterions and solvent. (c) Three-dimensional reaction-driven reversible volume variation (swelling/shrinking). Reprinted from ref. 3: T. F. Otero, in *Modern Aspects of Electrochemistry*, ed. R. E. White, J. O. Bockris and B. E. Conway, Springer US, New York, 1999, pp. 307–434, with the permission of Springer.

through hundreds of different random coil conformational states. When the polymer–polymer intramolecular interactions are stronger than the polymer–solvent interactions the chain gets a final folded, globular structure (Figure 6.1a).[3]

After oxidation by extraction of one electron, a polaron or radical cation unit is generated: a π-conjugated state extended along several monomeric units (Figure 1.1b) gives a planar, rod-like structure to the polaron chain section. Whatever the conformational coil structure of the monomeric units in the neutral state, the reaction drives the change to a new rod-like structure. The extraction of consecutive electrons from the same chain results in consecutive conformational movements, each giving a new rod-like polaronic segment. After the extraction of n electrons, n rod-like polaronic structures are uniformly distributed along the chain. The reaction-driven conformational movements together with electrostatic repulsions between neighbor radical cations (polarons) and the attraction of charge balance counterions from the solution bulk result in the transition from the globular (coil) structure (Figure 6.1a, top) of the neutral chain to a final linear (rod-like) chain structure (Figure 6.1a, bottom).[4,5]

By reduction at ambient temperature, consecutive electrons are injected into the chain. Polaronic conjugated structures are eliminated, recovering the free rotation between consecutive monomeric units. The chain's reduction results in opposed conformational movements, recovering a new folded, globular chain structure (Figure 6.1a). Reversible electrochemical oxidation/reduction reactions drive and control the reversible conformational movements of the chain, giving a basic molecular motor that moves under electrochemical (faradaic) reaction control from that shown in the top panel of

Figure 6.1a to the structure shown in the bottom panel of Figure 6.1a under oxidation, and then back again to the structure shown in the top panel of Figure 6.1a under reduction.

6.3 Oxidation/Reduction Reactions Drive Macroscopic Structural Changes

When the chains take up a part of actual polymeric films, the oxidation-driven conformational movements of the film's chains [eqn (4.1), forwards] must generate the free volume necessary to lodge anions, which are required for charge balance, and solvent (Section 5.2). Figure 6.1b shows a bi-dimensional representation of the oxidation-driven entrance of counterions. The presence of ions in the film drives the entrance of solvent molecules required for osmotic balance [eqn (7.4)]. As a consequence, during the oxidation of any film exchanging anions and solvent with the solution the film swells: the oxidation drives the film macroscopic swelling, which is a macroscopic structural change (Figure 6.1c, forwards reaction).

During film reduction [eqn (4.1), backwards] the reaction drives the expulsion of ions and solvent, and the film shrinks under the control of the number of injected electrons (Figure 6.1b, bottom and c, backwards reactions).[6]

In summary, the reversible oxidation/reduction of a polymeric film drives macroscopic and reversible structural changes in the form of film volume variations (swelling/shrinking). Reaction-driven volume changes have been followed by different methodologies (Section 6.16)[7–17] as they are the most efficient fully polymeric bending modes for artificial muscles or cantilevers.[18,19]

6.4 Reaction-Driven Structural Components

The presence of reaction-driven macroscopic swelling and shrinking structural processes is a direct result of the film reactions. The question now is if the reaction drives other structural changes in the reacting conducting polymer membrane and how each of the reaction-driven structural processes can be identified, characterized and quantified.

When the film oxidation/reduction is controlled by electrochemical methodologies, each of the electrochemical responses provides some structural information. Coulovoltammetric (Q/E) responses when the oxidation/reduction of any self-supported film electrode is controlled by potential cycles (cyclic voltammetry) result in one of the most efficient ways to identify the different macroscopic reaction-driven structural processes, *i.e.*, separating the potential domains for each structural process and quantifying the charge consumed by the reaction to promote every structural change.[20–23] These Q/E responses present some differences between conducting polymers exchanging anions [eqn (4.1)] and those exchanging cations [eqn (4.2)], and they require separate discussion (Sections 6.4.1 and 6.4.2).[21,23,24]

6.4.1 Reaction-Driven Anion Exchanges

Figure 6.2 presents a stationary voltammetric (*i*/*E*) response (Figure 6.2a) and the concomitant closed loop stationary coulovoltammetric (*Q*/*E*) response (Figure 6.2b) from a self-supported polypyrrole film in aqueous solution.[20,21] Positive charge increments in the *Q*/*E* response indicate film oxidation charges and negative increments indicate film reduction charges.

The closed *Q*/*E* loop guarantees that only reversible film oxidation/reduction processes are involved in the full potential range: the film oxidation charge equals the film reduction charge. The most remarkable aspect is that the redox charge follows a continuous evolution in the full studied potential range: film oxidation/reduction reactions occur at any potential between −2.5 and 0.6 V. This result contradicts most of the literature, where the film oxidation/reduction is restricted to the potential domain of the voltammetric (Figure 6.2a) oxidation/reduction maxima (−1.0 to 0.6 V here).

The presence of some overlapping irreversible reaction should give open *Q*/*E* responses (Section 4.8). Energy-dispersive X-ray (EDX) analysis and the bending movement of polypyrrole/tape [(Ppy)/tape] bilayer muscles (Section 6.13.2) confirm that the reaction drives the exchange of anions in the full potential range. The exchange of anions means that the forwards and backwards reactions shown in eqn (4.1) are the only driving reactions in the full potential range.

Any change of the *Q*/*E* slope (Figure 6.2b) indicates an abrupt change of the concomitant reaction rate for eqn (4.1):[21]

$$\frac{\Delta Q}{\Delta E} = \frac{\vartheta \Delta Q}{zFw\Delta E} \text{(note that mV s}^{-1}\text{ M mV}^{-1}\text{ g}^{-1} = \text{M g}^{-1}\text{ s}^{-1}\text{)}, \qquad (6.1)$$

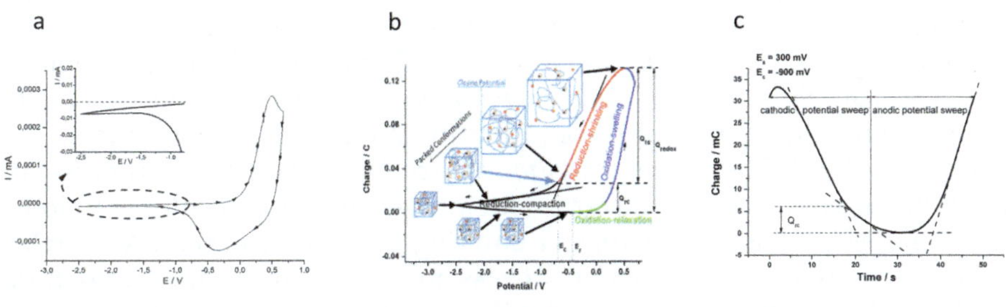

Figure 6.2 Stationary (a) voltammetric and (b) coulovoltammetric responses from a polypyrrole self-supported film, the electrolyte and the metal clamp contact separated by a transversal paint strip to avoid the electrolyte ascension by capilarity, subjected to consecutive potential sweeps between −2.5 and 0.7 V (*vs.* Ag/AgCl 3 M, reference electrode) at 1 mV s^{-1} in a 0.1 M $LiClO_4$ aqueous solution at ambient temperature. (c) Chronocoulogram attained from −900 to 300 mV at 50 mV s^{-1} using the potential sweep rate as a conversion constant, showing the continuity of the faradaic compaction processes (Q_{rc}) ahead of the cathodic potential sweep inside the anodic potential sweep. With kind permission from John Wiley and Sons, adapted from ref. 21: *Adv. Funct. Mater.*, 2014, **24**, 1265–1274.

where ϑ (in V s^{-1}) represents the potential sweep rate, z is the anion valence, F is the Faraday constant and w is the mass of the conducting polymer film reacting inside the solution. The Q/E maximum represents the most oxidized, swollen state of the membrane electrode under actual experimental conditions. From there, a fast reduction-driven shrinking process [*reduction–shrinking*, backwards reaction in eqn (4.1)] starts with expulsion of counterions and solvent molecules. The inflexion point, which correlates with the end of the voltammetric reduction maximum (Figure 6.2b), corresponds to *the closing potential*: decreasing average distances (by reduction–shrinking) between the chains fit the counterion unit diameter and the structure closes when a high percentage of the counterions (up to 40% in thick films, as detected by EDX analysis) still remain partially trapped inside the dense membrane. Ahead of the closing potential the Q/E slope decreases, but the consumption of reduction charge goes on. The polymer reduction with expulsion of anions continues, as confirmed by the bending movement of bilayer muscles, driving the film conformational compaction (*reduction–compaction*). The reduction–compaction rate is lower than the reduction–shrinking rate: before the closing potential, the counterions move towards the electrolyte under diffusion kinetic control; after the closing potential, the counterions must move the polymeric chains apart in front of them and the reaction occurs under kinetic control of the slow chain conformational movements. Behind each ion leaving the film the polymer chains occupy the empty volume, resulting in increasingly conformationally compacted states that yield slower reduction reactions.

The slow reduction–compaction of the film goes on through the cathodic potential limit at the beginning of the subsequent anodic potential sweep: the reduction overpotential is still high and a great number of counterions and balancing polarons are still present inside the membrane. The system presents a *great electrochemical inertia*: closed and packed partially reduced structures are *far from their electrochemical equilibrium* due to the decreasing reduction rate for increasingly conformationally compacted structures.

At the Q/E minimum, the *oxidation–relaxation* [forwards reaction in eqn (4.1)] of the compacted film starts generating the required volume to lodge balancing counterions and solvent entering from the solution. After relaxation, the film *oxidation–swelling* process goes on until the anodic potential limit. At the beginning of the cathodic potential sweep the oxidation goes on, showing great electrochemical inertia: the anodic overpotential is still high enough and the membrane has not attained the concomitant oxidation equilibrium state defined by the reactions in eqn (4.5)–(4.8).

Both film oxidation and reduction electrochemical inertial processes indicate that, even under steady state voltammetric conditions at a low sweep rate, the oxidation state of the film is out of its oxidation equilibrium state at any potential. The use of thinner films will produce faster structural processes, trapping lower percentages of counterions at the closing potential: individual chains should give the fastest and closest ones to equilibrium conditions.

By reduction beyond the closing potential, more reduced and conformationally packed energetic states of the membrane electrode are obtained: *increasingly conformationally compacted energetic states* of the partially reduced film are obtained by potential sweep until a high cathodic potential limit.

In summary, for conducting polymer film membrane electrodes exchanging anions, the coulovoltammetric responses indicate that the film oxidation/reduction expands beyond either the voltammetric maxima or the voltammetric potential limits, showing great electrochemical inertia. The different Q/E slopes identify four reaction-induced structural processes that, starting from the Q/E minimum, are named: oxidation–conformational relaxation (oxidation–relaxation or relaxation), oxidation–swelling (swelling), reduction–shrinking (shrinking) and reduction–conformational compaction (reduction–compaction or compaction). The reaction-induced structural changes produce a feedback influence on the reaction, modifying the reaction rate. So, by electrochemical reduction the membrane electrode shrinks and closes under a partially reduced state, promoting a drop of the reduction rate ahead of the closing potential, resulting in a slow polymeric reduction–compaction process.

Reduction–compaction and oxidation–relaxation are slow reaction-induced structural processes (the reduction–compaction being much slower and both are even slower as the film thickness increases). Despite the use of slow potential sweeps, the material is far from its electrochemical equilibrium defined by the reactions in eqn (4.4)–(4.7), resulting in great *electrochemical inertia*. The film oxidation–swelling goes on beyond the anodic potential limit at the beginning of the cathodic potential sweep, consuming oxidation charge (Figure 6.2a). The reduction–compaction inertia is stronger after the cathodic potential limit, going on during 2 V on the subsequent anodic potential sweep.

The tangent intersects allow good separation of the *potential domains* for the different structural processes (Figure 6.2b and c), defining the potential ranges. These potentials also allow good separation and quantification of *the structural charges* consumed by each of the reaction-induced structural processes.

6.4.2 Reaction-Driven Cation Exchanges

Figure 6.3 shows the coulovoltammetric response from blends of conducting polymers with macro-anions exchanging cations (Section 1.5.1) during p-doping and p-dedoping processes. After its most reduced–swollen state (Section 6.16) at the Q/E minimum, a fast *oxidation–shrinking* process due to the expulsion of cations and solvent is observed.[20,21,25] Following the fast growth of the oxidation charge, the inflexion point corresponds to the closing potential. After this point, a slow film *oxidation–compaction* starts. The film oxidation–compaction goes on at the beginning of the reduction branch until the Q/E maximum, confirming the great inertia of the electrochemical process. The slow *reduction–relaxation* of the film starts there, followed by fast *reduction–swelling*.

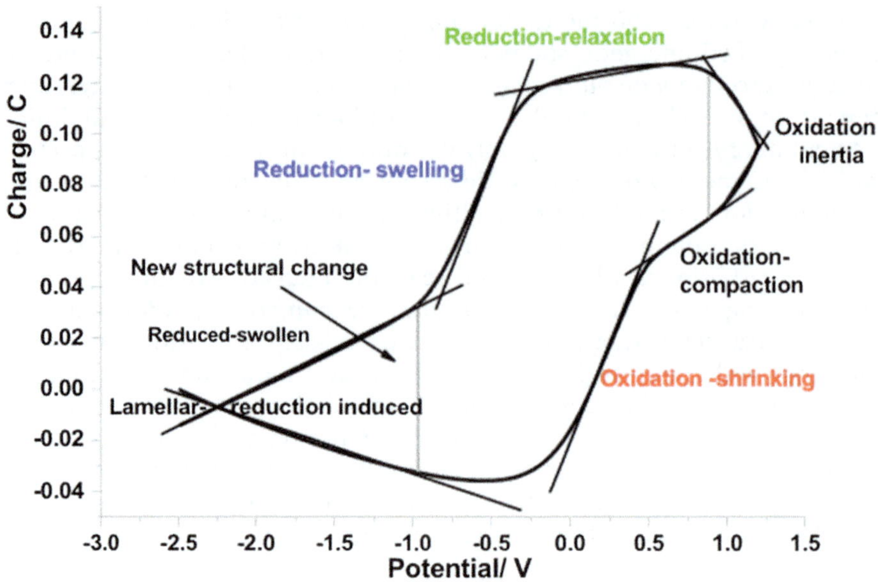

Figure 6.3 Coulovoltammetric responses from a self-supported polypyrrole (Ppy) film doped with dodecylbenzenesulfonate (DBS) in a 0.1 M NaCl aqueous solution at 6 mV s^{-1}, showing abrupt slope variations related to abrupt reaction rate changes, each due to the presence of a structural transition inside the film driven by reactions and described by the electrochemically stimulated conformational relaxation (ESCR) model: oxidation–shrinking, oxidation–compaction, reduction–relaxation and reduction–swelling. Moreover, the polymer blend presents a new, slow structural variation in the reduced and swollen state that could be due to micellar or lamellar organization of DBS and water. Reprinted from ref. 21: *Adv. Funct. Mater.*, 2014, **24**, 1265–1274, with the permission of Springer.

Bending movements of the Ppy-dodecylbenzenesulfonate/tape (Ppy-DBS/tape) bilayer muscles under potential cycling (Section 6.13.2) and EDX analysis of the oxidized and deeply reduced films confirm that the cation is the main exchanged ion in the full potential range, with the forwards and backwards reactions in eqn (4.2) being the only driving reactions.

It would be expected that deep reduction states could be attained very easily from these reduced, swollen membranes. Nevertheless, after a reduction threshold, a new slow reduction process emerges and goes on at the beginning of the next oxidation branch. The new, unexpected, slow reduction process makes completion of the film reduction very difficult and induces a great electrochemical inertia at the beginning of the next anodic potential sweep. For conducting polymers doped with large amphiphilic anions, the new, slow structural process initiated beyond the reduction threshold has been attributed to the formation of small vesicles (*reduction–vesicle*) or lamellar (*reduction–lamellar*) structural processes: *i.e.*, water and cations are surrounded by the amphiphilic membrane monolayer. A thin film of deeply reduced and conformationally compacted conducting polymer, which amphiphile ions have migrated to in order to form the vesicle membrane, should surround each vesicle.

In summary, for reaction-driven cation exchange membranes, the identified structural processes are named: oxidation–shrinking, oxidation–compaction, reduction–relaxation, reduction–swelling, and *reduction–vacuolar or reduction–lamellar formation.*

It appears clear that the complex blend family of conducting polymers (with different organic anions or different organic molecules adsorbed during electropolymerization; Section 1.5.1) constitutes an unexplored field for the identification and study of new reaction-driven structural biomimetic transformations.

The concomitant n-doping and n-dedoping processes from conducting polymer films exchanging cations [eqn (4.3)] have only been very superficially studied. Exploratory voltammetric (Figure 4.3) and chronoamperometric (Section 6.13.4) responses confirm the presence of some of the above-described basic reaction-driven structural processes: reduction–relaxation, reduction–swelling, oxidation–shrinking and oxidation–compaction. The oxidation–compaction traps a great percentage of the involved cations that can only be expelled at very high anodic overpotentials overlapping the polymer p-doping with entrance of the anions. An ionic crosscurrent exists inside the polymer during p-doping: *i.e.*, expulsion of cations to the electrolyte, and anion entrance from the electrolyte.

6.5 Erasing Structural and Chemical Memories: Steady State Responses

The history of the film membrane electrode (synthesis, drying, treatment in different solvents or electrolytes, pressure, stress, strain and so on) determines its internal conformationally compacted or relaxed structure. When the material is put inside an electrochemical cell under new physical, chemical or electrochemical conditions, several oxidation/reduction cycles are required under the new conditions in order to erase any previous three-dimensional structural memory inside the material. After cycling, the new steady state electrochemical response describes the structural electroactivity of the material under actual experimental conditions. In the 1980s the term *memory effect* was coined and it referred to anomalous electrochemical responses from conducting polymers, which are related to those here treated as structural and memory effects.[26–29]

Thus, when the conducting polymer is transferred inside a new electrolyte, several consecutive oxidation/reduction cycles are required to guarantee both removal of any structural memory and, if required, total removal and replacement of the old charge balancing ion by that present in the new electrolyte. During each cycle, the amount of material retaining the memory of the previous state (polymer–ion compacted structural memory) decreases by the progressive advance of conformational relaxation and the swelling front. The fraction of the old balancing counterion there included will be expelled during the subsequent reaction-driven shrinking process. The structural memory-erasing process goes on during consecutive cycles until total replacement of the old balancing ion by the new one.

6.6 Other Electrochemical Responses Reveal Reaction-Driven Structural Changes

In addition to coulovoltammetric responses when different electrochemical methodologies are applied to conducting polymer films, the obtained responses diverge from those described by electrochemical textbooks for the most usual electrochemical reactions. From the beginning of the 1980s, these unexpected responses were named *anomalous electrochemical responses*, *memory effects*, *relaxation effects*, *asymmetric effects* and so on.[26–28,30–42] Here, these anomalous responses will be presented and described according to the reaction-driven structural changes introduced previously.

The following descriptions always refer to electrochemical responses from films of conducting polymers after being previously subjected to consecutive oxidation/reduction cycles until stationary voltammetric responses were obtained in the same electrolyte: any previous structural or chemical (trapped ions) memory was erased (Section 6.5).

6.7 Voltammetric Responses

Self-supported membranes of conducting polymers exchanging anions during the reaction shown in eqn (4.1) were subjected to cyclic potential sweeps at rising cathodic potential limits or after reduction–compaction at the same potential for different reduction times (named *wait times* in the literature) in liquid electrolytes. The stationary anodic voltammetric responses (Figure 6.4) show that the membrane re-oxidation begins at rising anodic potentials for increasingly reduced–compacted initial states. Deeper reduced initial states of the film require rising anodic electrochemical energies (rising anodic overpotentials) to be re-oxidized. In the 1980s and 1990s these were quite surprising results: deeper reduced materials present rising difficulties with regards to their re-oxidization.[28–32,36,37,43,44] They diverge from voltammetric results described by electrochemical and chemical textbooks, where higher initial concentrations of any reduced species always give faster and easier subsequent oxidation reactions: they were named *anomalous voltammetric responses*.

The results are now rationalized by considering the reaction-induced structural changes. After the membrane reduction for rising reduction

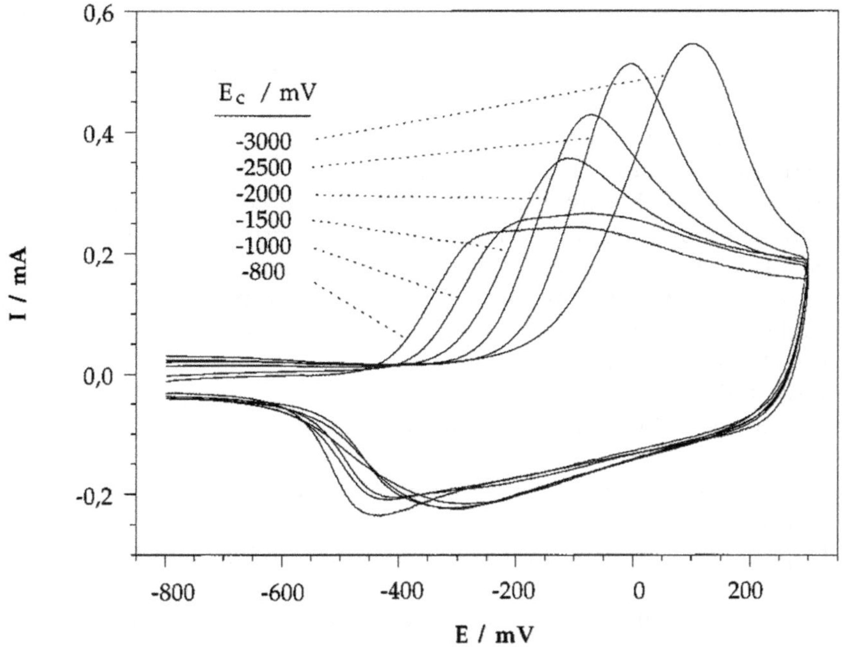

Figure 6.4 Stationary voltammetric responses from a polypyrrole film in 0.1 M $LiClO_4$/propylene carbonate solutions between different cathodic potentials (indicated on the figure) and 300 mV *vs.* a saturated calomel electrode (SCE) at 30 mV s^{-1} and 25 °C. Reprinted with permission from *J. Phys. Chem. B*, **101**, 8525–8533. Copyright (1997) American Chemical Society.

times or until more cathodic potential limits, increasingly reduced and conformationally packed states were attained, storing increasing chemo-conformational energies. During the subsequent anodic potential sweep, the beginning of the film membrane oxidation requires higher energies to relax these increasingly compacted energetic structures in order to generate the free volume required to lodge the anions entering from the solution for the film charge balance. Working under constant temperature (constant thermal energy) and a constant electrolyte concentration (constant chemical energy), the only available energetic source in the working electrochemical system is the rising applied anodic potential (electrochemical energy): more compacted initial structures require higher anodic overpotentials to start the oxidation–relaxation process.

Once relaxed, increasingly reduced–compacted initial structures give higher oxidation peaks. The oxidation state of the membrane at these high oxidation overpotentials is hindered in comparison to the standard oxidation potential of any polymer chain in solution or to that of any partially reduced and non-compacted film, as the initial state is more packed. As soon as the membrane relaxes, the film oxidation state corresponding to these high overpotentials must be recovered by consumption of higher charges, giving a higher current peak. After the maximum, the same oxidation–swollen state must be attained every time.

At the end of the oxidation maximum, the initial packed state was fully relaxed, oxidized and swollen: any previous conformational memory was erased. Whatever the initial reduced and packed conformational energetic state of the membrane, any conformational memory is erased by completion of its subsequent oxidation–swelling. If the attained final oxidation state is the same, whatever the initial reduced and conformationally packed state, the subsequent reduction–shrinking voltammetric branches overlap, as observed in Figure 6.4.

Any of the voltammetric responses from Figure 6.4 can be fully reproduced, starting from a fully oxidized, swollen film after:

(a) film reduction at the same potential each time for its specific reduction time;
(b) potential sweep until the concomitant cathodic potential limit.

This means that starting from the same oxidized, swollen state, the same reduced and conformationally compacted energetic state is attained by either of the two methods. In this way, starting from the same oxidized, swollen state of the film, any reduced, compacted state can be attained in a fully reproducible way, independent of the material's history of reversible processes (avoiding degradation processes) between both experiments. In summary, any reduced and conformationally compacted energetic state is the result of a reaction kinetic structural process. Rising reduction potentials result in higher reduction rates giving, for a constant reduction time, more reduced and packed final states. Film reductions at a constant potential give the same reduction rate: rising reduction times give more reduced and conformationally packed final states.

Otherwise, if the potential sweep is reversed at the beginning of the oxidation–relaxation process, the current gives a loop (Figure 6.5a and c): the oxidation–relaxation goes on even if the potential sweep is reversed. These voltammetric loops are characteristic responses from electrochemically induced nucleation processes (metal electrodeposition, pitting corrosion and so on). Here, the oxidation-driven conformational relaxation starts by formation of oxidized, relaxed nuclei in the reduced and conformationally packed matrix. If, starting from a reduced and conformationally compacted initial structure, the potential is reversed before completion of oxidation–relaxation–swelling (*i.e.*, at different potentials before the voltammetric maximum), the initial packed structure is only partially erased. The material keeps a fraction of the initial conformational energetic memory (the initial packed state) and the reduction voltammetric branches do not overlap.

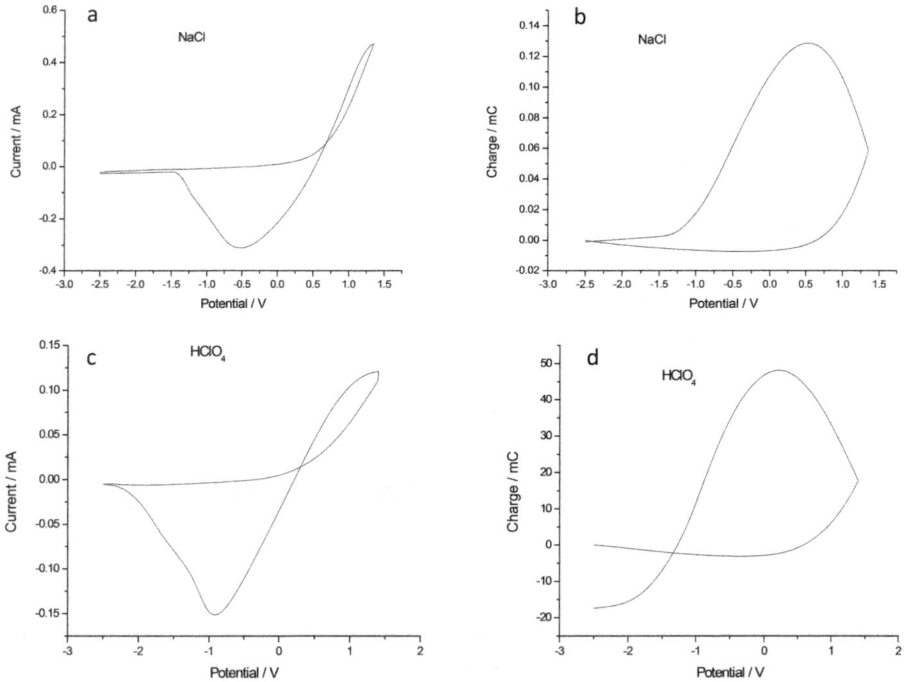

Figure 6.5 (a) Voltammetric and (b) coulovoltammetric responses of self-supported polypyrrole films in 0.1 M NaCl aqueous solutions from −2.5 to 1.4 V; (c) voltammetric and (d) coulovoltammetric responses in 0.1 M $HClO_4$ aqueous solutions from −2.5 to 1.4 V. The potential sweep rate was 3 mV s^{-1}. Reprinted from ref. 21: *Adv. Funct. Mater.*, 2014, **24**, 1265–1274, with the permission of John Wiley and Sons.

6.8 Chronoamperometric Responses

Any shrunk, open and partially reduced state (before the closing state) responds after application of an anodic potential step to give a high and narrow chronoamperometric oxidation peak, such as those described by electrochemical textbooks for electrochemical oxidation reactions (Figure 4.5, dashed line; and Figure 6.6 from −200 and −400 mV). Starting now from a reduced and conformationally compacted state, the subsequent chronoamperometric response to an anodic potential step presents a large oxidation–nucleation maximum (Figure 4.5, solid line; and Figure 6.6 for cathodic E_c values more than −400 mV). The current of the maximum decreases and the time required for its formation increases for increasingly reduced and conformationally compacted initial states.[38–42,45–47] These responses again illustrate the origin of the named anomalous electrochemical responses from conducting polymers: deeper reduced initial states require longer oxidation times with the flow of lower currents to be

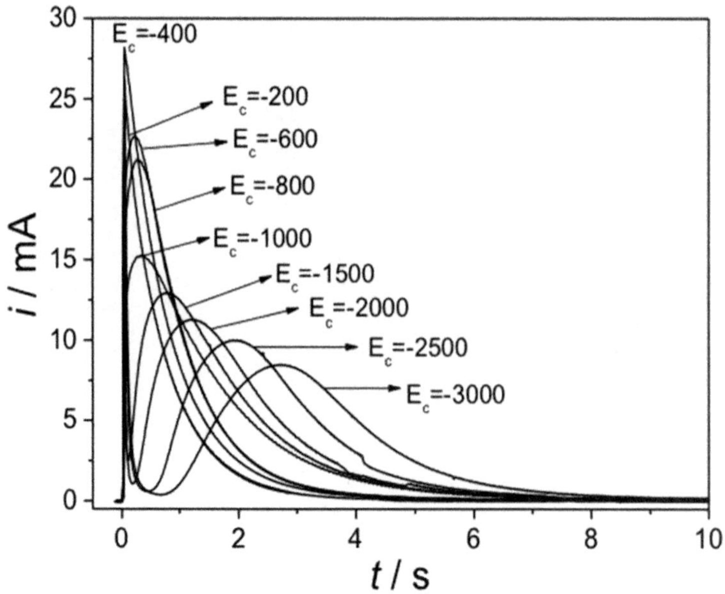

Figure 6.6 Chronoamperometric responses from a poly(3,4-ethylenedioxythiophene)-(PEDOT-) coated Pt electrode in a 0.1 M $LiClO_4$ acetonitrile solution subjected to potential steps from different cathodic potentials, kept at each step for 30 s, up to 500 mV. Reprinted from ref. 47: *Polym. Int.*, 2010, **59**, 329–336, with the permission of John Wiley and Sons.

oxidized at the same anodic potential. This should be equivalent to saying that higher initial concentrations of Fe^{2+} (*i.e.*, $[Fe^{2+}]$) give lower oxidation rates (r) to Fe^{3+}, contradicting the accepted oxidation rate from chemical kinetics: $r = k[Fe^{2+}]$.

After consideration of the reaction-induced structural changes, the "anomalous" chronoamperometric responses from conducting polymer membrane electrodes can be rationalized. A large chronoamperometric maximum is characteristic of any electrochemical reaction initiated by nucleation (corrosion pitting, metal electrodeposition or phase transformations). Here, nuclei of partially relaxed and oxidized polymer are formed and grow at the expense of the reduced and packed polymer membrane. The presence of a small and narrow initial peak is related to the capacitive charge of the electrical double layer (EDL) and that of the initial relaxation. The subsequent minimum corresponds to the beginning of nuclei formation. The following current increase is due to the nuclei expansion under conformational relaxation kinetic control. The inflexion point towards the maximum indicates the beginning of nuclei coalescence. After the maximum, nuclei coalescence is almost finished but continues until completion of the membrane oxidation–swelling under diffusion kinetic control of the counterions from the solution across the membrane.

After completion of oxidation, the reduction responses by potential step to the same initial potential, or the same reduction overpotential, overlap. Any structural memory from the previous conformationally packed state was erased by the completion of oxidation. If the membrane oxidation is interrupted before or around the large nucleation maximum, the membrane retains a partial structural memory of the initial reduction–compaction state.

6.9 Direct Visual Observation of the Oxidation–Relaxation–Nucleation Process

As indicated in Table 5.1, thin and uniform films of conducting polymers are electrochromic and change color in a reversible way during oxidation/reduction reactions: *i.e.*, reduced and oxidized states present a different color. During the potential steps, despite the nucleation shape of the chronoamperometric response, almost uniform color changes are observed according to most of the results described in the literature. This uniform color evolution may be due to a high density of nuclei initiated per unit of electrode area.

Nucleation points (including water or solvent condensation, the precipitation of salt crystals from solution, or the formation of a crystallographic structure of metals by melt cooling) used to be related to the presence of structural seeds or energetic irregularities. Mirror polishing diminishes the concentration of structural irregularities at the metal electrode surface. Very uniform and electrochromic thin films were then generated by consecutive potential steps.[48] A polypyrrole-coated electrode generated in this way was checked in a fresh electrolyte by consecutive square potential steps. After reduction at a potential selected in the reduction–compaction potential range, the film showed a yellow, transparent color. By stepping the electrode potential to an oxidation potential, only a few nuclei of the oxidized (dark blue to black) material were formed and expanded at the expense of the reduced yellow film during the oxidation time.[49] The black nuclei expansion and coalescence can be followed in parallel to the evolution of the chronoamperometric maximum. Consecutive square potential waves always show the oxidized nuclei starting at the same points of the membrane/electrolyte interface.

Each time, the reverse reduction, shrinking and conformational packing of the membrane during the cathodic potential step is visualised as a uniform color change. The membrane reduction–shrinking is a uniform three-dimensional process.

The procedure was repeated, generating and checking thin and uniform films of different conducting polymers, including polyaniline, polymethylaniline, polymethylpyrrole, poly-3-methylthiophene and polythiophene (Figure 6.7), following similar oxidation–relaxation–nucleation processes by subjecting every film to consecutive potential steps.[50,51]

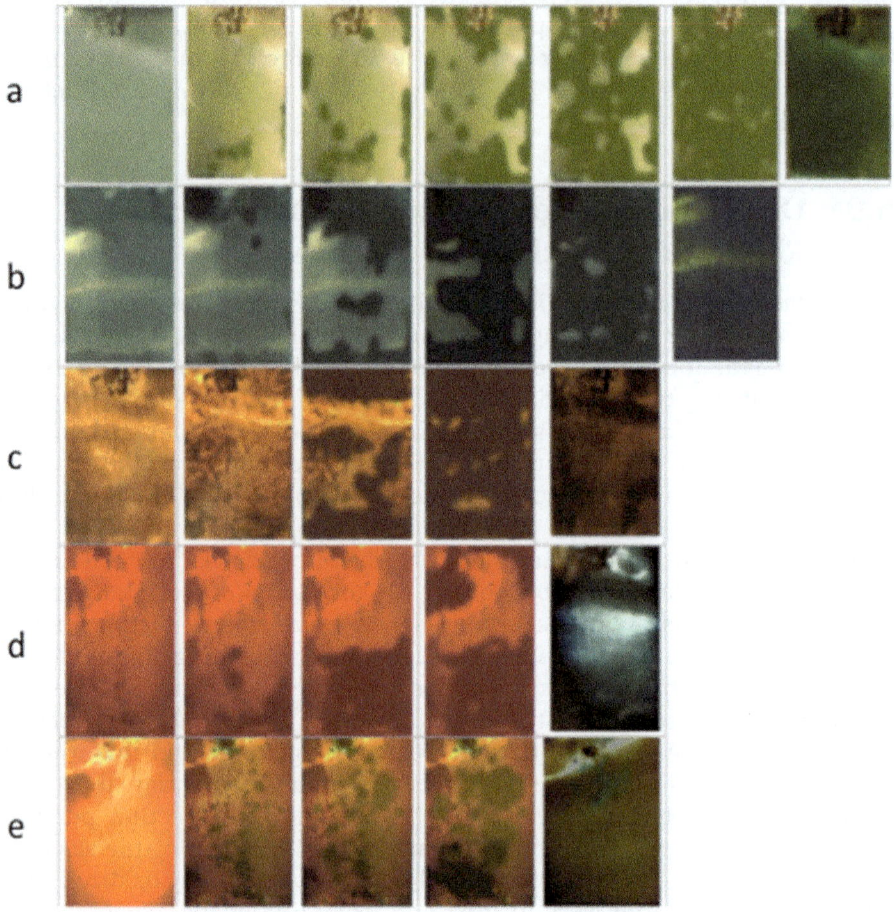

Figure 6.7 Formation and growth of oxidized nuclei during the oxidation of conducting polymer films, previously reduced and compacted by polarization at a high cathodic potential under a constant temperature (25 °C) in a 0.1 M $LiClO_4$ acetonitrile solution. (a) Polyaniline reduced at −850 mV, kept for 60 s (initial picture) and oxidized by potential step to 300 mV. (b) Polymethylaniline compacted at −1200 mV for 60 s and oxidized by potential step to 400 mV. (c) Poly-1-methylpyrrole compacted at −1800 mV for 60 s and oxidized by potential step to 300 mV. (d) Poly-3-methylthiophene compacted at −900 mV for 60 s and oxidized by potential step to 900 mV. (e) Polythiophene compacted at −600 mV for 60 s and oxidized by potential step to 800 mV. Pictures were taken at different oxidation times between 0 and 16 s. Reprinted from ref. 50: *ChemPhysChem*, 2003, **4**, 868–872, with permission from John Wiley and Sons.

6.10 Visual Tracing of the Giant Non-Stoichiometric Nature of Conducting Polymers

The possibility of seeing the formation and growth of oxidized nuclei offers a new, direct and visual way to follow and check the non-stoichiometric nature of materials.

In dealing with stoichiometric materials, *i.e.*, Fe, $Fe(OH)_2$ and $Fe(OH)_3$, it should be noted that local hydroxides produced by local corrosion can remain stable for decades in the absence of oxidizing or reducing agents (chemicals or currents). A great chemical potential gradient exists between the clean metal (μ_{Fe}) and the metal oxide ($\mu_{Fe(OH)2}$). This potential gradient cannot produce a partial $Fe(OH)_2$ reduction with simultaneous partial oxidation of Fe atoms to get a uniform surface coated by some intermediate and non-stoichiometric oxidized state because these non-stoichiometric oxidation states between 0 and 2 do not exist for Fe compounds.

However, conducting polymers are claimed to be giant non-stoichiometric compounds (Section 4.5). Following the same logic from the previous paragraph, the current flow can be switched off for any intermediate state of nuclei growth. As the oxidized polymer is a non-stoichiometric compound, the presence of oxidized and expanded nuclei in the reduced and conformationally packed film must result in instability. Two different energetic gradients act as driving forces to destroy this unstable situation. The first one is the chemical potential gradient between the oxidized nuclei and the reduced film fraction ($\Delta\mu = \mu_{ox} - \mu_{red}$). The second gradient is the mechanical stress between the oxidized and swollen nuclei and the surrounding reduced and conformationally packed film fraction. Taking into account the claimed existence of infinite intermediate and stable oxidation states, these gradients must drive the progressive reduction of the oxidized nuclei with parallel oxidation of the reduced film fraction to get a uniform final composition and color.

Experimental results corroborate these hypotheses. Each time, the procedure is initiated from the same oxidized state (a black, uniform color) of the polypyrrole film obtained by subjecting the electrode to a constant anodic potential for a constant time. The film is then reduced and packed by potential step to the same cathodic potential for the same reduction time for every experiment, yielding a uniform yellow and transparent packed film. The potential is then stepped back to the oxidation potential, taking pictures at different oxidation times during the chronoamperometric response (Figure 6.8A, points a–d, and Figure 6.8B). The full procedure was repeated and the cell electrical connection was switched off with interruption of the oxidation–nucleation growth at different oxidation times (points b–d of the chronoamperometric response in Figure 6.8A), which means a different stage of nuclei growth for every experiment. Whatever the attained nucleation evolution, after the current interruption the black oxidized nuclei progressively reduce and the yellow film fraction progressively oxidizes to give a final uniform green color (Figure 6.8B1 to 6.8B3), corresponding to an intermediate oxidized state.[50] The final uniform

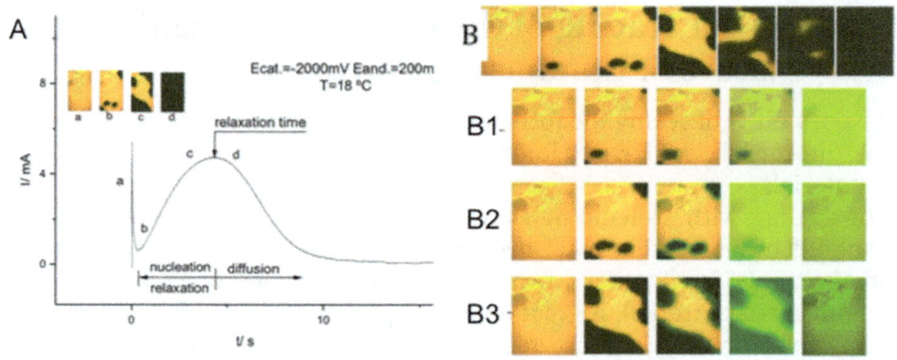

Figure 6.8 (A) Chronoamperometric response from polypyrrole films compacted by cathodic polarization at −1200 mV, maintained for 60 s and oxidized by a potential step to 100 mV at a constant temperature (25 °C) in a 0.1 M $LiClO_4$ acetonitrile solution, showing the formation and growth of nuclei (dark blue) on the compacted and reduced film (yellow) at times a, b, c and d. (B) Pictures showing the nucleation growth at different oxidation times. The procedure was repeated switching off the oxidation/nucleation after: (B1) 3 s, (B2) 5 s and (B3) 8 s. After interruption of the current, new photos were taken to follow the evolution of the system at intervals of 3 s. Reprinted from ref. 50: *ChemPhysChem*, 2003, **4**, 868–872, with permission from John Wiley and Sons.

color becomes darker as the oxidation–nucleation time increases before switching off the current flow (Figure 6.8B). By interrupting the oxidation–nucleation evolution at infinite intermediate times, infinite final stable and uniform oxidized states can be attained, as expected for a giant non-stoichiometric material.

6.11 Relaxation–Nucleation Starts at the Polymer/Electrolyte Interface

For decades, most researchers have supported the conducting/insulator transition model during the reduction of any conducting polymer.[52,53] This model requires the presence of a back metal supporting the conducting polymer film inside the electrolyte in order to allow the re-oxidation of the reduced and porous polymer insulator material. The oxidation can only start, according to this model, through the chains in direct contact with the metal at the polymer/metal interface. Then it advances towards the polymer/electrolyte interface. In order to support the model, electrochromic polymeric films casted from oligomeric solutions on a glass were used as electrodes. A metal contact (plate or sputtered) at the top allowed connection to the working electrode exit of the potentiostat. The system was immersed, including a fraction of the metal, in the electrolyte. After application of an anodic potential step, the polymer oxidation started, as was visualized by the color change, at the metal/polymer contact interface and advanced towards the film bottom.[54]

Despite so many experimental evidences of the high ionic content, electronic conductivity and normal re-oxidation of fully polymeric artificial muscles presented in the literature, the results of the electrochromic front border have been used as the foundation stone of the conductor/insulator transition model.

Recently, this experiment has been re-designed.[55] In order to simulate a transversal polymer ribbon from the metal to the solution, a fraction of the metal and most of the polymer film were protected from direct electrolyte contact by two para-film layers that were applied under pressure. Only 1 mm of the conducting polymer film remained uncoated at the electrode bottom. When used as a working electrode in an electrolyte, the color change after the anodic potential step always starts at the electrode bottom (far from the metal/polymer contact), advancing underneath the para-film towards the metal/polymer contact. This result unambiguously confirmed that the electronic conductivity of a deeply reduced polymer film is high enough to support its re-oxidation, starting at the polymer/electrolyte interface, as shown by the ESCR model. Using an unprotected reduced film, the minimum electrical resistance occurs at the metal/film contact and the color change starts there, advancing towards the film bottom, but these results do not mean that the reduced conducting polymer film is an insulator, as shown by the results from the new design.

In summary, we can conclude that the oxidation–nucleation of a reduced, packed film starts at the polymer/solution interface of the reduced and conformationally compacted film (Figure 6.9).[55] The electrons extracted from the chains flow through the reduced, compacted film towards the metal.

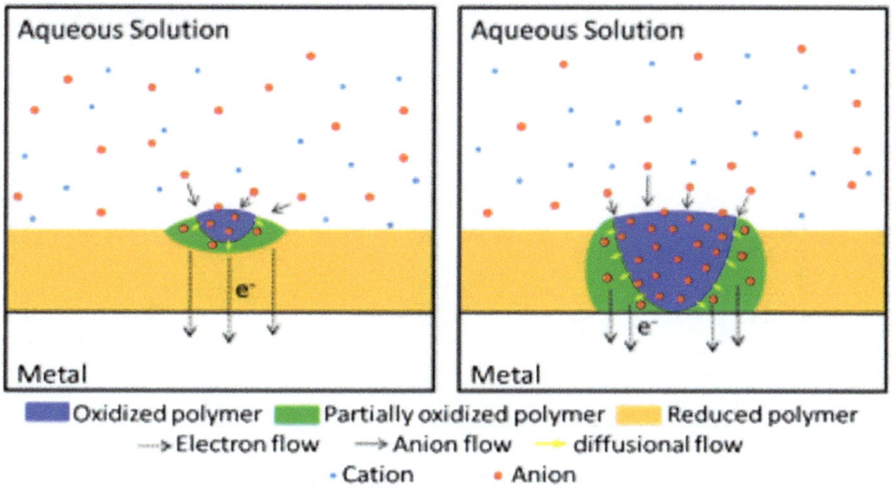

Figure 6.9 Scheme of the oxidation according to the ESCR model: the oxidation of conformationally packed conducting polymers starts by nucleation–relaxation at the polymer/solution interface. Reproduced from ref. 55 with permission.

6.12 Chronopotentiometric Responses

Increasingly reduced and conformationally packed states were attained by reduction at increasing cathodic potentials, each for a constant time. Then the electrode was subjected to consecutive square current waves: a constant anodic current, i_a, for time t, then a constant cathodic current, i_c, for the same time, t. The superposed chronopotentiometric responses during the film oxidation show rising initial peaks (Figure 6.10). Rising initial conformationally packed states act as increasing resistances to the conformational oxidation–relaxation and subsequent entrance of balancing anions.[56]

6.13 Ion Trapping by Structural Effects

According to the previous sections and Chapter 3, ions are trapped inside the polymer matrix during both electropolymerization and conformational compaction driven by reversible film oxidation/reduction processes. Here, for the moment, the best documented trapping effects will be presented.

6.13.1 Ion Trapping During Electropolymerization

The electropolymerization product is an oxidized polymer: the polymer synthesis takes place in an electrolyte and the electrogenerated polymeric chains [eqn (3.7)] immediately oxidize, generating positive charges and

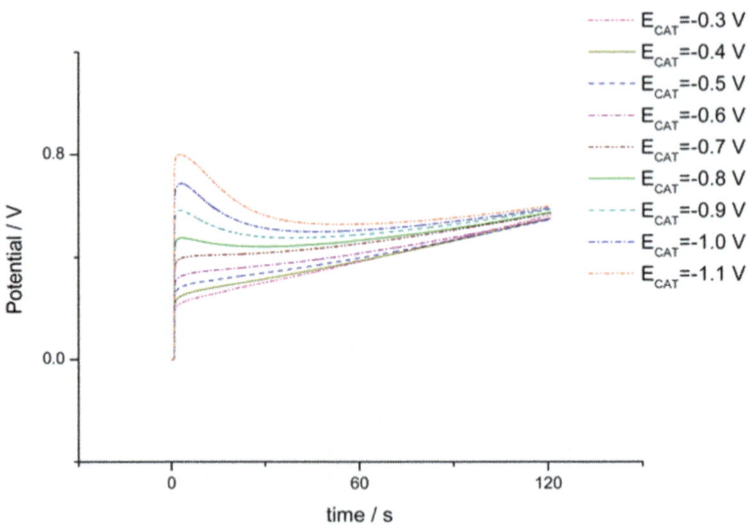

Figure 6.10 Experimental chronopotentiometric results attained from a polypyrrole film in a 0.5 M $LiClO_4$ aqueous solution when, after reduction at different cathodic potentials indicated on the figure for 60 s, a constant anodic current of 1 mA was applied over 100 s. Reprinted from ref. 56 with the permission of the Royal Society of Chemistry.

incorporating anions [eqn (3.8)] from the solution to the growing film for charge balance. When the diameter of the balancing anion is small, any subsequent oxidation/reduction of the material by consecutive potential cycles in a different electrolyte promotes the anion replacement by the new small anion present in the electrolyte [eqn (4.1)]. When the anion of the salt used during the electropolymerization process is too large (*i.e.*, polyelectrolyte, large organic macroions or large inorganic macroions) a conducting-polymer–macro-anion blend is formed.[57–61] The macro-anion remains trapped during any subsequent oxidation/reduction in different electrolytes, forcing the exchange of small cations for charge balance [eqn (4.2)].

6.13.2 Anion Trapping by Reduction–Compaction During p-Dedoping

For those conducting polymers exchanging small anions during the polymer oxidation/reduction [eqn (4.1)], the oxidized and swollen material shrinks during reduction and closes its structure. For thick films (some up to micrometres), up to 40% of the counterions involved in the full redox process are trapped at the closing potential.[22] But these are soft materials. From there, increasingly packed conformation structures are obtained by the slow reduction–compaction of the structure and a progressive volume decrease, as shown by the bending movement of bilayer muscles (Figure 6.11).[62] Increasingly packed conformational structures must correlate with either decreasing concentrations of trapped counterions (Section 6.10), a decreasing concentration of the balancing polaronic structures in the chains or the rising energetic conformationally packed states of chains.

6.13.3 Cation Trapping by Oxidation–Compaction During n-Dedoping or p-Doping

Blends of conducting polymers with organic macro-anions or anionic polyelectrolytes exchange cations during electrochemical reactions, thus the material shrinks and closes its structure, resulting in trapped cations and rising packed conformation structures by oxidation [eqn (4.2) and (4.3)]: cations are trapped by oxidized and conformationally packed states (Section 6.4.2).[22]

For n-doping materials, cations also penetrate in the film for charge balance during the film reduction and are trapped by the film oxidation–closing and subsequent oxidation–compaction. The conformational relaxation–nucleation is revealed now by the reduction chronoamperometric responses (Figure 6.12).

6.13.4 Low Band-Gap Polymers Trap Anions During p-Dedoping and Cations During n-Dedoping

Low band-gap conducting polymers experience either p-doping or n-doping processes in a potential range accessible from electrolytes using organic solvents. The cations that are trapped during n-dedoping by

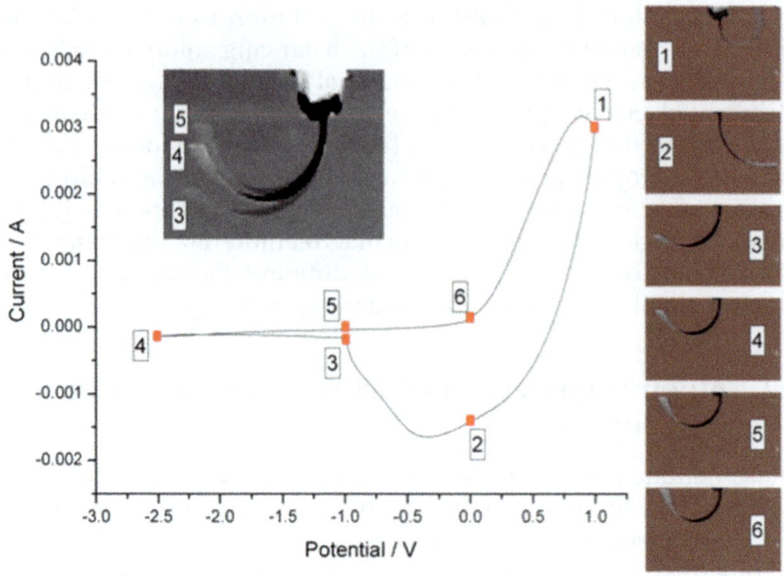

Figure 6.11 A bilayer polypyrrole/tape describes an anticlockwise bending movement during reduction–shrinking and compaction by the expulsion of anions from point 1 to point 6 of the polypyrrole film in a 0.1 M $LiClO_4$ aqueous solution. The movement between point 3 (closing potential) and point 6 confirms the presence of anions trapped at the closed potential. During oxidation–relaxation–swelling, the opposite movement occurs. Reprinted from ref. 62: T. F. Otero and J. G. Martinez, *Sens. Actuators, B*, 2014, **199**, 27–30, with permission from Elsevier.

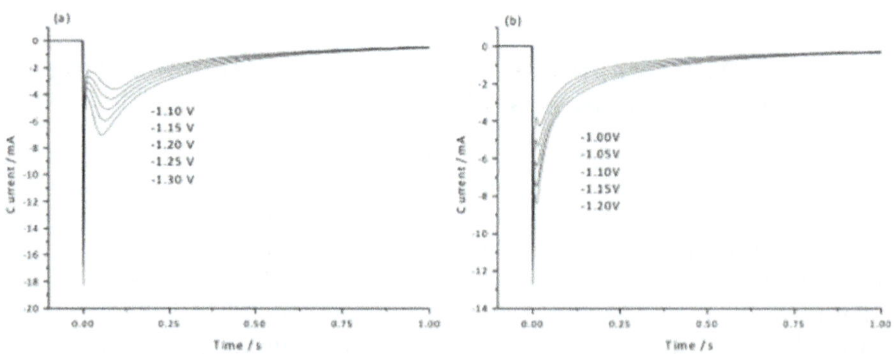

Figure 6.12 Chronoamperometric responses of a poly(2-ethylheptyl-DTCPD-*co*-3-methylthiophene)-coated Pt electrode in a 0.1 M tetrabutylammonium hexafluorophosphate ($TBAPF_6$) acetonitrile solution [where DTCPD is 2,5-dithienyl-3,4-(1,8-naphthylene)cyclopentadienone], subjected to potential steps from (a) 0.25 V and (b) 0.00 V anodic potentials, maintained for 10 s, to different cathodic potentials indicated on the figure. Different characteristic points are indicated. Reprinted with permission from ref. 63: *J. Phys. Chem. B*, **114**, 12777–12784. Copyright (2010) American Chemical Society.

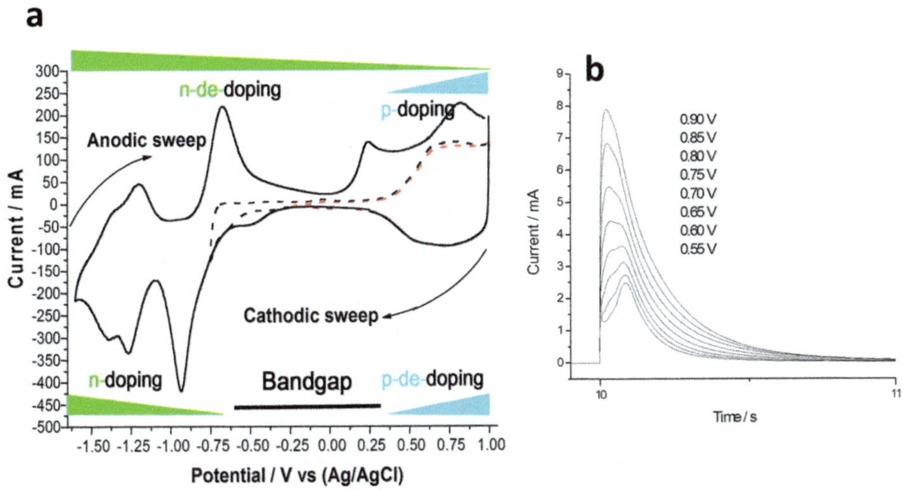

Figure 6.13 (a) Evolution of p-doping and n-doping processes along the anodic and cathodic potential sweeps from poly(2-ethylheptyl-DTCPD-*co*-3-methylthiophene). The dashed red voltammogram shows the p-doping/dedoping domain. The dashed black voltammogram indicates the beginning of the n-doping process, allowing determination of the band-gap. The solid black voltammogram shows all the p- and n-doping/dedoping processes. (b) Chronoamperometric responses subjected to potential steps, maintained for 10 s, from −0.50 V to different anodic potentials indicated on the figure. Reprinted with permission from ref. 63: *J. Phys. Chem. B*, 2010, **114**, 12777–12784. Copyright (2010) American Chemical Society.

oxidation–compaction are released during the material p-doping process that takes place at more anodic potentials and overlaps the oxidation–relaxation and oxidation–swelling with entrance of the anions (Figure 6.13a). The oxidation–compaction during n-dedoping gives reduction–relaxation–nucleation chronoamperometric responses (Figure 6.12). On the other hand, the anions trapped during p-dedoping by reduction–compaction are released during the material's n-doping, driving the membrane reduction–relaxation and reduction–swelling processes and overlapping with the entrance of cations,[63–65] giving oxidation–relaxation–nucleation chronoamperometric responses under potential step (Figure 6.13b). In both cases, the potential difference between the ion trapping and the ion release is close to 1 V (Figure 6.13a), which is suitable for the construction of chemical and chemo-ionic-conformational memories.

6.14 Analytical Evidences of the Ionic Content in Deeply Reduced Films

EDX analysis of closed, reduced or deeply reduced films exchanging anions during reactions show the important content of trapped counterions even after a deep reduction (Figure 6.14).[22] Electron paramagnetic resonance

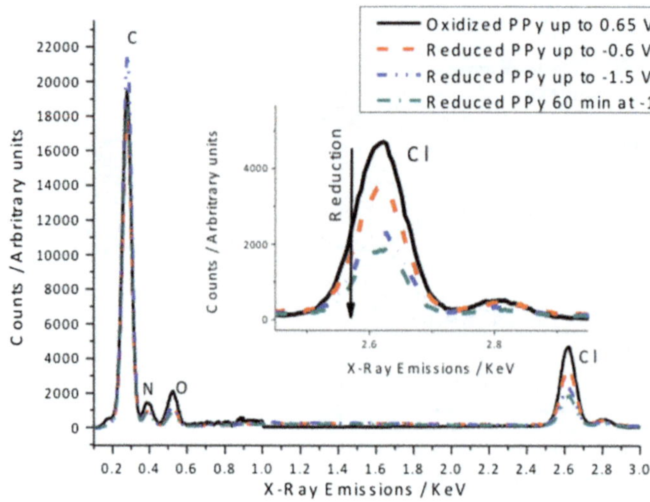

Figure 6.14 EDX spectra from four free-standing polypyrrole electrodes after two consecutive voltammetric cycles between −1.5 V and 0.65 V in a 0.1 M NaCl aqueous solution at 1 mV s^{-1}: the second cycle was stopped at 0.65 V (solid line; oxidized film), −0.70 V (red, dashed line; closing potential) or −1.50 V (blue, dash-point-point-dash line; reduced and packed) for each of three different films. A deeper reduction and packed state of the new film was attained by stopping the second cycle at −1.50 V (green, dash-point-dash line), keeping this reduction potential for 60 min. Reprinted from ref. 21: *Adv. Funct. Mater.*, 2014, **24**, 1265–1274, with the permission of John Wiley and Sons.

(EPR) analysis confirmed the presence of the charge compensating polarons in reduced films.[66–68] Spectroscopic analysis also confirmed the presence of the concomitant conjugated states corresponding to polaronic species.[69]

6.15 Electronic Conductivity of Deeply Reduced Films

Fully reduced films of any conducting polymer are expected to be, according to theoretical calculations from isolated chains, electronic insulators. The fact confirmed in the previous section states that, during reduction, the induced structural changes trap counterions and balancing polarons inside the conformationally packed reduced structure, providing electronic conductivity values to the deeply reduced films of higher than 10^{-4} S cm^{-1} measured in an inert atmosphere.[22,70–83]

As a consequence, any self-supported film can be reduced and compacted until cathodic potential limits as high as permitted by the electrolyte potential window (more cathodic than −3 V). Then the film is re-oxidized during the subsequent anodic potential sweep when the attained overpotential provides the extra energy required for the polymer oxidation–relaxation–nucleation.[21,84] These facts contradict the generally accepted insulator/conductor transition model that suggests the insulating nature of any deeply reduced conducting polymer, which should require a back metal support inside the electrolyte to be re-oxidized. Back metals are not required due to the relatively high conductivity of the reduced film. The back metal will reduce the lifetime of both the materials and electrochemical devices by supporting irreversible reactions that initiate polymer degradation (Section 4.8.1).

6.16 Hydrogen Inhibition from Aqueous Solutions

Despite the relatively high conductivity of the self-supported reduced films, the irreversible hydrogen evolution from aqueous solutions is fully inhibited (Figures 6.2 and 6.5a, b) by some families of conducting polymers up to overpotentials as high as −3 V.[21,84]

Conducting polymers thus become the champion material among the electrode materials inhibiting hydrogen evolution. This unparalleled property is waiting for analytical (competing with carbon powders, graphite, lead, mercury and other hydrogen-inhibiting materials) and technological applications.

In aqueous acid media, slow hydrogen evolution is recovered at high cathodic potentials, which is only revealed from open coulovoltammetric responses (Figure 6.5c and d).

6.17 *In situ* Monitoring of Reaction-Driven Dimensional Changes

Different *in situ* methodologies allow the oxidation/reduction of the polymer film with simultaneous determination of some of its dimensional variations, giving experimental evidences of the direct relationships between involved charges and macroscopic dimensional changes.

Under constant mechanical stress, length variations of the films were followed during oxidation/reduction reactions produced by subjecting the film to potential sweeps,[85–87] square potential steps[85,88,89] or square current steps.[90,91] *In situ* atomic force microscopy (AFM) methodologies allow the film thickness variation to be followed during reverse oxidation/reduction processes.[92–94] For bending bilayer or multilayer systems and low bending amplitudes, the bending beam method allows the characterization of mechanical stress across the multilayer produced by reactions.[95–97]

Electrochemical methodologies provide a very large amplitude (Figure 6.14) of the reverse macroscopic bending movement, beyond the limits of the bending beam method, during the conducting polymer swelling/shrinking driven by oxidation/reduction reactions.[98] Coulodynamic methods (charge consumed *versus* described angle) from bilayer artificial muscles offer a quantitative description of the charge–film volume variation relationships. In parallel, they allow identification of the presence of parallel physical and chemical processes (Sections 8.2.6 to 8.2.9), including the nature of the ionic exchanges during actuation.

6.18 *In situ* Monitoring of Reaction-Driven Mass Variations

The most used methodology to follow the film mass variation during reaction-driven entrance/expulsion of counterions and solvent is the quartz crystal microbalance. A thin gold electrode coating the quartz crystal is used as an electrode for the electrogeneration of a polymer film. Then, in a clean electrolyte, the voltammetric behavior of the film is characterized in parallel with the variation of the resonant frequency of the quartz crystal, which is proportional to the mass variation and to the viscoelastic behavior variation of the polymer film. With some assumptions for the viscoelastic behavior, the parallel evolution of the consumed charge and mass is determined, allowing the identification of the exchanged ions.[99–108]

After empirical evidence of the presence of parallel reactions taking place at the metal/solution interface (Section 4.8) from aqueous electrolytes and directly from the polymer with some families of conducting polymers (Section 4.8.2), results in the literature may require some revision.[109,110] The plot of the coulovoltammetric response allows quantitative determination of the charge consumed by irreversible reactions (not driving ionic exchanges) and its potential range during quartz crystal balance experiments.

6.19 Influence of the Charge Balancing Ion Dimensions

The factors that play a major role during reduction–compaction and oxidation–relaxation structural processes are the polymer film, exchanged anions and the solvent. For the same polymer film working under constant temperature and electrolyte concentration, the concomitant structural responses must be influenced by the dimensions of the counterion unit exchanged between the polymer and the electrolyte driven by the reaction. This unit can be constituted only by the anion with its crystallographic diameter or by the solvated anion (anion plus some solvent molecules). Any of the above-mentioned structural responses in the presence of electrolytes with the same cation and different anions must reveal the relative dimensions of the exchanged anion unit. Figure 6.15a shows the chronoamperometric responses when a polypyrrole film, after obtaining stationary voltammetric responses in each of the studied electrolytes, was reduced-compacted at the same potential for the same reduction time. Then the reduced-compacted film was oxidized by potential step to the same anodic potential in each electrolyte. Different monovalent anions give a different structural chronoamperometric response.

At the limit, when the dimensions of the solved counterion exceed the average inter-chain distance generated by oxidation–relaxation, the polymer oxidation does not occur due to steric hindrances. Figure 6.15b shows how,

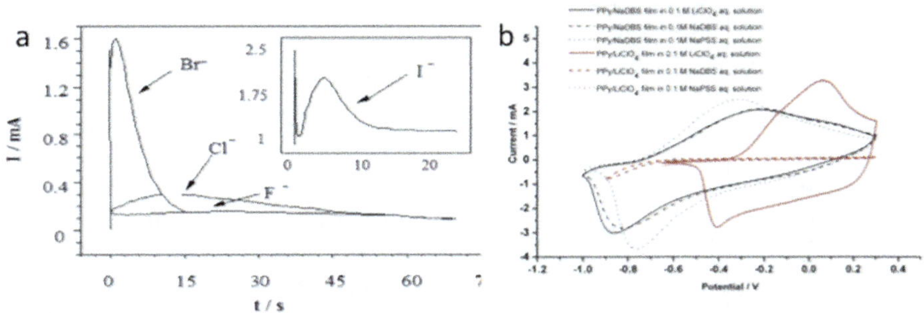

Figure 6.15 (a) Chronoamperometric responses from a Pt electrode coated with a polypyrrole film subjected to a potential step, kept for 120 s, from −2000 mV to 300 mV *vs.* SCE in 0.1 M solutions of different lithium halides (indicated on the figure) in propylene carbonate. The figure shows the influence of the solvated counterion diameter on the position and intensity of the maxima (from *Synth. Met.*, 1996, **83**, 205). (b) Stationary voltammetric responses from polypyrrole films coating Pt electrodes electrogenerated from a 0.1 M pyrrole and 0.1 M $LiClO_4$ acetonitrile solution (red lines) or from a 0.1 M pyrrole and 0.1 M NaDBS aqueous solution (black lines) and checked in a 0.1 M $LiClO_4$ aqueous solution (solid lines), a 0.1 M NaDBS aqueous solution (pH = 5.46; dashed lines) or in a 0.1 M sodium polystyrene sulfonate (NaPSS) aqueous solution (pH = 8.45; dotted lines). Scan rate of 50 mV s^{-1}. Reprinted from ref. 109: *Adv. Funct. Mater.*, 2013, **23**, 3929–3940, with the permission of John Wiley and Sons.

in the presence of a polyelectrolyte (polymeric anion), the polypyrrole voltammetric response is flat, while the same film presents, before and after this control, the usual voltammetric oxidation/reduction response in the presence of small ClO_4^- anions.

6.20 Solvent Influence

Let us keep in mind here that the solvent, despite being included in the reactions shown in eqn (4.1)–(4.3), cannot be considered as a reactant: reactions also occur (slowly) in the presence of solid electrolytes, allowing the movement of the same anion. Different parts of this book underline the central role of the solvent for the membrane osmotic balance. Moreover, the solvent acts as a plasticizer, assisting the different structural processes and influencing structural reaction rates. In addition, the solvent acts on the dissociation degree of the salt and on the actual ionic concentration inside the polymer film and in the electrolyte. Figure 6.16 illustrates the result of the solvent influence on the structural chronoamperometric responses from a polypyrrole film in electrolytes with the same salt and the same salt concentration in different solvents.[111] The film is reduced-compacted every time at the same cathodic potential for the same reduction time and then it is oxidized by potential step to the same anodic potential.

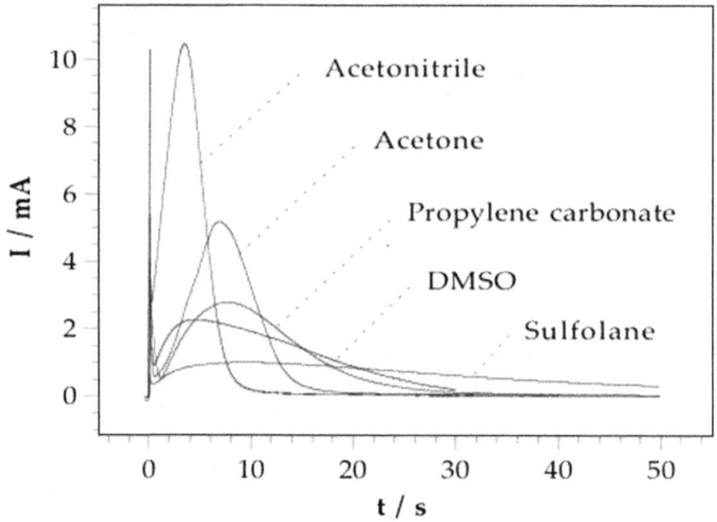

Figure 6.16 Chronoamperometric responses to potential steps carried out using a polypyrrole electrode from −2000 to 300 mV *vs.* SCE, each kept for 50 s, in 0.1 M $LiClO_4$ solutions of different solvents (DMSO, dimethylsulfoxide). Reprinted from ref. 111: *J. Non-Cryst. Solids*, 1998, **235–237**, 619–622, with the permission of Elsevier.

6.21 Other Reaction-Driven Conformational, Allosteric and Structural Responses from Different Artificial and Biological Materials

Electrochemical responses such as those presented in this chapter have been observed in different redox materials, such as carbon nanotubes or porphyrin monolayers. Parallel reaction-driven conformational, allosteric, folding or mis-folding processes are ubiquitous during biological reactions taking place in living cells. Some complementary aspects of artificial and natural reaction-driven structural processes will be presented in Chapter 10.

Now the challenge is in developing the most advanced (compared to the current) chemical models including the structural aspects of the reaction in order to get quantitative and predictive descriptions of both artificial and biological reaction-driven structural processes. This is the aim of the next chapter, where the ESCR model and the structural chemical kinetic (SCK) model will be presented.

6.22 Physical-Driven Conformational Changes

Ultraviolet-visible (UV-vis) spectroscopy has revealed thermochromic, solvatochromic or barochromic effects that have also been attributed to reversible conformational changes from rod-like to coil-like conformations of the conducting polymer chains.[112]

Increasing interest is being shown in the intermediate (between chemically and physically) induced conformational adaptation of the conducting polymer chains to proteins, nucleic acids, polysaccharides and different biological compounds.[113] The reciprocal and synergic conformational influence is opening new artificial borders (see Section 8.10) mimicking biosensing and biocatalytic processes.

References

1. T. F. Otero, *J. Mater. Chem. B*, 2013, **1**, 3754–3767.
2. F. Carpi and E. Smela, *Biomedical Applications of Electroactive Polymer Actuators*, John Wiley & Sons, Chichester, 1st edn, 2009.
3. T. F. Otero, in *Modern Aspects of Electrochemistry*, ed. R. E. White, J. O. Bockris and B. E. Conway, Springer US, New York, 1999, pp. 307–434.
4. T. F. Otero and J. Rodriguez, in *Intrinsically Conducting Polymers: An Emerging Technology*, ed. M. Aldissi, Kluwer Academic Publishers, Dordrecht, 1993, vol. 246, pp. 179–190.
5. T. F. Otero, in *Modern Aspects of Electrochemistry*, ed. R. E. White, J. O. Bockris and B. E. Conway, Springer US, New York, 1999, pp. 307–434.
6. T. F. Otero, in *From Non-Covalent Assemblies to Molecular Machines*, ed. J.-P. Sauvage and P. Gaspard, Wiley-VCH, Weinheim, 2011, pp. 443–452.
7. E. Smela and N. Gadegaard, *J. Phys. Chem. B*, 2001, **105**, 9395–9405.
8. E. Smela and N. Gadegaard, *Adv. Mater.*, 1999, **11**, 953–957.
9. P. R. Singh, S. Mahajan, S. Raiwadec and A. Q. Contractor, *J. Electroanal. Chem.*, 2009, **625**, 16–26.
10. M. F. Suarez and R. G. Compton, *J. Electroanal. Chem.*, 1999, **462**, 211–221.
11. M. Pyo and C. H. Kwak, *Synth. Met.*, 2005, **150**, 133–137.
12. L. Lizarraga, E. M. Andrade and F. V. Molina, *J. Electroanal. Chem.*, 2004, **561**, 127–135.
13. L. Lizarraga, E. M. Andrade, M. I. Florit and F. V. Molina, *J. Phys. Chem. B*, 2005, **109**, 18815–18821.
14. K. Kaneto, Y. Sonoda and W. Takashima, *Jpn. J. Appl. Phys.*, 2000, **39**, 5918–5922.
15. P. Chiarelli, D. Derossi, A. Dellasanta and A. Mazzoldi, *Polym. Gels Networks*, 1994, **2**, 289–297.
16. X. Chen and O. Inganas, *Synth. Met.*, 1995, **74**, 159–164.
17. D. S. H. Charrier, R. A. J. Janssen and M. Kemerink, *Chem. Mater.*, 2010, **22**, 3670–3677.
18. Q. Pei and O. Inganas, *J. Phys. Chem.*, 1992, **96**, 10507–10514.
19. T. Otero, E. Angulo, J. Rodriguez and C. Santamaria, *J. Electroanal. Chem.*, 1992, **341**, 369–375.
20. T. F. Otero, M. Alfaro, V. Martinez, M. A. Perez and J. G. Martinez, *Adv. Funct. Mater.*, 2013, **23**, 3929–3940.
21. T. F. Otero, J. G. Martinez, M. Fuchiwaki and L. Valero, *Adv. Funct. Mater.*, 2014, **24**, 1265–1274.

22. T. F. Otero and J. G. Martinez, *Adv. Funct. Mater.*, 2014, **24**, 1259–1264.
23. M. Fuchiwaki, J. G. Martinez and T. F. Otero, *Adv. Funct. Mater.*, 2015, **25**, 1535–1541.
24. L. Valero, J. G. Martinez and T. F. Otero, *J. Solid State Electrochem.*, 2015, **19**, 2683–2689.
25. M. Fuchiwaki and T. F. Otero, *J. Mater. Chem. B*, 2014, **2**, 1954–1965.
26. A. Abd-Elwahed and R. Holze, *Synth. Met.*, 2002, **131**, 61–70.
27. M. Lapkowski, M. Zagorska, I. Kulszewiczbajer and A. Pron, *Synth. Met.*, 1993, **55**, 1570–1573.
28. H. Randriamahazaka, *Smart Mater. Struct.*, 2011, **20**, 124010.
29. H. Tang, A. Kitani and M. Shiotani, *J. Electroanal. Chem.*, 1995, **396**, 377–383.
30. C. Odin, M. Nechtschein and P. Hapiot, *Synth. Met.*, 1992, **47**, 329–350.
31. K. Aoki, J. Cao and Y. Hoshino, *Electrochim. Acta*, 1994, **39**, 2291–2297.
32. A. Buzarovska and L. Arsov, *Polym. Bull.*, 2002, **48**, 99–104.
33. H. Grande and J. Rodríguez, *Synth. Met.*, 1997, **85**, 1077–1078.
34. M. Lapkowski and E. Vieil, *Synth. Met.*, 2000, **109**, 199–201.
35. K.-M. Mangold, K. Morgenschweis and K. Jüttner, *Electrochim. Acta*, 1999, **44**, 1865–1869.
36. V. Noel, H. Randriamahazaka and C. Chevrot, *J. Electroanal. Chem.*, 2003, **542**, 33–38.
37. M. J. R. Presa, D. Posadas and M. I. Florit, *J. Electroanal. Chem.*, 2000, **482**, 117–124.
38. B. J. West, T. F. Otero, B. Shapiro and E. Smela, *J. Phys. Chem. B*, 2009, **113**, 1277–1293.
39. K. Fraoua, M. Delamar and C. P. Andreiux, *Synth. Met.*, 1998, **97**, 91–95.
40. M. Kalaji, L. Nyholm and L. Peter, *J. Electroanal. Chem.*, 1992, **325**, 269–284.
41. M. R. Nateghi and B. Savabieh, *Electrochim. Acta*, 2014, **121**, 128–135.
42. T. F. Otero, H. Grande and J. Rodriguez, *Electrochim. Acta*, 1996, **41**, 1863–1869.
43. M. Lapkowski, M. Zagorska, I. Kulszewiczbajer and A. Pron, *J. Electroanal. Chem.*, 1992, **341**, 151–163.
44. R. Mazeikiene and A. Malinauskas, *Synth. Met.*, 1998, **92**, 259–263.
45. C. Odin and M. Nechtschein, *Synth. Met.*, 1993, **55**, 1281–1286.
46. J. Baudoin, F. Chao and M. Costa, *J. Chim. Phys. Phys.-Chim. Biol.*, 1990, **87**, 1125–1141.
47. T. F. Otero and M. Caballero Romero, *Polym. Int.*, 2010, **59**, 329–336.
48. T. Otero and E. Delarretaazelain, *J. Chim. Phys. Phys.-Chim. Biol.*, 1989, **86**, 131–141.
49. T. F. Otero, H. J. Grande and J. Rodriguez, *J. Phys. Chem. B*, 1997, **101**, 3688–3697.
50. T. F. Otero and I. Boyano, *ChemPhysChem*, 2003, **4**, 868–872.
51. T. F. Otero, I. Boyano, M. T. Cortes and G. Vazquez, *Electrochim. Acta*, 2004, **49**, 3719–3726.
52. L. Groenendaal, G. Zotti, P. H. Aubert, S. M. Waybright and J. R. Reynolds, *Adv. Mater.*, 2003, **15**, 855–879.

53. J. Heinze, B. A. Frontana-Uribe and S. Ludwigs, *Chem. Rev.*, 2010, **110**, 4724–4771.
54. Y. Tezuka, S. Ohyama, T. Ishii and K. Aoki, *Bull. Chem. Soc. Jpn.*, 1991, **64**, 2045–2051.
55. T. F. Otero, J. G. Martínez and B. Berrueco, *Front. Bioeng. Biotechnol.*, 2015, **3**, 15.
56. J. G. Martinez and T. F. Otero, *RSC Adv.*, 2014, **4**, 29139–29145.
57. L. Buckley, D. Roylance and G. Wnek, *J. Polym. Sci., Part B: Polym. Phys.*, 1987, **25**, 2179–2188.
58. D. S. Maddison and J. Unsworth, *Synth. Met.*, 1989, **30**, 47–55.
59. W. Wernet, M. Monkenbusch and G. Wegner, *Makromol. Chem., Rapid Commun.*, 1984, **5**, 157–164.
60. W. Wernet, M. Monkenbusch and G. Wegner, *Mol. Cryst. Liq. Cryst.*, 1985, **118**, 193–197.
61. M. Depaoli, R. Waltman, A. Diaz and J. Bargon, *J. Chem. Soc., Chem. Commun.*, 1984, 1015–1016.
62. T. F. Otero and J. G. Martinez, *Sens. Actuators, B*, 2014, **199**, 27–30.
63. J. Arias-Pardilla, W. Walker, F. Wudl and T. F. Otero, *J. Phys. Chem. B*, 2010, **114**, 12777–12784.
64. J. Arias-Pardilla, T. F. Otero, R. Blanco and J. L. Segura, *Electrochim. Acta*, 2010, **55**, 1535–1542.
65. T. F. Otero, J. Arias-Pardilla, H. Herrera, J. Luis Segura and C. Seoane, *Phys. Chem. Chem. Phys.*, 2011, **13**, 16513–16515.
66. A. Zykwinska, W. Domagala, A. Czardybon, B. Pilawa and M. Lapkowski, *Chem. Phys.*, 2003, **292**, 31–45.
67. A. Petr and L. Dunsch, *J. Phys. Chem.*, 1996, **100**, 4867–4872.
68. A. Osterholm, B. Meana-Esteban, C. Kvarnstrom and A. Ivaska, *J. Phys. Chem. B*, 2008, **112**, 6331–6337.
69. A. Osterholm, A. Petr, C. Kvarnstroem, A. Ivaska and L. Dunsch, *J. Phys. Chem. B*, 2008, **112**, 14149–14157.
70. A. F. Diaz and J. I. Castillo, *J. Chem. Soc., Chem. Commun.*, 1980, 397.
71. G. Tourillon and F. Garnier, *J. Electroanal. Chem. Interfacial Electrochem.*, 1984, **161**, 407–414.
72. P. M. McManus, R. J. Cushman and S. C. Yang, *J. Phys. Chem.*, 1987, **91**, 744–747.
73. H. Mao and P. G. Pickup, *J. Am. Chem. Soc.*, 1990, **112**, 1776–1782.
74. A. Berlin, G. Zotti, S. Zecchin, G. Schiavon, B. Vercelli and A. Zanelli, *Chem. Mater.*, 2004, **16**, 3667–3676.
75. E. M. Genies and P. Noel, *Synth. Met.*, 1993, **57**, 4192–4197.
76. G. Zotti, S. Zecchin, G. Schiavon, A. Berlin, G. Pagani, M. Borgonovo and R. Lazzaroni, *Chem. Mater.*, 1997, **9**, 2876–2886.
77. W. W. Focke, G. E. Wnek and Y. Wei, *J. Phys. Chem.*, 1987, **91**, 5813–5818.
78. U. Olgun and M. Gülfen, *RSC Adv.*, 2014, **4**, 25165.
79. D. Ofer, R. M. Crooks and M. S. Wrighton, *J. Am. Chem. Soc.*, 1990, **112**, 7869–7879.

80. M. D. Levi, Y. Gofer, D. Aurbach, M. Lapkowski, E. Vieil and J. Serose, *J. Electrochem. Soc.*, 2000, **147**, 1096.
81. J. R. Reynolds, A. D. Child, J. P. Ruiz, S. Y. Hong and D. S. Marynick, *Macromolecules*, 1993, **26**, 2095–2103.
82. K. Mamma, K. Siraj and N. Meka, *J. Polym. Eng.*, 2013, **33**, 785–792.
83. A. B. Kaiser, *Adv. Mater.*, 2001, **13**, 927–941.
84. T. F. Otero and M. J. Ariza, *J. Phys. Chem. B*, 2003, **107**, 13954–13961.
85. R. Kiefer, P. A. Kilmartin, G. A. Bowmaker and J. Travas-Sejdic, in *Electroactive Polymer Actuators and Devices (EAPAD) 2008*, ed. Y. Bar-Cohen, Spie-Int Soc Optical Engineering, Bellingham, 2008, vol. 6927.
86. L. Bay, K. West, P. Sommer-Larsen, S. Skaarup and M. Benslimane, *Adv. Mater.*, 2003, **15**, 310–313.
87. P. Chiarelli, D. Derossi, A. Dellasanta and A. Mazzoldi, *Polym. Gels Networks*, 1994, **2**, 289–297.
88. H. L. Wang, J. B. Gao, J. M. Sansinena and P. McCarthy, *Chem. Mater.*, 2002, **14**, 2546–2552.
89. K. Kaneto, Y. Sonoda and W. Takashima, *Jpn. J. Appl. Phys.*, 2000, **39**, 5918–5922.
90. T. F. Otero, J. J. Lopez Cascales and G. Vazquez Arenas, *Mater. Sci. Eng., C*, 2007, **27**, 18–22.
91. T. F. Otero, G. Vazquez Arenas and J. J. Lopez Cascales, *Macromolecules*, 2006, **39**, 9551–9556.
92. E. Smela and N. Gadegaard, *Adv. Mater.*, 1999, **11**, 953–957.
93. E. Smela and N. Gadegaard, *J. Phys. Chem. B*, 2001, **105**, 9395–9405.
94. M. F. Suarez and R. G. Compton, *J. Electroanal. Chem.*, 1999, **462**, 211–221.
95. Q. Pei and O. Inganas, *J. Phys. Chem.*, 1992, **96**, 10507–10514.
96. G. Alici, P. Metz and G. M. Spinks, in *2005 IEEE/ASME International Conference on Advanced Intelligent Mechatronics*, IEEE, New York, 2005, vol. 1 and 2, pp. 1029–1034.
97. Q. Pei and O. Inganas, *Adv. Mater.*, 1992, **4**, 277–278.
98. T. F. Otero and M. T. Cortes, *Chem. Commun.*, 2004, 284–285.
99. K. Naoi, *J. Electrochem. Soc.*, 1991, **138**, 440.
100. H. Daifuku, T. Kawagoe, N. Yamamoto, T. Ohsaka and N. Oyama, *J. Electroanal. Chem. Interfacial Electrochem.*, 1989, **274**, 313–318.
101. S. Cordoba-Torresi, C. Gabrielli, M. Keddam, H. Takenouti and R. Torresi, *J. Electroanal. Chem. Interfacial Electrochem.*, 1990, **290**, 269–274.
102. A. R. Hillman, M. J. Swann and S. Bruckenstein, *J. Phys. Chem.*, 1991, **95**, 3271–3277.
103. G. Inzelt, V. Kertész and A.-S. Nybäck, *J. Solid State Electrochem.*, 1999, **3**, 251–257.
104. V. M. Schmidt and J. Heitbaum, *Electrochim. Acta*, 1993, **38**, 349–356.
105. S. Skaarup, *Solid State Ionics*, 2003, **159**, 143–147.
106. R. M. Torresi, S. I. Córdoba de Torresi, T. Matencio and M.-A. De Paoli, *Synth. Met.*, 1995, **72**, 283–287.
107. X. Wang, B. Shapiro and E. Smela, *J. Phys. Chem. C*, 2009, **113**, 382–401.
108. X. Wang and E. Smela, *J. Phys. Chem. C*, 2009, **113**, 369–381.

109. T. F. Otero, M. Alfaro, V. Martinez, M. A. Perez and J. G. Martinez, *Adv. Funct. Mater.*, 2013, **23**, 3929–3940.
110. T. F. Otero, J. G. Martinez, M. Fuchiwaki and L. Valero, *Adv. Funct. Mater.*, 2014, **24**, 1265–1274.
111. H. Grande, T. F. Otero and I. Cantero, *J. Non-Cryst. Solids*, 1998, **235**, 619–622.
112. M. Leclerc, *Adv. Mater.*, 1999, **11**, 1491–1498.
113. R. C. Quintanilha, E. S. Orth, A. Grein-Iankovski, I. C. Riegel-Vidotti and M. Vidotti, *J. Colloid Interface Sci.*, 2014, **434**, 18–27.

CHAPTER 7

Conformational, Allosteric and Structural Chemistry: Theoretical Description

7.1 Introduction

Chapter 6 collected multiple experimental evidences of the reaction-driven conformational, allosteric and structural changes from different materials, molecules, polymers and macromolecules, be they biological or artificial. The list could be extended for pages: any surprising, anomalous or non-quantifiable result from most of the studied biochemical processes and functions are attributed in the literature to the conformational changes of the involved biological molecules, enzymes, proteins or nucleic acids. The pending point now is the development of a self-consistent physical–chemical model able to integrate and quantify these reaction-driven conformational and structural changes by chemical kinetic models.

When the reaction-driven conformational and structural movements are ignored, the conducting-insulator transition model can attempt a description of the anomalous electrochemical responses from conducting polymers (Figure 7.1).[1–3] At decreasing temperatures, conducting polymers experience a conductor–insulator transition.[4–8] In addition, theoretical calculations related to the conductivity of an isolated polymeric chain give, as expected from the molecular structure, including C–C σ single bonds between consecutive monomeric units (Figure 1.1a), theoretical conductivities lower than 10^{-9} S cm^{-1}.[9–16] In this context, most electrochemists have considered that the potential ranges of the oxidation–reduction voltammetric maxima correspond to the electronic conducting and electroactive regions of the material.

RSC Smart Materials No. 19
Conducting Polymers: Bioinspired Intelligent Materials and Devices
By Toribio Fernández Otero

Published by the Royal Society of Chemistry, www.rsc.org

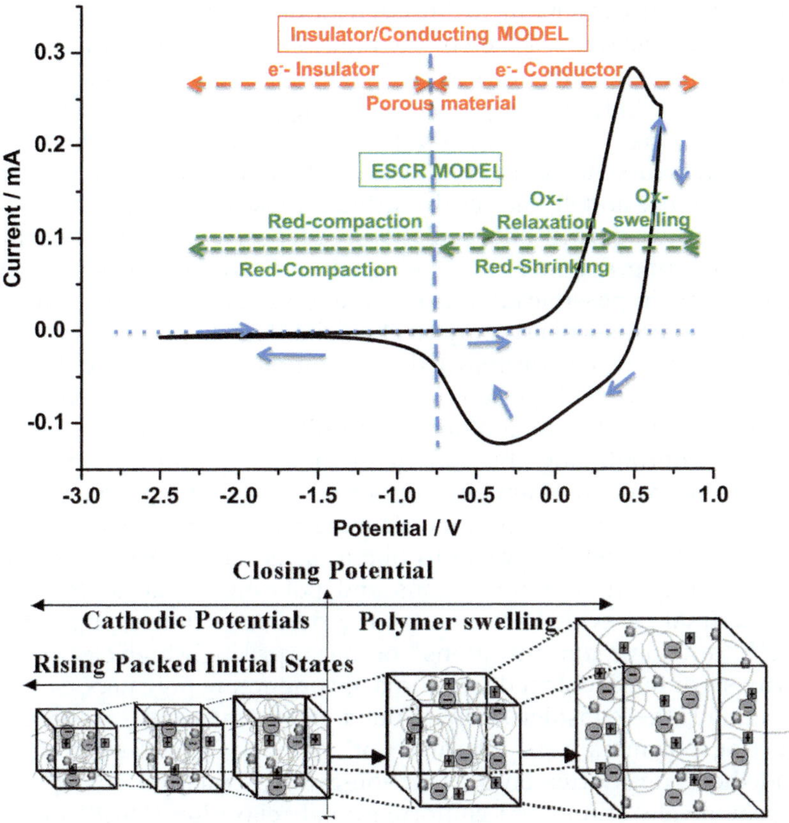

Figure 7.1 Stationary voltammetric responses from a self-supported polypyrrole (Ppy) film in a $LiClO_4$ aqueous solution and potential domains for the different processes or structures according to the conductor/insulator transition model (red) and the electrochemically stimulated conformational relaxation (ESCR) model (green). Reversible structural (volume) variations of a cubic material unit stimulated by the reaction-driven exchanges of the counterion and solvent with the electrolyte are shown below the graph.

Beyond the voltammetric reduction maximum, the current drop during a cathodic potential sweep has been attributed to completion of the polymer reduction, resulting in the conductor–insulator material transition, as referred to in most reviews and books.[2,17–19] The polymer re-oxidation during the subsequent anodic potential sweep only occurs if the polymer is a porous material (porous model, Figure 7.1) supported by a metal electrode. The polymeric re-oxidation can only be initiated, according to this model, through the polymeric chains in direct contact with the metal electrode at the metal–polymer interface, advancing towards the polymer–electrolyte interface through channels. The electrolyte from inside the pores provides the required counterions for charge balance. The material conductivity is explained by percolation of the conducting (oxidized) regions.[20–27] The porous-percolation name of the model was inspired by the multiple works at that time involving composites of conducting polymers with non-conducting polymers, showing percolation thresholds for very low conducting polymer contents.[28–30]

The conducting-insulator transition model has dominated the literature for the last 25 years. According to this theoretical model, any fully polymeric device (Chapters 8 and 9), such as the artificial muscles described since 1992,[31] could only work during one oxidation/reduction cycle. The re-oxidation of the conducting polymer film taking part of any fully polymeric device should become impossible after its first reduction until high cathodic potentials, contradicting very early empirical results.[31]

Experimental results contradict the insulating nature of deeply reduced self-supported films. Fully polymeric artificial muscles work in a reversible way for thousands of cycles. In addition, coulovoltammetric results (Figure 6.2b) show that, contradicting the conducting-insulator transition model, the reduction–compaction of any self-supported conducting polymer film goes on beyond the reduction voltammetric maximum, being re-oxidized during the next anodic potential sweep to obtain steady responses. The polymer reduction–compaction trapping ions and polarons is supported by multiple experimental evidences, such as the determination of high spin,[32–34] charge[33,35] and ionic concentrations[36] or the surprisingly high electronic conductivities[36] of the deeply reduced films of conducting polymers.

In parallel, the considered anomalous electrochemical responses from conducting polymers (*e.g.*, relaxation or memory effects) were attributed to, and theoretically treated as, reaction-driven structural changes by the electrochemically stimulated conformational relaxation (ESCR) model.[37–66] The polymer shrinks (for conducting polymers exchanging anions) during the reduction maximum. The current drop after the reduction maximum is a structural effect due to the reduction–closing of the polymeric structure. The polymer film traps a high percentage of the counterions (and its balancing polarons), corroborating analytical results from the literature, and this supports the surprisingly high electronic conductivity of deeply reduced films. In fact, after the reduction maximum, the current never drops to zero (Figure 6.2a): the material reduction goes on, as revealed by coulovoltammetric responses, driving the slow film reduction–compaction up to very high cathodic potential limits. Empirical determinations of the concentration of counterions (energy-dispersive X-ray spectroscopy, EDX) and polarons (electron paramagnetic resonance, EPR)[67] explain the resulting relatively high electronic conductivities ($>10^{-4}$ S cm^{-1}) of deeply reduced self-supported films. This conductivity is high enough to support polymeric re-oxidation during the subsequent anodic potential sweep: the polymer film doesn't require any back metal support (as claimed by the conductor–insulator transition supporters and followed by most of the device designers) in the electrolyte to guarantee film re-oxidation.

Consequently, this chapter is concentrated on the theoretical treatments of the reaction-driven structural movements, the theoretical description of the above-presented structural electrochemical responses and the new structural chemical kinetics merging with classical chemical kinetic models included in the reaction-driven structural changes. The possible extension of the attained quantitative results to quantify enzymatic, allosteric and other biochemical reactions will be presented.

7.2 The ESCR Model

The model will be described for reaction-driven structural changes in conducting polymer films exchanging anions during the reaction shown in eqn (4.1). Similar reaction-driven structural changes are described in Chapters 6 and 10 for conducting polymers exchanging cations, with coverage of the model regarding those materials, reactions and structural changes.

The ESCR model was based on the following hypotheses:

- The reversible oxidation–reduction of a conducting polymer chain promotes reversible reorganization of its double bonds, resulting in reversible conformational movements: reaction-driven conformational movements (Section 6.1).
- The electrochemical reduction of any oxidized, swollen conducting polymer film [the backward reaction in eqn (4.1)] consecutively drives through the conformational movements shrinking, closing and conformational compaction.
- The film reduction–shrinking process occurs under diffusion kinetic control of the counterions through the polymer film towards the solution.
- The reaction drives a progressive decrease of the intra-chain average distances in the film. When those intra-chain distances approach the counterion unit diameter, the structure closes and the escaping ions drag solvent (electro-osmosis) towards the solution, accelerating the membrane closing–compaction process.
- The structure closes, trapping a high percentage of counterions inside the closed and still partially reduced structure.
- Being a soft material, the reduction reaction shown in eqn (4.1) (backwards), with expulsion of counterions, goes on after closing, but now at a lower reduction rate and under kinetic control of the slow conformational movements. Ions advancing towards the solution must push apart the polymer chains in front of them, as a swimmer does in seawater filled with dense algae.
- The chains occupy the volume that the escaping ions leave behind them, producing increasingly reduced and conformationally packed structures that induce decreasing reduction rates.
- Producing a fully reduced state (free of counterions) from a very thin conducting polymer film (nanometers) becomes a difficult and time-consuming task.[68] This is essentially impossible (for practical

reduction times) from thick (over 10 μm) self-supported films due to the decreasing reduction rates from the increasingly compacted states.

- Starting from the same oxidized and swollen state, stable, reproducible, reduced and compacted states (with a decreasing concentration of trapped counterions) with increasing conformational energies are obtained by reduction at a rising cathodic overpotential (η_c, where $\eta_c = E_c - E$; E_c is the closing potential and E the applied cathodic potential of reduction–compaction) every time, each for a constant time.
- The conformational energy required to compact 1 mole of polymeric segments by electrochemical reduction is $z_c\eta_c$, where z_c is the charge consumed to compact 1 mole of polymeric segments, q_c ($z_c = q_c = it_c$, where i is the current and t_c is the compaction time) under reduction at the compaction overpotential η_c.
- To be re-oxidized, any reduced and compacted energetic state needs (related to any partially reduced, shrunken conducting polymer) some extra energy (the conformational relaxation energy) and time, τ (the relaxation time), in order to relax its structure, generating the required free volume to lodge counterions and solvent needed for charge and osmotic balance, respectively.
- The energy required to relax by oxidation 1 mole of compacted polymeric segments is $z_r\eta$, where z_r is the charge required to relax 1 mole of compacted polymeric segments, q_r ($z_r = q_r = it_r$, where t_r is the relaxation time) and η is the oxidation overpotential: $\eta = E - E_{0,}$ where E is the actual polymer oxidation potential and E_0 is the standard oxidation potential of the polymer (without compaction).
- The amorphous, reduced and packed film surface at the film–electrolyte interface is not energetically uniform: the membrane oxidation–relaxation of any conformationally packed membrane begins by nucleation at the points of the interface with higher conformational mobility of the chains.
- The oxidized, relaxed nuclei advance very fast towards the film–metal interface due to the rising potential gradient between the base of the advancing nuclei front and the metal, forming oxidized and swollen cylinders across the reduced and compacted polymer film.
- The film oxidation advances then by lateral growth and coalescence of oxidized cylinders under conformational relaxation kinetic control.
- Behind the relaxed front, the completion of oxidation–swelling goes on under diffusion control of the counterions and solvent molecules from the solution across the oxidized cylinder.
- The entrance of balancing counterions is a faradaic process controlled by the forwards reaction in eqn (4.1): the entrance of one anion requires the simultaneous extraction of z electrons, z being the ion charge, from the chains.
- The entrance of solvent for osmotic balance is a physical process that must adapt to the structural conditions occurring in parallel to, or following after some time, the faradaic ionic exchange.

The quantitative definition of the molar conformational energy ($z_c\eta_c$) suggests that different methodologies can be used to obtain very reproducible reduced and compacted states of rising conformational energy. The initial state is a reproducible, oxidized and swollen state of the film attained each time by polarization at the same oxidation overpotential, η, for a long enough oxidation time. Then, rising conformational energetic states are attained:

- by reduction at rising compaction overpotentials, η_c, each for the same reduction–compaction time (same z_c);
- by reduction at the same compaction overpotential, η_c, for rising compaction times, t_c, which means rising z_c;
- under potential cycles, switching off the cathodic potential sweep at a higher reduction–compaction overpotential, η_c, every time;
- by flow of a constant cathodic current (i_c) during rising reduction times (rising $z_c = i_c t_c$).

7.2.1 Conformational Relaxation–Nucleation: Relaxation Time, Conformational Energy and Relaxation Energy

Two key concepts support the quantitative description of reaction-driven structural changes: *i.e.*, the film's conformational reduction–compaction and the subsequent oxidation–relaxation of any reduced and conformationally compacted state. The theoretical problem involves polymers following electrochemical reactions. Any theoretical treatment must integrate electrochemical (macroscopic) models and polymeric (molecular) models.

The integration of similar macroscopic and molecular theoretical components has been solved in the literature by developing relaxation models (magnetic, mechanical and relaxation polarization currents): we can now try a new relaxation model that includes chemical reactions, *i.e.*, the ESCR model.

As for any other relaxation model, the relaxation time (τ) is related to the change of the system molar energy (ΔH) by the Arrhenius expression, with R as the gas constant and T as the temperature:

$$\tau = \tau_0 \, \mathrm{e}^{-\left(\frac{\Delta H}{RT}\right)}. \tag{7.1}$$

The molar energy for reaction-driven structural polymeric changes under constant pressure (molar enthalpy) includes, according to the above hypotheses, three terms:[53,54]

$$\Delta H = -\Delta H^* - \Delta H_c + \Delta H_r, \tag{7.2}$$

where ΔH_c represents the increment of the polymer molar conformational energy by polymeric reduction–compaction at the cathodic overpotential η_c (related to the closing potential):

$$\Delta H_c = z_c \eta_c, \tag{7.3}$$

where z_c is the conformational compaction coefficient: the charge required to compact, by reduction, 1 mole of polymeric segments (in C M^{-1}). ΔH_r represents the decrease of the polymer molar conformational energy by its oxidation–relaxation at the anodic overpotential η (related to the standard polymeric oxidation potential):

$$\Delta H_r = z_r\eta. \tag{7.4}$$

ΔH_r is the energy required to relax 1 mole of reduced, compacted polymeric segments, with z_r being the conformational relaxation coefficient, *i.e.*, the charge required to relax 1 mole of polymeric segments (in C M^{-1}). The anodic overpotential η is the driving magnitude of any electrochemical oxidation (Butler–Volmer equation). Through ΔH_r, the ESCR model includes electrochemical kinetics, electrochemical models or electrochemical methodologies. The term $\Delta H*$ collects the rest of the energetic components of the system, such as the polymer–polymer, polymer–solvent, polymer–ion or ion–solvent interactions inside the polymer film for each of the oxidation states and in the absence of electric fields. Let us keep in mind that this term can also include any other energetic interaction with the polymer chain: photo-physical or photo-chemical, magneto-physical or magneto-chemical, piezoelectrical, electrostrictive, electromagnetic interactions and so on. Some of these processes, such as electroporation, allow the penetration of drugs through living cell membranes that would otherwise be blocked. Unless otherwise specified, these energetic interactions are not considered (*i.e.*, they are kept constant) in this book.

Thus, through the energetic addition of $\Delta H* + \Delta H_c$, the ESCR model can include polymeric models and polymeric methodologies. The relaxation time becomes:

$$\tau = \tau_0\, e^{-\left(\frac{\Delta H* + z_c\eta_c - z_r\eta}{RT}\right)}. \tag{7.5}$$

7.2.2 Structure and Chemical Reactions

Through the electrochemistry of conducting polymers and other electrochemically active materials, scientists have at their disposal reactive dense gels, the polymeric chains of which act as reactants experiencing molecular (conformational) and macroscopic (swelling, shrinking, compaction, relaxation) structural changes driven by electrochemical reactions. These reactive dense gels, mimicking the intracellular matrix of living cells, can be used as new materials and reaction models to check the limits of present chemical kinetic models. The resulting models could be applied to describe biological reactions and for the quantification of biological functions.

The ESCR model offers different experimental ways to get, in a very reproducible way, different reduced and shrunken, and different reduced and compacted states, as described at the end of Section 7.1. Now the empirical chemical kinetics can be attained for the same reaction but starting the reaction from initial states storing a different conformational energy (different reduction–compaction state) each time.

7.2.3 Structural Chemical Kinetic (SCK) Model

The study of the empirical chemical kinetics of a reaction gives, through any of the chemical kinetic methodologies described by present textbooks, the reaction kinetic magnitudes: the activation energy (E_a), the reaction coefficient (k) and the reaction orders (a, b) related to each reactant concentration. For the reactions shown in eqn (4.1)–(4.3), the reactants are active chain sites present inside the polymer film (most of them) and ions present in solution. The new reaction-driven structural processes offer the, until now non-existent, opportunity of repeating the full kinetic methodology using, each time, a partially reduced and swollen state (before the structure closing) or one of the increasingly reduced and conformationally packed states.[37,56,59–62,64,69–75] From each of these initial states, a new set of E_a, k, a and b values will be attained. According to current chemical kinetic models, this is a pointless, difficult and time-consuming task: being the same reaction, the same values of the kinetic magnitudes should be expected from each of the studied initial states. Let us see.

From the reaction in eqn (4.1), the oxidation empirical kinetics are given by the rate (r_{ox}) at which the active sites from [Pol*] chains disappear. From eqn (2.3), this reaction rate is:

$$r_{ox} = -\frac{d[\mathrm{Pol}*]}{dt} = -\frac{d\left(\frac{Q}{zwF}\right)}{dt} = -\frac{1}{zwF}\frac{dQ}{dt} = -\frac{1}{zF}I = k[\mathrm{A}^-]^a[\mathrm{Pol}*]^b. \quad (7.6)$$

By subjecting the conducting polymer film to a potential step from an initial reduced (shrunk or compacted) state to the same final state (the same anodic potential) gives, every time, the empirical polymeric oxidation rate (r_{ox}) evolution as chronoamperometric, i–t, responses (Figure 7.2a); chronocoulometric, Q–t, responses (Figure 7.2b); or variation of the concentration of the film active sites with time, [Pol*]–t (Figure 7.2c). The chronoamperometric responses from any conformationally compacted initial state present a nucleation–coalescence maximum (Figure 7.2a). The chronocoulometric (Figure 7.2b) and the concentration–time (Figure 7.2c) responses present an inflexion point: these are reactions with an induction time.

The kinetic methodology of the initial reaction rates can be selected to get the empirical kinetics. The initial oxidation rate from partially shrunk initial states is [eqn (7.6)] the initial slope of the concentration of active sites with time (d[Pol*]/dt; Figure 7.2c, green line). This can also be determined [eqn (7.6)] from the initial slope of the concomitant chronocoulometric (dQ/dt) response (Figure 7.2b, green line) as: $(1/zwF)(dQ/dt)$. The initial chronocoulometric slope is the current ($dQ/dt = I$) at the chronoamperometric maximum (Figure 7.2a, blue and green lines), thus the initial oxidation rate can also be calculated with the current at the maximum as: $[(1/zwF)I = (1/zF)I]$.

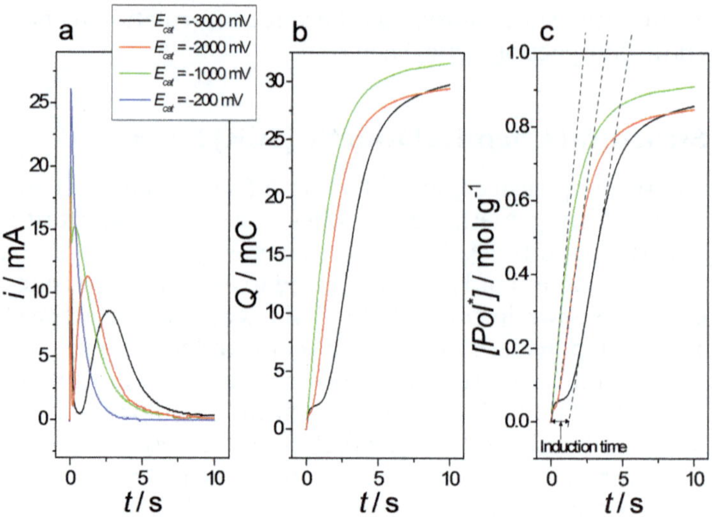

Figure 7.2 (a) Chronoamperometric responses of a poly(3,4-ethylenedioxythiophene)- (PEDOT-) coated Pt electrode in a 0.1 mol L^{-1} $LiClO_4$ acetonitrile solution subjected to potential steps from different cathodic potentials kept for 30 s to 500 mV. (b) Chronocoulograms obtained by integration of these experimental chronoamperograms. (c) Variation of the concentration of the active centers in the polymeric film with time, calculated using eqn (4.11). Reprinted from ref. 59: *Polym. Int.*, 2010, **59**, 329–336, with the permission of John Wiley and Sons.

Using initial conformationally compacted states, the empirical kinetics present an induction time (Figure 7.2c, red and black lines).[59] The initial oxidation rate is the slope at this induction time. This can also be determined $[(1/zwF)(dQ/dt)]$ from the slope of the concomitant chronocoulometric responses (Figure 7.2b, red and black lines). Taking into account the correlation between the induction points (Figures 7.2 and 7.3) and the second chronoamperometric maxima (Figure 7.2), the initial oxidation rate can also be calculated from the current at that maximum: $[(1/zF)I]$.

Once the initial oxidation rate has been obtained, the empirical polymeric oxidation kinetics can be calculated from each of the chosen initial shrunk or conformationally compacted states following the usual kinetic methodology described by present textbooks of chemical or biochemical kinetics.

By including the activation energy in eqn (7.6) the oxidation empirical kinetics become:

$$r_{ox} = k_0 e^{-\frac{E_a}{RT}} [A^-]^a [Pol*]^b \tag{7.7}$$

and taking logarithms gives:

$$\ln r_{ox} = \ln k_0 - E_a/RT + a\ln[A^-] + b\ln[Pol*]. \tag{7.8}$$

By subjecting the polymer film to potential steps from the same initial (relaxed or conformationally packed) state to the same oxidation potential [constant concentration of active sites and final concentration of counterions, eqn (4.11)] under a constant electrolyte concentration $[A^-]$ and at a different temperature (T, °K) every time, eqn (7.8) under these experimental conditions becomes:

$$\ln r_{ox} = \ln k' - E_a/RT, \tag{7.9}$$

where the constant terms from eqn (7.8) are collected in a new constant k':

$$\ln k' = \ln k_0 + a\ln[A^-] + b\ln[\mathrm{Pol}*]. \tag{7.10}$$

By plotting the logarithm of the attained (as above described) empirical initial oxidation rates (r_{ox}) *vs.* $1/T$, the chemical nature of the studied process is confirmed if eqn (7.9) is followed. The activation energy is obtained from the slope (E_a/RT). The origin ordinate ($\ln k'$) will allow, after obtaining the reaction orders a and b, the reaction coefficient k_0 to be obtained from eqn (7.10).

By now subjecting the polymer to potential steps from the same initial (relaxed or conformationally packed) state to the same oxidation potential (a constant concentration of active sites) in an electrolyte of a different $[A^-]$ concentration each time and under constant temperature (T), eqn (7.8) becomes:

$$\ln r_{ox} = \ln k'' + a\ln[A^-], \tag{7.11}$$

where the new constant terms from eqn (7.8) are collected in a new constant k'':

$$\ln k'' = \ln k_0 - E_a/RT + b\ln[\mathrm{Pol}*]. \tag{7.12}$$

Using the new empirical oxidation reaction rates to plot $\ln r_{ox}$ *vs.* $\ln[A^-]$, the chemical nature of the studied process is confirmed if the empirical results follow eqn (7.11). The reaction order a is obtained from the slope. The origin ordinate ($\ln k''$) will allow, once E_a is known and after obtaining b, the reaction coefficient k_0 to be obtained from eqn (7.12).

By now subjecting the polymer to potential steps from the same initial (relaxed or conformationally packed) state to a different oxidation potential every time [different concentrations of the active sites in the film consume different oxidation charges, eqn (4.11)] under a constant temperature (T) and constant electrolyte concentration $[A^-]$, eqn (7.8) becomes:

$$\ln r_{ox} = \ln k''' + b\ln[\mathrm{Pol}*], \tag{7.13}$$

where all the new constant terms from eqn (7.8) are collected in a new constant k''':

$$\ln k''' = \ln k_0 - E_a/RT + a\ln[A^-]. \tag{7.14}$$

Using the new empirical oxidation reaction rates to plot $\ln r_{ox}$ *vs.* $\ln[\mathrm{Pol}*]$, the chemical nature of the studied process is confirmed if the empirical results follow eqn (7.13). The slope is the reaction order b. The origin ordinate ($\ln k'''$) will allow, using the above obtained values of E_a and a, the reaction coefficient k_0 to be obtained by using eqn (7.14).

Up to here, we have confirmed the universal validity of the chemical kinetic model described by current chemical, electrochemical or biochemical textbooks when applied to conducting polymers. Now the new open front can be explored: the full procedure described from eqn (7.8) until the previous paragraph is repeated, as many times as possible, using a different reduced, shrunken state every time or a different reduced, compacted state, as indicated by Figure 7.3.

Experimental results are depicted by Figure 7.4 for the oxidation of PEDOT. Figure 7.4c–e confirm that, whatever the initial shrunken or conformationally compacted states used for the oxidation, eqn (7.9), (7.11) and (7.13) are followed, verifying the chemical nature [eqn (4.1)] of each of the studied oxidation processes. Figure 7.4f, g and i present unexpected chemical kinetic results: the activation energy, the reaction coefficients and each of the reaction orders of the polymer oxidation reaction are no longer constant and they change as a function of the initial conformationally packed state used for the kinetic study.

This conclusion contradicts what is stated in textbooks. In addition, the attained shifting values of E_a, k, a and b as a function of the initial conformationally packed state used for the reaction mean that those kinetic magnitudes include (and can provide) quantitative structural information related to both the initial conformationally packed state and its evolution during the film reaction until the final swollen conformational state. The activation energy, the reaction coefficient and the reaction orders become the structural activation energy, the structural reaction coefficient and the structural reaction orders. The chemical kinetics have become structural chemical kinetics.

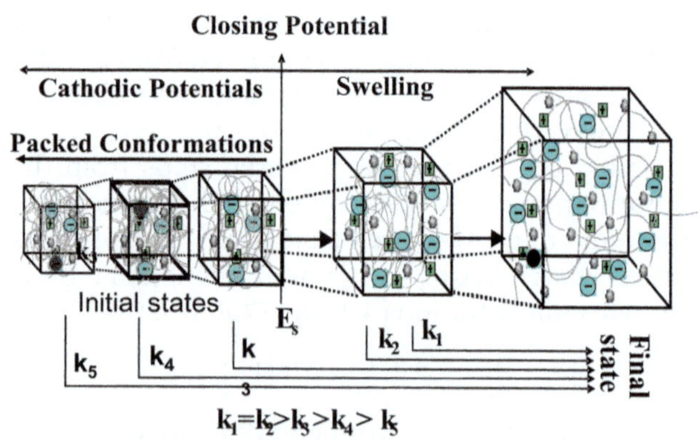

$$(p3\text{-}MTh)_s + n(A^-)_{aq} + mS \rightarrow [(p3\text{-}MTh^{n+})(A^-)_n\,(S)_m]_{gel} + (n\,e^-)_{metal}$$

Figure 7.3 Chemical kinetic methodology followed to attain the oxidation rate coefficient of poly-3-methylthiophene (p3-MTh) from different shrunken or conformationally packed initial states to the same final state every time, S being the solvent. Reprinted from ref. 69: *J. Electroanal. Chem.*, 2007, **610**, 96–101, with the permission of Elsevier.

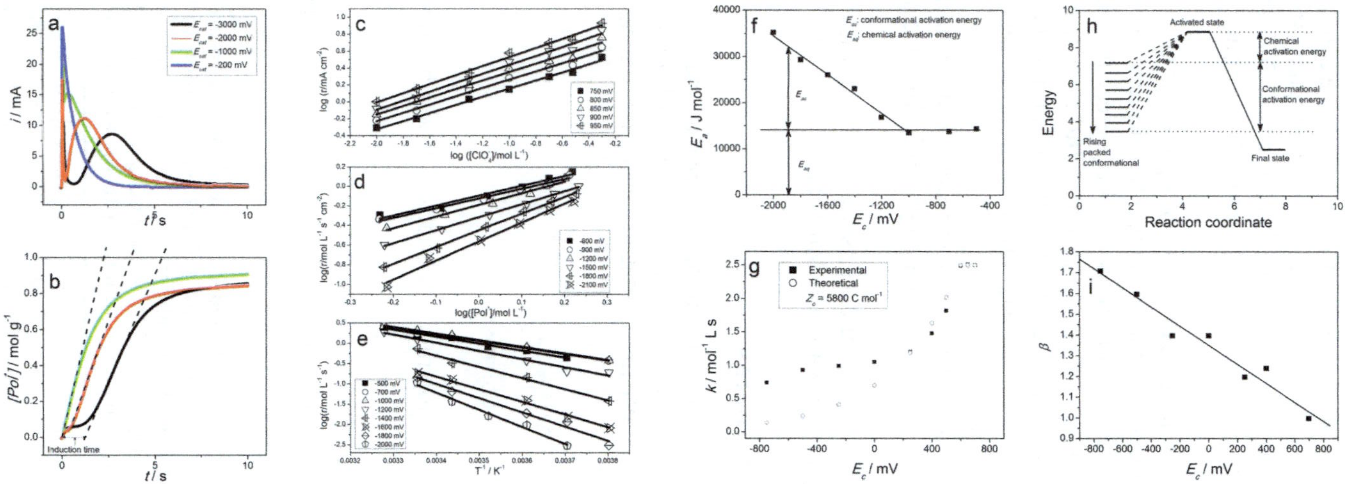

Figure 7.4 (a) Chronoamperometric responses of a PEDOT-coated Pt electrode in a 0.1 mol L^{-1} $LiClO_4$ acetonitrile solution subjected to potential steps from different cathodic potentials kept for 30 s to 500 mV. (b) Evolution of the concentration of the active centres in the film obtained from (a). Dashed lines represent the initial reaction rate. (c) Each line represents the evolution of the initial oxidation rate as a function of the electrolyte concentration at different final oxidation states. (d) Each line represents the initial oxidation rate for different concentrations of active centres in the film. The influence of the initial packed conformational state was investigated by reduction at different cathodic potentials for the same time. (e) Each line represents the initial oxidation rate evolution for different temperatures. The influence of the initial packed conformational state is investigated. Evolution of (f) the activation energy, (g) the reaction coefficient and (i) the reaction order related to the concentration of the active centers as a function of the packed conformations attained by reduction at rising cathodic potentials. (h) Schematic representation of the chemical and conformational activation energies following (f). Reprinted from ref. 65 with the permission of the Royal Society of Chemistry.

As a summary, either the activation energy, reaction coefficient or the reaction orders of any reaction involving conformational, allosteric or folding/unfolding processes include quantitative information about those conformational, allosteric and structural reaction-driven changes. Now scientists can start to explore, separate, quantify and decode each of the reaction-driven structural components to determine the physicochemical, biochemical and functional meanings.

As a direct consequence, a new scientific door can be opened as any material property or biological function related to these reaction-induced conformational changes (muscular action, haptic muscles, biological molecular motors, enzymatic reactions, allosteric effects in proteins and nucleic acids, brain memory, and proprioception or consciousness, to mention just a few) could be quantified and described by self-sustained structural chemical kinetic models in the coming times.

Table 7.1 illustrates the empirical increase of the reaction order and the parallel decrease of the origin ordinate ($\ln k''$) for a polypyrrole film in a 0.1 M $LiBF_4$ acetonitrile solution.

Now some preliminary quantifications and consequences from the new structural chemical kinetic magnitudes can be advanced.

7.2.4 Structural Activation Energy

The energetic components of the reaction-driving conformational changes are defined by eqn (7.2)–(7.4). Taking into account eqn (7.5) and some basic thermodynamic relationships, the polymer oxidation activation energy presented in eqn (7.7) becomes:

$$E_a = RT + \Delta H* + z_c\eta_c - z_r\eta, \qquad (7.15)$$

which is constituted by three different terms:

- The activation molar energy of polymeric relaxation driven by the electrochemical oxidation, E_{relax}:

$$E_{relax} = -z_r\eta. \qquad (7.16)$$

Table 7.1 Values of the reaction order β obtained from the slope, origin ordinate (to get k) and correlation coefficient for polypyrrole oxidation in $LiBF_4$ obtained by applying different cathodic potentials for 30 s each before oxidation by potential step to 0.25 V. Reproduced from *Smart Mater. Struct.*, 2013, **22**, 104019, with permission from Taylor & Francis Group.

Potential (V)	−0.30	−0.50	−0.70	−0.90	−1.10	−1.50	−1.70
β	0.91	1.39	1.92	2.00	1.86	1.93	3.87
Origin ordinate	−1.31	−1.82	−2.52	−2.82	−2.68	−3.14	−4.71
R^2	0.99	1.00	0.95	0.99	1.00	0.98	1.00

- The molar energy of the polymer–electrolyte ($E_{\text{P-E}}$) system in the absence of any flowing current or applied electric field:

$$E_{\text{P-E}} = RT + \Delta H*. \tag{7.17}$$

- And the molar conformational energy of the initial compacted polymeric state for the oxidation [E_{conf}; eqn (7.3)]:

$$E_{\text{conf}} = z_c \eta_c.$$

As a summary, eqn (7.15) quantifies the structural activation energy and describes the empirical results presented by Figure 7.4f. The activation energy of the polymer oxidation reaction when the reaction starts from partially reduced and shrunken initial states ($\eta_c = 0$) is constant ($E_a = RT + \Delta H* - z_r\eta$). When the oxidation is initiated from increasingly reduced and compacted initial states, the activation energy of the oxidation reaction increases linearly with the reduction–compaction overpotential (η_c): $z_c\eta_c$. This term is the conformational energy of the initial state obtained by reduction for a constant time at rising cathodic overpotentials (η_c).

Thus, under constant electrochemical conditions, the structural activation energy of the reaction includes two components: the constant electro-chemo-polymeric activation energy ($E_a = RT + \Delta H* - z_r\eta$) and the conformational energy ($z_c\eta_c$) of the initial state. Ultimately, the conformational energy involved in the chemical or biochemical reactions becomes a number that could be empirically attained and theoretically described. Figure 7.4h represents the structural activation energy going from initial states with different conformational energetic states to the same reaction product: the oxidized and swollen polymer.

This opens the way for subsequent quantifications of biochemical reactions and biological functions based on chemically induced conformational movements of any reactant macromolecule. Most biochemical reactions occur in the absence of electric fields and/or currents, but the chemical reaction still drives conformational compaction processes involving the molar conformational energy (E_{conf}) or requires the molar conformational relaxation (E_{relax}) energy to occur. Similar to eqn (7.15) for electrochemically driven processes, the activation energy for reversible conformational, allosteric, folding or unfolding structural processes driven by biochemical reactions should become:

$$E_a = RT + E_{\text{P-E}} + E_{\text{conf}} - E_{\text{relax}}. \tag{7.18}$$

This equation will be used in Chapter 10 to attempt a quantitative description of some reversible and irreversible biochemical reaction rates.

Moreover, eqn (7.15) states the way to include any new energetic component of the activation energy for structural biochemical reactions induced by photons, electric fields, magnetic fields, electromagnetic fields and so on, inducing simultaneous conformational movements.

7.2.5 Structural Reaction Coefficient

By comparing eqn (7.6) and (7.8), and considering the activation energy from eqn (7.15), the reaction coefficient becomes:

$$\ln k = \ln k_0 + \left(\frac{z_r\eta - \Delta H*}{RT}\right) - \left(\frac{z_c\eta_c}{RT}\right), \tag{7.19}$$

describing that, under constant chemical and thermal conditions, the reaction coefficient decreases exponentially for rising conformational energies ($z_c\eta_c$) of the reaction initial state. That means that the reaction coefficient includes and can give quantitative information about the conformational energetic state of the reactant: reaction coefficients become structural reaction coefficients.

Eqn (7.19) fits the empirical results presented in Figure 7.4g. The reaction coefficients attained for the oxidation of any partially reduced and shrunken initial state ($\eta_c = 0$) keep a constant value. The reaction coefficient of the polymer oxidation under constant chemical and thermal conditions decreases exponentially for increasingly conformationally packed initial states obtained by reduction for a constant time at rising cathodic potentials (*i.e.*, rising values of η_c) beyond the closing potential. Thus, the structural reaction coefficient, eqn (7.19), is a quantitative description of the anomalous electrochemical responses: rising reduced, packed initial structures give, under constant chemical and electrochemical conditions (constant thermal and chemical energies), slower subsequent oxidation–relaxation reaction rates.

For biochemical, reaction-driven, reversible conformational compaction–relaxation, irreversible compaction or irreversible relaxation processes in the absence of electrical fields or currents, eqn (7.19) can be written as:

$$\ln k = \ln k_0 + \frac{E_{\mathrm{relax}} - \Delta E_{\mathrm{P-C}}}{RT} - \frac{E_{\mathrm{conf}}}{RT}, \tag{7.20}$$

which describes that, under constant chemical conditions, the biochemical reaction rate decreases exponentially for the increasing conformational compaction energy (E_{conf}) of the initial state. Related to the second term, the reaction rate increases exponentially for increasing relaxation energies (E_{relax}) or for decreasing interaction of the polymeric reactant ($\Delta E_{\mathrm{P-C}}$) with the ambient (water, other molecules and ions) when the second reactant arrives. Eqn (7.20) can provide some support for a quantitative approach to biochemical, reaction-driven allosteric effects, such as the hemoglobin–oxygen reaction discussed in Chapter 10.

7.2.6 Structural Reaction Orders

From eqn (7.10) and (7.17), the reaction coefficient k' is given by:

$$\ln k' = \ln k - \left(\frac{z_r\eta - \Delta H*}{RT}\right) - \left(\frac{z_c\eta_c}{RT}\right) + a\ln\left[A^-\right] + b\ln[\text{Pol}*].$$

And the reaction order a becomes:

$$a = \frac{\ln k' - \ln k}{RT\ln[A^-]} + \left(\frac{z_r\eta - \Delta H*}{RT\ln[A^-]}\right) + \left(\frac{z_c\eta_c}{RT\ln[A^-]}\right) - \frac{b\ln[\text{Pol}*]}{\ln[A^-]}. \quad (7.21)$$

Working under constant chemical and thermal conditions by oxidation to the same anodic potential and changing only the cathodic potential limit required for getting the initial state of conformational compaction:

$$a = K + \frac{z_c\eta_c}{RT\ln[A^-]}. \quad (7.22)$$

This equation describes the reaction order as a linear function of the molar conformational energy of the partially reduced and conformationally packed initial state. A similar expression is attained for the reaction order b, related to the concentration of the active polymeric sites in the membrane. The reaction orders become a function of the reactant conformational energy: they are structural reaction orders.

Eqn (7.19) describes the empirical results presented in Figure 7.4i. The reaction orders of the polymer oxidation increase linearly with the conformational energy of the reactant ($z_c\eta_c$) obtained by reduction for a constant time at rising cathodic potentials (rising values of η_c).

As a final comment, the structural kinetic magnitudes defined by eqn (7.15), (7.19) and (7.21) also include, through terms like $\Delta H*$, the evolution of the intra- and inter-molecular interactions of the molecules and ions, whether reactants or not, present in the biochemical reaction arena. Scrutiny of the possible biochemical implications constitutes an exciting, pending task.

Now some of these structural energetic components can be used to describe the structural electrochemical responses attained from conducting polymers and other electrochemically active materials.

7.2.7 The SCK Model Includes Chemical Kinetic Models

The SCK model is a result of classical chemical kinetic methodologies and theoretical treatments [eqn (7.7)–(7.14)] applied to the different shrunken or conformationally compacted initial states of reactive gels. In the absence of any kinetic control by conformational movements driven by the reaction, eqn (7.18), (7.19) and (7.22) give the constant values of the reaction order, reaction coefficient and reaction activation energy, respectively, described by present chemical and biochemical kinetic models. As a conclusion, the structural chemical kinetic model may be considered an advanced chemical kinetic model that includes present chemical kinetic models when the reaction doesn't occur under any conformational kinetic control of the reactants.

7.3 Structural Chronoamperometric Responses: Theoretical Simulation

After partial reduction and conformational compaction by polarization at compaction potential E_c, the film oxidation gives, by potential step to E, a chronoamperometric response with a nucleation–relaxation maximum (Figures 7.2a and 7.5). The ESCR model allows formulation of the following hypotheses:

1. A number of nuclei per square centimeter is formed by oxidation–relaxation when the oxidation of the conformational compacted membrane starts.
2. Each nucleus forms a cylinder of oxidized polymer going from the electrolyte–membrane interface to the membrane–metal interface, the diameter of which expands during oxidation–relaxation until their coalescence with neighboring expanding cylinders.
3. During this time, the expanding lateral area of the cylinders separates reduced and shrunken, and oxidized and swollen membrane fractions.
4. At the oxidized/reduced borders, the oxidation of 1 mole of polymeric segments involves the loss of q_r electrons, with formation of the same number of positive charges on chains and the simultaneous entrance of the same number of monovalent anions from the solution for charge balance.
5. The oxidation–relaxation of a polymeric segment having a length λ takes time τ [eqn (7.1)].
6. Behind the relaxation border, the membrane oxidation is completed by oxidation–swelling under diffusion kinetic control of the counterions from the solution through the oxidized cylinder core.
7. The involved energy (ΔH) includes three components, defined by eqn (7.2)–(7.4).
8. The oxidation charge (Q_t) includes, at any oxidation time, two components: the oxidation–relaxation charge (Q_r) consumed to relax the conformationally compacted cylinder borders and the oxidation–diffusion charge (Q_d) consumed to complete the oxidation–swelling of the just relaxed front.

$$Q_t = Q_r + Q_d. \tag{7.23}$$

The relaxation charge is attained under geometrical considerations of the expanding cylinders and the coalescence (Avrami's treatment) to get the theoretical descriptions for the evolution of the current $I_r(t)$, charge $Q_r(t)$ and area $A_r(t)$ during oxidation–relaxation:

$$I_r(t) = 2atQ_r e^{(-at^2)} \tag{7.24}$$

$$Q_r(t) = Q_r[1 - e^{(-at^2)}] \tag{7.25}$$

$$A_r(t) = A[1 - e^{(-at^2)}], \tag{7.26}$$

where the constant a is given by:

$$a = \frac{\pi N_0 \lambda^2}{\tau^2 A} e^{\left[-\frac{2\Delta H}{RT}\right]}, \tag{7.27}$$

where N_0 is the number of nuclei, λ is the length of a polymeric segment, τ is the relaxation time and A is the polymer film surface area.

After oxidation–relaxation, the oxidation–swelling of this relaxed material is completed by consumption of the oxidation–diffusion charge. At any time, the flow of ions towards the partially oxidized material is assumed to be proportional to the gradient between the actual charge density in the relaxed material, $\delta_d(t)$, and that of the steady oxidation state at the applied oxidation potential, δ_d. The theoretical equations describing the evolution of the diffusion charge and that of the diffusion current are attained as follows:

$$Q_d(t) = Q_d\left[1 - e^{-at} - 2ae^{-bt}\int_0^{t'} t' e^{(bt' - at'^2)}\,dt'\right] \tag{7.28}$$

$$I_d(t) = 2abQ_d e^{(-bt)}\int_0^{t'} t' e^{(bt' - at'^2)}\,dt', \tag{7.29}$$

where the constant b is given by:

$$b = 2D/h^2, \tag{7.30}$$

and D is the diffusion coefficient of ions through the swelling membrane, h is the thickness of the dry membrane, t is the oxidation time (after the potential step) and t' accounts for the infinitesimal relaxation times. If the initial state for the membrane oxidation is an open and shrunken, partially reduced state, then $t' = 0$ and eqn (7.28) and (7.29) become:

$$Q_d(t) = Q_d[1 - e^{-bt}] \tag{7.31}$$

$$I_d(t) = bQ_d e^{(-bt)}. \tag{7.32}$$

By combining eqn (7.23), (7.25) and (7.28), the theoretical chronocoulometric (Q/t) responses are attained:

$$Q(t) = Q_r + Q_d\left[1 - e^{-at^2}\right] - 2aQ_d e^{-bt}\int_0^{t'} t' e^{(bt' - at'^2)}\,dt'. \tag{7.33}$$

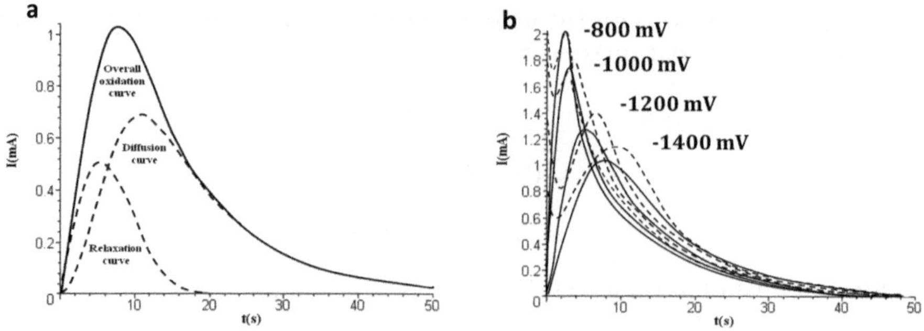

Figure 7.5 (a) The dashed lines present the oxidation–relaxation [eqn (7.21)] and oxidation diffusion [eqn (7.29)] components of the theoretical [eqn (7.34)] chronoamperometric response (solid line). (b) Overlapping of theoretical (solid lines) and experimental (dashed lines) chronoamperometric responses in 0.1 M $LiClO_4$ acetonitrile solutions by potential steps from different cathodic potentials of reduction–compaction (−800, −1000, −1200 and −1400 mV *vs.* saturated calomel electrode, SCE) maintained for 60 s, each to the same oxidation potential of 450 mV. Electrode, 1 cm^2 Pt foil coated with a 0.22 μm thick polyaniline film; temperature, 25 °C. Reprinted from ref. 54 with permission: *J. Phys. Chem. B*, 1997, **101**, 3688–3697. Copyright 1997 American Chemical Society.

By combining eqn (7.24) and (7.29), the theoretical chronoamperometric (I/t) responses are attained:

$$I(t) = 2aQ_r t e^{-at^2} + 2abQ_d \left[e^{(-bt)} \int_0^{t'} t' e^{(bt' - at'^2)} dt' \right]. \tag{7.34}$$

Figure 7.5a shows the two theoretical components (dashed lines), the oxidation–relaxation current [eqn (7.24)] and the oxidation–diffusion current [eqn (7.29)] of the chronoamperometric response [eqn (7.34); full line] when a conformationally compacted polymeric film is oxidized by a potential step.

Figure 7.5b shows the good agreement between theoretical [eqn (7.34); full lines] and experimental (dashed lines) chronoamperometric responses when a polyaniline film was oxidized by potential step to the same anodic potential every time under a constant electrolyte concentration and working temperature after its reduction–compaction at a different cathodic potential (indicated on the figure) for a constant time.[38,48,50,54,76,77]

Similar results were attained from different electrolyte concentrations at different working temperatures from different conducting polymers and in different electrolytes (solvents or salts).[48,50,54,75–77]

7.4 Structural Voltammetric Responses: Theoretical Description

After reduction compaction by potential sweep until rising cathodic potential limits, the membrane attains a different compaction degree of the polymeric entanglement (υ). The subsequent polymer oxidation by relaxation–nucleation starts at rising anodic potentials: *i.e.*, the nucleation potential, $\eta_n = E_n - E_0$. For simplicity, N_0 nuclei are supposed to be initiated when the potential sweep attains η_n. The closure degree is zero for a relaxed structure: $\upsilon = 0$, $\eta_n = 0$ and $E_n = E_0$; and $\upsilon = 1$ when both the conformational polymeric compaction and the nucleation overpotential attain saturation, η_n^0. Assuming a linear dependence between both extremes:

$$\eta_n = \eta_n^0 \upsilon. \tag{7.35}$$

On the other hand, the probability (P) of occurrence of a conformational movement for any polymeric segment after compaction until η_c at the nuclei border is the inverse of the relaxation time. Including all the constant terms from eqn (7.5) in P gives:

$$P = \mathrm{e}^{-\frac{z_c}{RT}\eta_c}. \tag{7.36}$$

When $\eta_c = 0$ and $\upsilon = 0$, the structure is relaxed and $P = 1$, and beyond the standard oxidation potential E_0 a planar oxidation front is initiated and advances from the polymer–solution interface of the membrane electrode to the polymer–metal interface. When η_c is high, υ is close to 1 and P nears zero. Thus, P and υ are linearly related:

$$\upsilon = 1 - \mathrm{e}^{-\frac{z_c}{RT}\eta_c} \tag{7.37}$$

and by combining eqn (7.35) and (7.37):

$$\eta_n = \eta_n^0 \left[1 - \mathrm{e}^{-\frac{z_c}{RT}\eta_c}\right] \text{ and } \ln\left[1 - \frac{\eta_n}{\eta_n^0}\right] = -\frac{z_c}{RT}\eta_c. \tag{7.38}$$

Eqn (7.38) can be checked empirically.

As described for chronoamperometric responses, the overall voltammetric charge at any potential during the anodic voltammetric response includes two components: the oxidation–relaxation charge and the oxidation–diffusion charge [eqn (7.24)]. From geometrical considerations for the nuclei growth and applying Avrami's treatment to the nuclei coalescence, the evolution during the potential sweep of both the relaxation charge $Q_r(E)$ and that of the relaxation current $I_r(E)$ are attained:

$$Q_r(E) = Q_r\left[1 - \mathrm{e}^{\left\{-\frac{\pi N_0 \lambda^2}{\upsilon^2 A \tau_0^2}\left(\frac{RT}{z_r}\right)^2\left[\mathrm{e}^{-\frac{\Delta H}{RT}} - \mathrm{e}^{-\left(\frac{\Delta H_n}{RT}\right)}\right]^2\right\}}\right] \tag{7.39}$$

$$I_r(E) = \frac{2\pi N_0 \lambda^2}{\upsilon A \tau_0^2} Q_r \frac{RT}{z_r} \mathrm{e}^{-\frac{\Delta H}{RT}} - \mathrm{e}^{\left\{-\frac{\pi N_0 \lambda^2}{\upsilon^2 A \tau_0^2}\left(\frac{RT}{z_r}\right)^2\left[\mathrm{e}^{-\frac{\Delta H}{RT}} - \mathrm{e}^{-\left(\frac{\Delta H_n}{RT}\right)}\right]^2\right\}}. \tag{7.40}$$

Eqn (7.39) and (7.40) describe the evolution of the charge (coulovoltammetric response, Q/E) or the current (voltammetric response, I/E), respectively, during the film oxidation along the anodic potential sweep. The reaction occurs under conformational relaxation kinetic control initiated by nucleation of the oxidized polymer in the reduced and conformationally packed film.

The relaxation component (the first term), despite being marginally related to the diffusion term, will define and give information about the characteristic voltammetric peak potentials as a function of the experimental variables. Thus, if the potential of the voltammetric maximum (peak, E_p) is much higher than the nucleation overpotential ($\eta_p \gg \eta_n$), from the eqn (7.40) maximum:

$$\eta_p = \frac{\Delta H^* + z_c \eta_c}{z_r} + \frac{RT}{z_r}\ln\left[\frac{RT\lambda v}{z_r \tau_0}\sqrt{\frac{\pi N_0}{A}}\right]. \tag{7.41}$$

From this equation, under constant temperature (T) and sweep rate (v), a linear dependence of the voltammetric maximum overpotential (η_p) with the cathodic voltammetric potential limit of reduction–compaction (η_c) is predicted:

$$\eta_p = a + b\eta_c. \tag{7.42}$$

When only the temperature is changed, from eqn (7.41) the temperature influence on η_p is attained:

$$\frac{\eta_p}{T} = a' + b'\ln T. \tag{7.43}$$

From eqn (7.41) the influence of the sweep rate on η_p is also predicted under constant temperature and for the same cathodic potential limit every time:

$$\eta_p = a'' + b''\ln v \tag{7.44}$$

Eqn (7.42)–(7.44) have been attained empirically by some pioneering electrochemical studies on conducting polymers.[78,79]

Behind the relaxation front, the film oxidation–swelling is completed under diffusion kinetic control of the ions through the film from the solution. Following a similar procedure to the above-described for the diffusion chronoamperometric component, the diffusion charge becomes:

$$Q_d(\eta) = Q_{ds}(\eta)\frac{2e^{(-c\eta)}}{v\eta}\int_{\eta_n}^{\eta}\frac{I_r(\eta')}{Q_r}\left[\int_{\eta'}^{\eta} c\eta e^{c\eta}\,d\eta\right]d\eta', \tag{7.45}$$

where Q_{ds} represents the diffusion charge consumed until a given potential η is reached in the considered area, and $c = D/h^2v$ is a constant including potential-dependent diffusion coefficient D. Eqn (7.45) describes the coulovoltammetric (Q/E) response of the film during the anodic potential sweep.

The current flowing at any potential due to the film oxidation–swelling under diffusion control will be $I_d(\eta) = v\frac{dQ_d(\eta)}{d\eta}$ and by substitution of the diffusion charge evolution from eqn (7.45):

$$I_d(\eta) = \frac{2D}{h^2}\left[Q_{ds}(\eta)\frac{A(\eta)}{A} - Q_d(\eta)\right]. \tag{7.46}$$

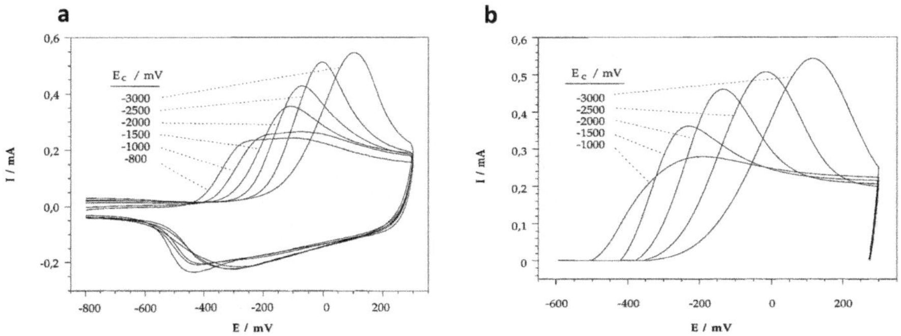

Figure 7.6 (a) Experimental and (b) theoretical, following eqn (7.46), stationary voltammetric responses from a polypyrrole film in 0.1 M $LiClO_4$/propylene carbonate solutions from different initial cathodic potentials (indicated on the figure) and 300 mV *vs.* SCE at a scan rate of 30 mV s^{-1} and a temperature of 25 °C. Reprinted with permission from ref. 40: *J. Phys. Chem. B*, 1997, **101**, 8525–8533. Copyright 1997 American Chemical Society.

Eqn (7.40) and (7.46) describe the voltammetric responses (I/E) from conducting polymer films starting under conformational relaxation kinetic control and going on under diffusion kinetic control of the counterions across the oxidized film from the solution. Experimental results from different cathodic potentials (Figure 7.6a) were simulated from eqn (7.46), as depicted by Figure 7.6b. Similar experimental and theoretical results were attained from different variables.[40]

7.5 Structural Coulovoltammetric Responses: Theoretical Description

Eqn (7.39) and (7.45) describe, during a potential sweep, the evolution of the oxidation–relaxation and oxidation–swelling charge, respectively. Figure 7.7 shows both the experimental (solid lines) and theoretical (dotted lines) coulovoltammetric responses from different cathodic potential limits. Similar results were attained at different sweep rates under different temperatures or electrolyte concentrations. A good agreement exists between experimental and simulated results.[42]

7.6 Structural Chronopotentiometric Responses: Theoretical Description

Two new hypotheses were added here:

- The oxidation/reduction of any polymeric chain occurs through n consecutive steps defined by the reactions shown in eqn (4.4)–(4.7).

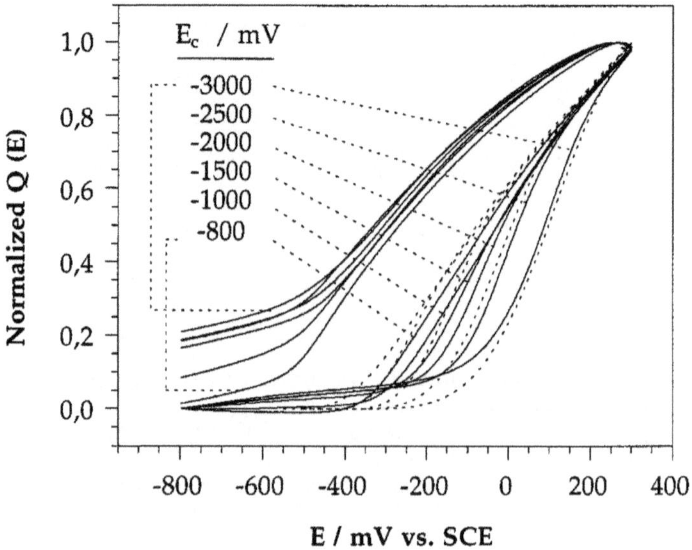

Figure 7.7 Experimental (solid lines) and theoretical [dotted lines; from eqn (7.45)] normalized coulovoltammetric responses from a polypyrrole film in a 0.1 M $LiClO_4$ propylene carbonate solution from different departure potentials, indicated in the figure, to 300 mV *vs.* SCE at 30 mV s^{-1} at room temperature. Reprinted with permission from ref. 42: *J. Phys. Chem. B*, 1998, **102**, 7535–7540. Copyright 1998 American Chemical Society.

- In addition to the oxidation–relaxation and oxidation–swelling charges some capacitive charge will be also expended. It can be considered as some kind of electrical double layer (EDL).

Thus, under the flow of a constant anodic current through a reduced and compacted conducting polymer electrode, the evolution of the electrode potential with time $E(t)$ during its oxidation is described by a stair function (Section 8.9):[80]

$$E(t) = \sum E_n(t)p_n(t), \tag{7.47}$$

where $p_n(t)$ is the unitary pulse function, taking a unitary value in the nth interval of time (*i.e.*, the time required to extract the nth electron from each of the film polymeric chains) and being zero outside this interval;[81] being the evolution of the film potential during that time, $E_n(t)$ is described by eqn (9.6). The anodic current applied to oxidize a conformationally compacted film must be consumed by the two structural oxidation components: the oxidation–relaxation current described by eqn (7.24) and the oxidation–swelling current described by eqn (7.32).

$$i_a(t) = i_r(t) + i_d(t) + i_{EDL}(t) = 2aQ_r t\exp(-at^2) + bQ_d\exp(-bt) + i_0[1 - \exp(t/\tau)]. \tag{7.48}$$

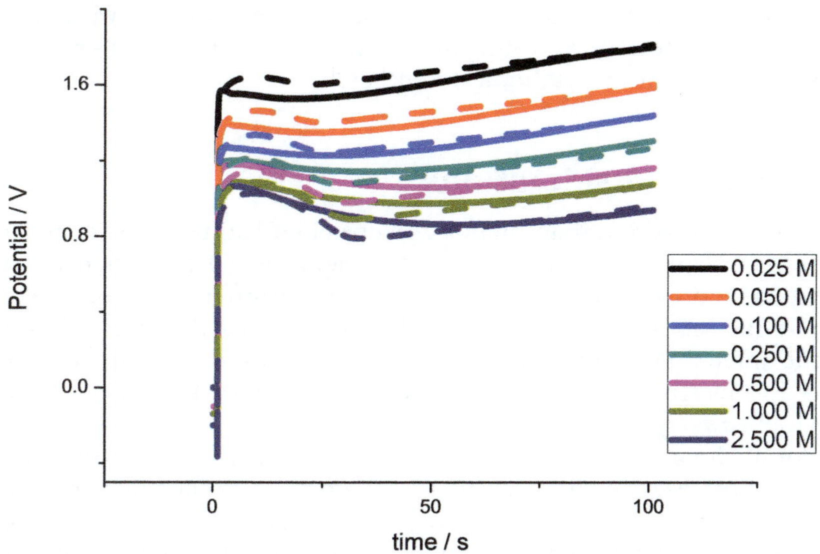

Figure 7.8 Experimental and theoretical chronopotentiometric results obtained from a freestanding polypyrrole film electrode in $LiClO_4$ aqueous solutions with different salt concentrations. The material was prearranged following the same procedure every time: by reduction at −0.5 V for 60 s, then oxidized each time at 0.5 V for 60 s and then reduced-compacted at −0.9 V for 60 s. Then a constant current of 1 mA was applied over 100 s to obtain the chronopotentiometric results (solid lines). Theoretical results from eqn (7.48) (dashed lines) overlap with the experimental findings. Reproduced from ref. 44 with permission from the Royal Society of Chemistry.

By combining eqn (8.5) and (7.48) the evolution of the membrane potential during oxidation by flow of a constant anodic current is obtained:

$$E_n(t) = E_0 + E_z + (n-1)\Delta E + \frac{RT}{(1-\alpha)F}\left\{\ln\left(\frac{2aQ_r t\exp(-at^2) + bQ_d\exp(-bt) + i_0\left[1-\exp(t/\tau)\right]}{FV}\right) - d\ln[A^-] - e\ln\left([\mathrm{Pol}*]_{\mathrm{initial}} - \frac{Q}{FV}\right) - \ln k_{a0}\right\}, \tag{7.49}$$

including the influence of each experimental variable: the cathodic and anodic potential limits [through the constant a, eqn (7.27)], the diffusion coefficient of the counterions in the swelling polymer (D) and its dependence on the film potential and structure, the working temperature (T), and the electrolyte concentration $[A^-]$. By changing one of the experimental variables each time, while keeping the other variables constant, Figure 7.8 shows the experimental (solid lines) and theoretical (dashed lines) results obtained from a compacted, freestanding polypyrrole film electrode in $LiClO_4$ aqueous solutions with different salt concentrations. Similar responses were attained at different temperatures starting the current flow after the film's reduction–compaction at different cathodic potentials.[44]

7.7 Final Considerations

Most of the considered anomalous electrochemical responses, up to the development of the ESCR model, were theoretically described by inclusion of reaction-driven conformational relaxation–nucleation and conformational compaction processes. So, this means they were described by the inclusion of structural aspects driven by electrochemical reactions in electrochemical models: electrochemistry becomes structural electrochemistry.

A lot of theoretical work is still required in order to unify the hypotheses stated to describe and quantify the different electrochemical responses. Then, new structural responses, such as those already detected during the reduction–swelling of conducting polymer–organic macro-anion blends discussed in Section 6.4.2, will emerge from the study of the different conducting polymer families, requiring new expansions of the ESCR model for their description.

By inclusion of the molar energy of conformational compaction (the compaction conformational energy or conformational energy) and the molar energy of conformational relaxation (the relaxation energy), the chemical kinetics became conformational chemical kinetics or structural chemical kinetics. The structural activation energy includes the constant chemical (or electrochemical) activation energy plus the conformational energy of the initial state, which increases linearly with the compaction overpotential. The structural reaction coefficient decreases exponentially for rising molar conformational energies of the reactant, describing in a quantitative way why more reduced initial states of the conducting polymer exchanging anions during the reaction give slower subsequent oxidation rates (anomalous electrochemical responses). The structural reaction order also results in a linear function of the molar conformational energy of the partially reduced and conformationally packed initial state.

The attained quantitative expressions for these structural and conformational electrochemical-driven reactions have been translated to parallel theoretical expressions for reactions driving conformational changes taking place in the absence of both electric fields and the flow of electrical currents. This is the case with conformational and allosteric biochemical reactions. Most of these reactions support life and originate life functions that remain outside of the scope of chemical models described by present textbooks. In Chapter 10 some predictions of the attained structural equations when applied to different biochemical reactions, such as enzymatic catalysis or allosteric effects, will be presented.

References

1. K. Aoki, J. Cao and Y. Hoshino, *Electrochim. Acta*, 1994, **39**, 2291–2297.
2. J. Heinze, B. A. Frontana-Uribe and S. Ludwigs, *Chem. Rev.*, 2010, **110**, 4724–4771.
3. A. Diaz and J. Castillo, *J. Chem. Soc., Chem. Commun.*, 1980, 397–398.
4. A. Epstein, J. Ginder, F. Zuo, R. Bigelow, H.-S. Woo, D. Tanner, A. Richter, W.-S. Huang and A. MacDiarmid, *Synth. Met.*, 1987, **18**, 303–309.
5. J. M. Ginder, A. F. Richter, A. G. MacDiarmid and A. J. Epstein, *Solid State Commun.*, 1987, **63**, 97–101.
6. M. Reghu, K. Vakiparta, C. O. Yoon, Y. Cao, D. Moses and A. J. Heeger, *Synth. Met.*, 1994, **65**, 167–171.
7. C. O. Yoon, M. Reghu, D. Moses and A. J. Heeger, *Phys. Rev. B*, 1994, **49**, 10851–10863.
8. C. O. Yoon, M. Reghu, D. Moses, A. J. Heeger, Y. Cao, T.-A. Chen, X. Wu and R. D. Rieke, *Synth. Met.*, 1995, **75**, 229–239.
9. O. Lhost and J. L. Brédas, *J. Chem. Phys.*, 1992, **96**, 5279.
10. E. M. Genies and E. Vieil, *Synth. Met.*, 1987, **20**, 97–108.
11. C. Y. Lee, H. M. Kim, J. W. Park, Y. S. Gal, J. I. Jin and J. Joo, *Synth. Met.*, 2001, **117**, 109–113.
12. V. Coropceanu, J. Cornil, D. A. da Silva Filho, Y. Olivier, R. Silbey and J.-L. Brédas, *Chem. Rev.*, 2007, **107**, 926–952.
13. M. Remmers, D. Neher, J. Grüner, R. H. Friend, G. H. Gelinck, J. M. Warman, C. Quattrocchi, D. A. dos Santos and J.-L. Brédas, *Macromolecules*, 1996, **29**, 7432–7445.
14. J. L. Brédas, D. Beljonne, J. Cornil, J. P. Calbert, Z. Shuai and R. Silbey, *Synth. Met.*, 2001, **125**, 107–116.
15. W. R. Salaneck and J. L. Brédas, *Solid State Commun.*, 1994, **92**, 31–36.
16. J. Brédas, D. Beljonne, Z. Shuai and J. Toussaint, *Synth. Met.*, 1991, **43**, 3743–3746.
17. G. Inzelt, M. Pineri, J. Schultze and M. Vorotyntsev, *Electrochim. Acta*, 2000, **45**, 2403–2421.
18. N. Gospodinova and L. Terlemezyan, *Prog. Polym. Sci.*, 1998, **23**, 1443–1484.
19. A. Pron and P. Rannou, *Prog. Polym. Sci.*, 2002, **27**, 135–190.
20. J. S. Andrade, N. Ito and Y. Shibusa, *Phys. Rev. B*, 1996, **54**, 3910–3915.
21. K. Aoki, J. Cao and Y. Hoshino, *Electrochim. Acta*, 1994, **39**, 2291–2297.
22. K. Aoki and M. Kawase, *J. Electroanal. Chem.*, 1994, **377**, 125–129.
23. K. Aoki, *J. Electroanal. Chem.*, 1994, **373**, 67–73.
24. J. Cao and K. Aoki, *Electrochim. Acta*, 1996, **41**, 1787–1792.
25. Y. Y. Suzuki, A. J. Heeger and P. Pincus, *Macromolecules*, 1990, **23**, 4730.
26. Y. Tezuka, T. Kimura, T. Ishii and K. Aoki, *J. Electroanal. Chem.*, 1995, **395**, 51–55.
27. K. Xu, L. Zhu, J. Li and H. Tang, *Electrochim. Acta*, 2006, **52**, 723–727.
28. P. Banerjee and B. M. Mandal, *Macromolecules*, 1995, **28**, 3940–3943.

29. A. Fizazi, J. Moulton, K. Pakbaz, S. D. D. V. Rughooputh, P. Smith and A. J. Heeger, *Phys. Rev. Lett.*, 1990, **64**, 2180–2183.
30. K. Levon, A. Margolina and A. Z. Patashinsky, *Macromolecules*, 1993, **26**, 4061–4063.
31. T. Otero, E. Angulo, J. Rodriguez and C. Santamaria, *J. Electroanal. Chem.*, 1992, **341**, 369–375.
32. A. Petr and L. Dunsch, *J. Phys. Chem.*, 1996, **100**, 4867–4872.
33. A. Osterholm, A. Petr, C. Kvarnstroem, A. Ivaska and L. Dunsch, *J. Phys. Chem. B*, 2008, **112**, 14149–14157.
34. A. Zykwinska, W. Domagala, A. Czardybon, B. Pilawa and M. Lapkowski, *Chem. Phys.*, 2003, **292**, 31–45.
35. A. Osterholm, B. Meana-Esteban, C. Kvarnstrom and A. Ivaska, *J. Phys. Chem. B*, 2008, **112**, 6331–6337.
36. T. F. Otero and J. G. Martinez, *Adv. Funct. Mater.*, 2014, **24**, 1259–1264.
37. T. Otero and E. Angulo, *Solid State Ionics*, 1993, **63–65**, 803–809.
38. T. Otero, H. Grande and J. Rodriguez, *J. Electroanal. Chem.*, 1995, **394**, 211–216.
39. T. F. Otero, H. Grande and J. Rodriguez, *Synth. Met.*, 1996, **76**, 285–288.
40. T. F. Otero, H. Grande and J. Rodriguez, *J. Phys. Chem. B*, 1997, **101**, 8525–8533.
41. H. Grande, T. F. Otero and I. Cantero, *J. Non-Cryst. Solids*, 1998, **235**, 619–622.
42. H. Grande and T. F. Otero, *J. Phys. Chem. B*, 1998, **102**, 7535–7540.
43. H. Grande and T. F. Otero, *Electrochim. Acta*, 1999, **44**, 1893–1900.
44. J. G. Martinez and T. F. Otero, *RSC Adv.*, 2014, **4**, 29139–29145.
45. T. F. Otero and M. J. Ariza, *J. Phys. Chem. B*, 2003, **107**, 13954–13961.
46. T. F. Otero and I. Boyano, *ChemPhysChem*, 2003, **4**, 868–872.
47. T. F. Otero and I. Boyano, *Electrochim. Acta*, 2006, **51**, 6238–6242.
48. T. F. Otero, M. T. Cortes and I. Boyano, *J. Electroanal. Chem.*, 2004, **562**, 161–165.
49. T. F. Otero, H. Grande and J. Rodriguez, *J. Phys. Org. Chem.*, 1996, **9**, 381–386.
50. T. F. Otero, H. Grande and J. Rodriguez, *Electrochim. Acta*, 1996, **41**, 1863–1869.
51. T. F. Otero, H. Grande and J. Rodriguez, *Synth. Met.*, 1996, **83**, 205–208.
52. T. F. Otero, H. Grande and J. Rodriguez, *Synth. Met.*, 1997, **85**, 1077–1078.
53. T. F. Otero and H. Grande, *J. Electroanal. Chem.*, 1996, **414**, 171–176.
54. T. F. Otero, H. J. Grande and J. Rodriguez, *J. Phys. Chem. B*, 1997, **101**, 3688–3697.
55. T. F. Otero and J. Padilla, *J. Electroanal. Chem.*, 2004, **561**, 167–171.
56. T. Otero, J. Rodriguez and H. Grande, *J. Braz. Chem. Soc.*, 1994, **5**, 179–181.
57. T. F. Otero, in *From Non-Covalent Assemblies to Molecular Machines*, ed. J.-P. Sauvage and P. Gaspard, Wiley-VCH, Weinheim, 2011, pp. 443–452.
58. T. F. Otero, M. Alfaro, V. Martinez, M. A. Perez and J. G. Martinez, *Adv. Funct. Mater.*, 2013, **23**, 3929–3940.
59. T. F. Otero and M. Caballero Romero, *Polym. Int.*, 2010, **59**, 329–336.

60. T. F. Otero and J. M. Garcia de Otazo, *Synth. Met.*, 2009, **159**, 681–688.
61. T. F. Otero and J. G. Martinez, *J. Solid State Electrochem.*, 2011, **15**, 1169–1178.
62. T. F. Otero and J. G. Martinez, *Adv. Funct. Mater.*, 2013, **23**, 404–416.
63. T. F. Otero and J. G. Martinez, *Sens. Actuators, B*, 2014, **199**, 27–30.
64. T. F. Otero and F. Santos, *Electrochim. Acta*, 2008, **53**, 3166–3174.
65. T. F. Otero, *J. Mater. Chem. B*, 2013, **1**, 3754–3767.
66. B. J. West, T. F. Otero, B. Shapiro and E. Smela, *J. Phys. Chem. B*, 2009, **113**, 1277–1293.
67. A. Zykwinska, W. Domagala, A. Czardybon, B. Pilawa and M. Lapkowski, *Chem. Phys.*, 2003, **292**, 31–45.
68. C. Odin and M. Nechtschein, *Phys. Rev. Lett.*, 1991, **67**, 1114–1117.
69. T. F. Otero and R. Abadias, *J. Electroanal. Chem.*, 2007, **610**, 96–101.
70. T. F. Otero and R. Abadias, *J. Electroanal. Chem.*, 2008, **618**, 39–44.
71. J. Arias-Pardilla, T. F. Otero, R. Blanco and J. L. Segura, *Electrochim. Acta*, 2010, **55**, 1535–1542.
72. T. F. Otero, J. Arias-Pardilla and E. Chermak, *Synth. Met.*, 2010, **160**, 425–431.
73. J. Arias-Pardilla, W. Walker, F. Wudl and T. F. Otero, *J. Phys. Chem. B*, 2010, **114**, 12777–12784.
74. T. F. Otero, J. Arias-Pardilla, M. I. Roca and J. G. Martinez, *Mol. Cryst. Liq. Cryst.*, 2012, **555**, 295–305.
75. J. G. Martinez, T. Sugino, K. Asaka and T. F. Otero, *ChemPhysChem*, 2012, **13**, 2108–2114.
76. T. F. Otero and I. Boyano, *J. Phys. Chem. B*, 2003, **107**, 6730–6738.
77. T. F. Otero and I. Boyano, *J. Phys. Chem. B*, 2003, **107**, 4269–4276.
78. C. Odin, M. Nechtschein and P. Hapiot, *Synth. Met.*, 1992, **47**, 329–350.
79. C. Odin and M. Nechtschein, *Phys. Rev. Lett.*, 1991, **67**, 1114–1117.
80. J. G. Martinez and T. F. Otero, *J. Phys. Chem. B*, 2012, **116**, 9223–9230.
81. *Handbook of Mathematical Functions with Formulas, Graphs, and Mathematical Tables*, ed. M. Abramowitz and I. A. Stegun, Dover, New York, 1964.

CHAPTER 8

Electro-Chemo-Biomimetic Devices

8.1 Introduction

Each of the composition-dependent biomimetic properties listed in Table 5.1, driven by the membrane electrochemical reactions, mimics a biological function that is the result of biochemical reactions in the organs of living beings.[1–10] The reversible variation of the value of each property under electrochemical control is being used to develop one-tool electro-chemo-biomimetic devices:

- Reversible volume variations give artificial muscles.
- The opening/closing of inter-chain distances control transversal ionic conductivity, diffusivity and the flow/non-flow of different ions to give smart membranes.
- Storage/liberation of ions result in smart drug delivery devices, smart chemical delivery devices (mimicking glands), electron/ion transducers for artificial chemical synapses, or pre-concentration analytical and de-contamination systems by ionic trapping–storage.
- Conformational energetic states can result in new chemo-ionic-conformational memories (that can mimic brain memory).
- Charge storage in the chains yields organic batteries and supercapacitors (mimicking electric organs).
- Control of the intermolecular forces gives smart surfaces (wettability control or self-cleaning control).

RSC Smart Materials No. 19
Conducting Polymers: Bioinspired Intelligent Materials and Devices
By Toribio Fernández Otero

Published by the Royal Society of Chemistry, www.rsc.org

- Color changes in the visible or infrared (IR) regions result in smart windows, glasses, mirrors, filters or camouflage systems.
- The material potential and its variation give sensors, biosensors and sensing devices.

All of these devices are reactive and biomimetic electrochemical devices, so the electrochemical reactions of the device constitutive materials drive each device actuation. Actuation here means a change of the material's working property that results from the material's working function. The present world market only offers, among millions of technological devices, a few reactive ones, such as batteries, smart mirrors, some chemical and electrochemical sensors, and some photoelectrochemical devices.

In parallel, control of the electronic conductivity of conducting polymers (CPs) is contributing to the new explosion of transparent and flexible microelectronics, photoelectronics and optoelectronics.[11–18] These are physical devices and their fields allow, from the point of view of this book, traditional physical technologies based on the physical properties of the device constitutive materials to be developed, supported by solid theoretical models developed during the past century. However, these dry physical devices, the compositions of which remain constant while working, that are of such importance in current technologies are outside the scope of this book.

By contraposition, the full development and control of new reactive devices requires the parallel development of new theoretical models, such as those presented in Chapter 7. These models must progressively include more precise and quantitative information related to the volume and structural variations of the device constituent materials during actuation, the influence of the reaction-driven structural processes, the kinetics of each of the reaction-driven properties, the influence of different additives for specific control of each structural kinetic component, or the material degradation and stabilization kinetics. In addition to empirical and theoretical difficulties for developing the new electrochemical devices, our present technology, based on dry physical devices, pushes engineers and designers to produce solid state reactive devices: solid polymeric electrodes, and solid ionic conducting and electronic insulating membranes. Thus, the new theoretical models will require the parallel development of properties and devices that work in the presence of liquid electrolytes (solvent, even in very small amounts, is present as a plasticizer) or that work as fully solid state devices. Despite these difficulties, in particular the low effort devoted nowadays to the study of basic theoretical aspects, some technological products, mainly based on empirical results, are emerging. Here, the state of the art of these empirical biomimetic one-tool devices, together with some basic theoretical supports, will be presented.

8.2 Artificial Muscles

In striated mammal muscles, an electrical order arrives from the brain as a nervous pulse, which liberates calcium ions inside the cell sarcomere and adenosine triphosphate (ATP) ions immediately link actin and myosin chains. The ATP reaction (hydrolysis), giving adenosine diphosphate (ADP), provides the energy required for the conformational movements of the chains driving the cell contraction.[19,20] The consecutive events are electrical pulse, calcium exchange, chemical reaction, reaction-driven conformational movement and anisotropic volume change, generating muscle contraction and producing mechanical energy. Natural muscles are electro-chemo-mechanical transducers. The basic component is a polymeric chemo-conformational motor.

Those named *artificial muscles* in the literature are closer to natural muscles, as the working device reproduces most of these consecutive events. At present, almost any transducer using whatever original energy (electrical, chemical, optical, magnetic and so on) and giving mechanical energy is named an artificial muscle.[21–31] Most of the electrical artificial muscles based on polymeric materials are electromechanical (physical) actuators or electromechanical artificial muscles, *i.e.*, the mechanical deformations are induced by electric field gradients.

Here, we are concerned with electro-chemo-mechanical artificial muscles, so muscles driven by the electrochemical reactions of the constitutive polymers.[2,3,32–44]

Good physical and mathematical models developed over a century ago for electromechanical actuators (piezoelectric, ferroelectric, coulombic, electro-osmotic or electrostrictive) have pushed most physicists, engineers and even some electrochemists involved in this field to expand such physical and mathematical models to try to describe chemical-driven actuators.[45–50] This book is focused on electrochemically (faradaic) driven actuation and models.

In dense reactive gel electrodes from CPs, electrical pulses that arrive from the generator under a computer order promote electrochemical reactions, driving the conformational movements of the chain reactant (basic molecular motor), together with ionic and aqueous exchanges with the surroundings.[49,51–54] The final result is the generation of quasi-isotropic dimensional variations of the material. The consecutive events overlap step-by-step those events taking place in natural muscles, except, at the moment, for the anisotropic volume variation and the reversibility of the CP electrochemical reaction. The parallel ATP hydrolysis in natural muscles is an irreversible reaction, requiring the glucose cycle to recover ATP from the generated ADP: natural muscles only work by contraction, while artificial muscles based on the electrochemistry of CPs work under contraction as well as under expansion.

From the isotropic (in fact quasi-isotropic) volume variation under electrochemical control (Figure 6.1c), different artificial muscles give linear,[53,55–57] bending[51,58,59] and origami[60] movements (among others), producing mechanical energy, and these artificial muscles are being developed as an extension of molecular, microscopic and macroscopic devices.[61–72] In order to describe the general, basic aspects that can be translated to any other structure or shape, here we will choose bending, fully polymeric (not including any metallic component) systems.

8.2.1 Bilayer and Tri-Layer Bending Devices

Small volume variations driven by the exchange of ions and solvent during the CP reaction (Figure 6.1c) are transduced into large bending movements by the construction of a bilayer or triple layers using a second, inactive layer, such as a tape,[51] cellulose or paper,[73] a metal,[74] a membrane[75] and others. Most of the claimed triple layer actuators are in fact CP/metal/membrane/metal/CP five-layer actuators. The membrane, or the central material, is usually metal sputtered in order to provide it with a very high surface electronic conductivity: this is a consequence of the accepted conducting-insulator transition model (Chapter 7). During actuation, the metal can give parallel irreversible electrochemical reactions, such as water electrolysis (Section 4.8.1), resulting in subsequent actuator degradation. Here, we will describe bilayers constituting the CP/tape form, not including any metallic component, *i.e.*, fully polymeric devices. Figure 8.1 describes the construction of a bilayer (a) and (b) and a triple layer (a)–(e) of a fully polymeric artificial muscle using a two-sided tape and a metal electrode coated with two CP films electropolymerized on either side of the metal electrode.

When a CP/tape bilayer muscle is used as the working electrode (WE) in a three electrode electrochemical cell (Figure 8.2a), the flow of an anodic current through the CP layer promotes the polymer oxidation driving, through the reaction shown in eqn (4.1), the entrance of anions and solvent: the CP film swells. Across the CP/tape interface the swelling CP generates expanding stress gradients (Figure 8.2b)[54] and the bilayer bends, keeping the CP in the convex part and the tape in the concave part. The bottom of the tape that is free to move (unclamped) shows a large angular movement that exceeds, in some cases, 360°. During the CP film reduction by the flow of cathodic currents, opposite processes and bending movements (Figure 8.2a) occur.

Figure 8.2c shows the structures of the electro-mechanical and electro-chemo-mechanical triple-layer actuators. In electro-mechanical actuators, the two electrodes are passive elements required for the generation of the electric field gradient: volume variations only occur across the electroactive polymer, driven by the electric field gradient. In electro-chemo-mechanical

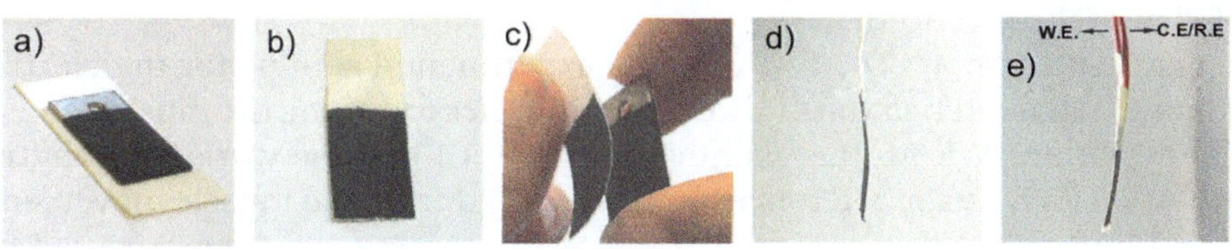

Figure 8.1 Construction of a three-layer fully polymeric muscle from an electrogenerated polymer film coating the metal electrode: (a) a double-sided tape was adhered and the bilayer (b) was peeled off from the electrode. (c) The free side of the tape was adhered to the second film of the CP and (d) the three layers were peeled off from the metal. Each CP film was connected with conducting glue to a metal wire (e), where WE is the working electrode, CE is the counter-electrode and RE is the reference electrode.

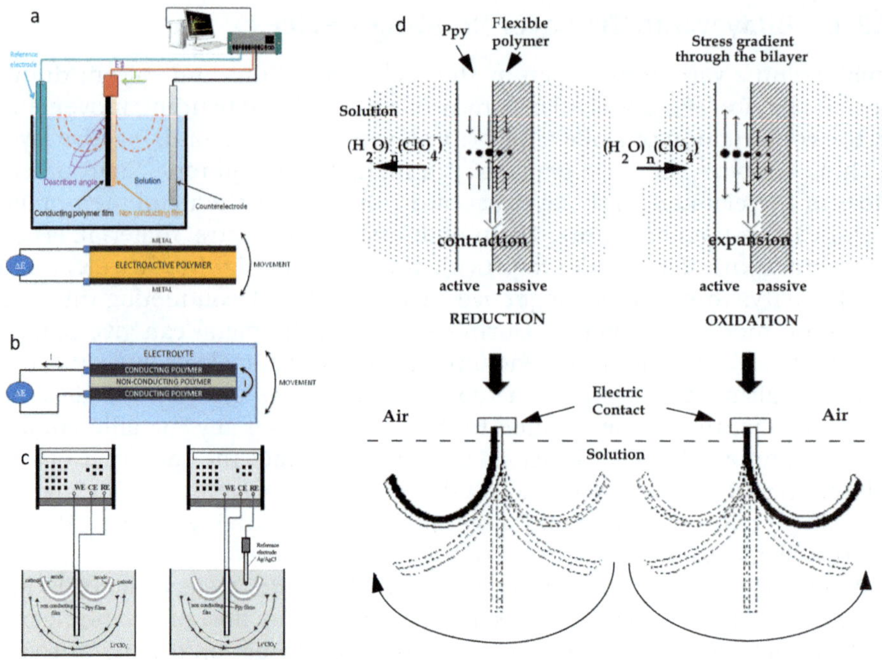

Figure 8.2 (a) Electrochemical characterization of a bilayer muscle showing the experimental set-up. (b) Scheme of the electro-mechanical (top) and electro-chemo-mechanical (bottom) actuators (reprinted from ref. 8: *Electrochim. Acta*, 2012, **84**, 112–128, with the permission of Elsevier). (c) Experimental set-ups for three-layer actuators following the muscle potential (left) or the WE potential (right) during actuation by current flow. (d) Scheme of the ionic exchanges, generated stress gradients and bending movements during the CP film reduction (left) or oxidation (right). Reprinted from ref. 1: *Modern Aspects of Electrochemistry*, with the permission of Springer.

actuators, the two electrodes respond to the current flow with electrochemical reactions (oxidation or reduction) resulting in swelling and shrinking electrode processes. The intermediate tape is a passive film (from the point of view of the reaction).

Figure 8.2d shows the two possible experimental set-ups for the electrochemical characterization of three-layer muscles following the muscle potential evolution (WE *vs.* the counter-electrode, CE) or the working potential evolution (WE *vs.* the reference electrode, RE) during the muscle movement.

8.2.2 Electro-Chemo-Dynamic Characterization of Artificial Muscles

Any CP-based artificial muscle (bilayers here) can be subjected to consecutive potential cycles, square potential waves, square current waves or any other electrochemical methodology allowing the reversible oxidation/reduction of the CP film. In parallel, the reversible bending movement can be followed and video-recorded. In this way, electrical magnitudes obtained from the electrochemical responses can be plotted *versus* mechanical-dynamic magnitudes (described angles, movement rate and so on) to get electro-chemo-dynamic relationships.[61,76–78]

Voltammetric and coulovoltammetric responses up to different potential limits give unequivocal information (Chapters 5 and 6) about the potential domains for the reversible oxidation/reduction of the CP layer, the potential ranges for the different reaction-driven structural processes, the presence of irreversible oxidation or reduction reactions overlapping the reversible film oxidation/reduction, at which potential the irreversible reaction begins, whether the CP degradation (and the muscle degradation) beyond these limits starts, and (the key point) the maximum reversible charge the muscle can support during actuation without risking either chemical degradation by overoxidation or other parallel electrochemical reactions.

Coulodynamic (charge/angle) responses are attained by correlation between the coulovoltammetric responses and the amplitude of the angle described at each potential when the CP film taking part of the muscle is cycled between the studied initial and final potential limits.

When the muscle is characterized by square current waves, the coulodynamic representation correlates the consumed charge and described angle at every time of the current flow. The characterization can be repeated now using rising currents for constant oxidation and reduction charges, and represents the angular rate as a function of the applied current (ampero-dynamic responses).

8.2.3 The Driving Current Controls the Angular Movement of the Polymeric Motor

The resulting empirical ampero-dynamic responses indicate that the angular rate of the bending muscular movement is a linear function of the driving current (Figure 8.3A and B):

$$\omega = kI, \tag{8.1}$$

where ω (in rad s^{-1}) is the angular rate of the movement, I (in A g^{-1}) is the driving specific current and k (in rad g A^{-1} s^{-1}) is the ampero-dynamic constant of the CP/tape/electrolyte system.

Driven by a constant current, the reaction shown in eqn (4.1) controls the entrance in the film of the same number of counterions and water molecules per unit of time [eqn (4.9), where $Q = it$]. This means that there is a constant volume variation rate for the CP film (inside the oxidation range of the CP, determined by the voltammetric control).

Eqn (8.1) describes the control of the muscle movement by the imposed current: the movement can be accelerated, retarded or stopped at any time by increasing, decreasing or switching off, respectively, the flowing current.

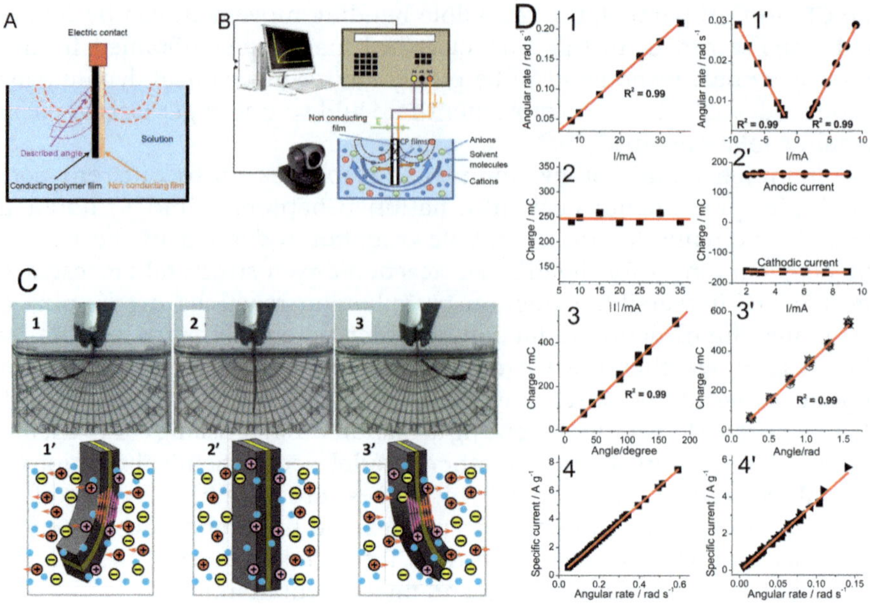

Figure 8.3 (A) Scheme of the electrochemical cell and configuration of the electrodes used to study the electrochemical actuation and sensing behavior of the triple-layer actuators (CP film/tape/CP film). (B) A vision system was used for video-recording and to quantify the angular movement. (C) Anti-clockwise from (a) to (c), or clockwise from (c) to (a), bending movements and concomitant ionic exchanges and stress gradients, (c') and (a'), from a Ppy-DBS/tape/Ppy-DBS triple-layer in a 0.1 M $LiClO_4$ aqueous electrolyte (Ppy, polypyrrole; DBS, dodecylbenzenesulfonate). (D) Artificial muscles are electrochemical devices with linear relationships between the electrical-driven magnitudes and angular movements: (a)–(d) for muscles exchanging anions (Ppy/ClO_4^-; reproduced from ref. 77 with the permission of the Royal Society of Chemistry); (a')–(d') for muscles exchanging cations (Ppy-DBS/tape; reproduced from ref. 84: *Electrochim. Acta*, 2011, **56**, 3721–3726, with the permission of Elsevier). The same driving charge per unit of reduced polymer mass gives the same angular movement whatever the mass of CP in the device (d and d'). Adapted from ref. 83: *J. Mater. Chem.*, 2011, **21**, 17265–17272, with the permission of the Royal Society of Chemistry.

Furthermore, by reversing the sense of the current flow, from anodic to cathodic or *vice versa*, the sense of the bending movement can be reversed at any time. Artificial muscles based on CPs, or any other reactive material, are faradaic polymeric motors: the imposed current controls the movement rate and the sense of the movement. The current flow (charge per unit of time) and its variation determine the number of ions exchanged at any moment per unit of time between the CP film and the electrolyte. Consequently, it controls the CP film volume variation per unit of time, the stress gradient increment generated by this volume variation at the

CP/tape interface per unit of time and the subsequent angular displacement per unit of time. Artificial muscles constituted by CPs exchanging anions and muscles constituted by CPs exchanging cations follow eqn (8.1): the ampero-dynamic constant k is different for each system (CP/tape/electrolyte) and the sense of the angular movement driven by the same anodic current is opposed because the CP oxidation promotes opposed swelling or shrinking volume variations in CPs exchanging anions [eqn (4.1)] or cations [eqn (4.2)], respectively.

8.2.4 The Consumed Charge Controls Displacement and Relative Position

The empirical coulodynamic (charge/angle, Q/α) responses (Figure 8.3D panels c and c') confirm that the charge controls the angular displacement through the CP film volume increment defined by the number of exchanged ions [eqn (2.3), (4.11) and (5.1)] and solvent molecules (see Section 8.2.8). Working inside the reversible oxidation/reduction charge of the CP layer, the same consumed charge must promote the same angular displacement whatever the initial angular position (α_0) of the muscles and the way (constant current, current steps, potential steps) to promote the flow of this charge:

$$\alpha = \alpha_0 + k'Q, \tag{8.2}$$

where α (in rad) represents the new angular position of the muscle, α_0 (in rad) is the initial angular position, Q (in C) is the consumed charge and k' (in rad C^{-1}) is the coulodynamic constant, characteristic of the CP/tape/electrolyte system.

Eqn (8.2) indicates that artificial muscles based on CPs (or any other electrochemically active material) are polymeric motors in which the relative (angular) position of the muscle base is under faradaic (linear) control of the consumed charge. From any original position, α_0, a new position at a constant angular distance, α, is attained by consumption of charge Q. Artificial muscles are electro-chemo-positioning devices (Figure 8.3D panels c and c').[2,8,79–85]

The sense of the angular displacement (clockwise or anticlockwise) is determined by the relative tape/CP position, the nature of the anodic/cathodic charge and the anion/cation exchanged by the CP layer during reaction. Thus, for a tape located on the left-hand side and a CP film exchanging anions during a reaction on the right-hand side, anodic charges result in polymer oxidation–swelling, giving clockwise movements, and the cathodic charges promote polymer reduction–shrinking, giving anticlockwise angular displacements. For the same structure and CP films exchanging cations during reaction, anodic charges result in polymer oxidation–shrinking, giving anticlockwise movements, and cathodic charges promote polymer reduction–swelling, giving clockwise angular displacements.

8.2.5 Artificial Muscles are Robust, Reproducible, Reliable and Faradaic Polymeric Motors

Eqn (8.2) indicates that whatever the initial position of the muscles the same charge always results in the same angular displacement: the same charge results in the same increment of the ionic composition [eqn (4.11)] of the CP film. This means that the CP composition variation must control either the CP volume variation, the concomitant stress gradient variation across the muscle or the angular displacement resulting from that stress gradient.

Only the amount of CP taking part of the muscle and reacting inside the electrolyte influences the muscle's displacement. There, the same ionic-specific composition variation in the CP should give the same transversal stress gradient and the same angular displacement, whatever the shape or dimensions (from nanometric to macroscopic) of the muscle, the thickness of the constituent CP film or the amount of CP taking part of the muscle. In summary, bending bilayer or multilayer artificial muscles manufactured following the same procedures (the same CP synthesis conditions and tape, and checked in the same electrolyte) in different laboratories or in the same laboratory with different geometric surface areas, different shapes and with different CP film thicknesses must produce the same angular rate by consumption of the same specific current, i (current per unit of actuating CP mass, in A g^{-1}) and must describe the same angle by consumption of the same specific charge, q (charge per unit of actuating CP mass, in C g^{-1}). In this way, eqn (8.1) and (8.2) become the general expressions:

$$\omega = ki \text{ and } \alpha = \alpha_0 + k'q, \tag{8.3}$$

confirmed by empirical α/i relationships, as depicted by Figure 8.3D (panels d and d') for different muscles, including CP films exchanging anions[79,81] or cations.[84] Each of these figures presents results from different muscles, each having a different shape, different surface area, different thickness and a different mass of the constitutive CP film. Of course, the same relative CP/tape position in the muscle gives opposite bending movements during the same reaction: during oxidation, the polypyrrole–ClO_4^- film swells, while the polypyrrole–DBS film shrinks.

When the faradaic nature of these artificial muscles is ignored and the electro-chemo-mechanical devices are treated as electro-mechanical devices (responding to the electric potential instead of being driven by current and charge), quite complex equations are attained to describe both the movement rate and relative position in order to develop control systems.[53,58,86–103]

8.2.6 Dynamic Hysteresis and Creeping Effects Under Cycling

Eqn (8.3) describes and quantifies how robust the fabrication procedure is, giving very reproducible devices and good control of both the motor movement rate and of its relative position at any time. The remaining problems to obtain fully confident devices for engineering and robotic applications are hysteresis dynamic effects and creeping effects under cycling.

Creeping effects: Some fully polymeric muscles produce the same angular amplitude per oxidation/reduction-driven cycle but the initial position shifts by a small, constant angle per cycle. Creeping effects are not present in freestanding, fully polymeric bilayer muscles giving closed coulovoltammetric responses: the CP layer exchanges small anions. The use of CP–organic macroion blends in bilayer muscles gives stationary voltammetric (Figure 8.4a), open coulovoltammetric and dynamo-voltammetric responses (Figure 8.4b), and almost linear coulodynamic (charge/angle) responses (Figure 8.4c). Nevertheless, the muscle's angular position at the

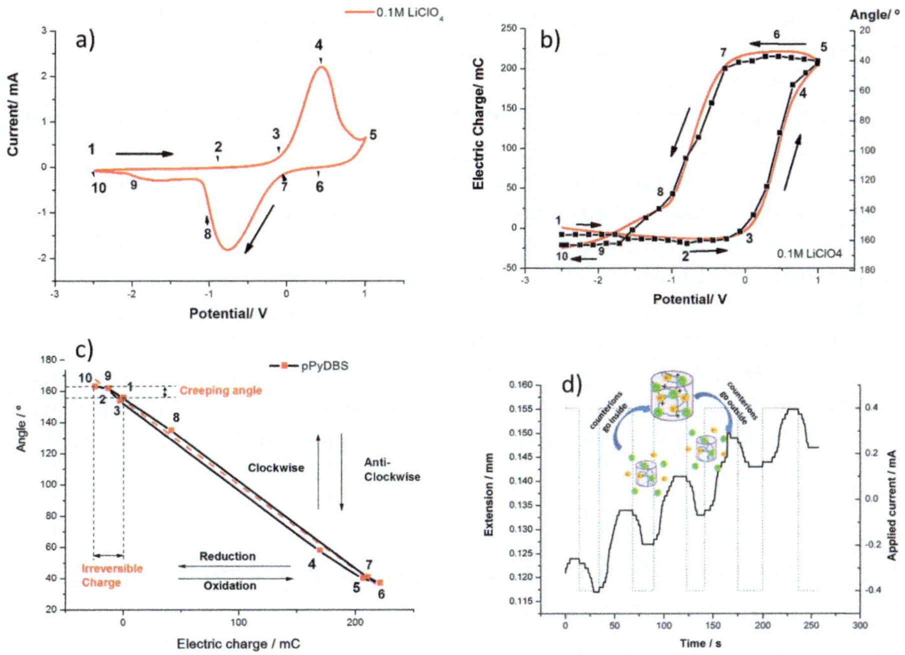

Figure 8.4 (a) Stationary voltammetric response from a tape/Ppy-DBS bilayer muscle subjected to consecutive potential sweeps between −2.50 V and 1.0 V at 6 mV s^{-1} in a 0.1 M $LiClO_4$ aqueous solution. (b) Coulovoltammetric (charge–potential; red lines) and dynamo-voltammetric (charge–angle; black dots) responses obtained by integration of the voltammogram, and parallel evolution of the video-recorded angle described by the muscle. (c) Evolution of the angular displacement of the muscle with the consumed charge (the coulodynamic response) along the potential cycle, indicating irreversible charge and a creeping angle (adapted from *J. Solid State Electrochem.*, 2015, **19**, 2683–2689, Copyright Springer). (d) Ppy film length variations when, under a constant force of 0.1 N, it was oxidized and reduced by applying five consecutive square current waves (±0.4 mA; dotted line) in a 0.1 M $LiClO_4$ aqueous solution showing the creeping effect. The dimensions of the Ppy film were 7.32 mm × 1.94 mm × 34 µm. Reprinted from ref. 78: *Sens. Actuators, B*, 2014, **1990**, 27–30, with the permission of Elsevier.

end of the potential cycle (point 10 in Figure 8.4) is shifted related to the muscle's angular position at the beginning of the potential cycle (point 1): this is the creeping displacement per cycle. Irreversible charge and creeping displacement per cycle seem linearly related.[104] The irreversible charge was attributed to the presence of the organic macroions and to hydrogen formation by discharge of the concomitant organic acid by cycling in the presence of water. The creeping displacement should correspond with a progressively deep reduction of the CP layer: anodic polarizations at a constant potential reverse most of the creeping angle. In addition, under different mechanical working conditions (trailed weights or under mechanical stress), a second physical creeping effect appears by progressive elongation of the CP for the same oxidized state (Figure 8.4d). Regardless of whether the chemical creeping originated from irreversible charges, there is a physical creeping due to mechanical deformation, reversible or not, which still remains an open question.

Dynamic hysteresis: During cyclic actuation by oxidation/reduction (potential sweep, square current or square potential waves), the muscle presents, for the same oxidized state of the CP, a different angular position for the oxidation or reduction branches. This is a dynamic hysteresis effect resulting from osmotic and electro-osmotic physical effects following or driven by, respectively, the faradaic exchange of ions.

Both creeping effects and dynamic hysteresis require a good clarification and quantification in order to obtain, in addition to very reproducible artificial muscles, a good (as perfect as possible) control of the movement rate and position. This means reliable actuators and motors for different tools and robots. In this way, dynamic hysteresis will be more deeply considered in Section 8.2.9.

8.2.7 Artificial Muscles as a Tool to Clarify Reaction-Driven Ionic Exchanges

Most of the *in situ* methodologies used to determine electrochemical reaction-driven ionic exchanges (quartz crystal microbalance, QCM; atomic force microscopy, AFM; bending beam and so on) use CP films on metal (plates, foils or sputtered) electrodes. They require the parallel determination of the consumed charge and a property varying with the ionic exchange (dimensions, frequency–mass, *etc.*). The presence of the metal electrode can result in parallel irreversible reactions (such as water electrolysis) that consume charge. These irreversible reactions can now be detected and quantified, as indicated in Section 4.5. The consumed irreversible charge corrupts empirical determinations from the literature when the non-detected irreversible charge is attributed to faradaic ionic exchanges between the film and the electrolyte.

Fully polymeric bilayer muscles are the most efficient tools to translate *in situ* small volume variations into large bending movements (up to ±360°).

A transversal paint strip around the muscle and below the clamp contact outside the electrolyte avoids the possible contact by capillarity of the electrolyte and the metal clamp, and the subsequent parallel reactions during muscle characterization. The paint strip is kept just above the electrolyte meniscus.

In this way, the CPs synthesized in the presence of small anions give, during subsequent oxidation/reductions, anion exchanges (Section 4.1), as revealed by voltammetric, coulovoltammetric and parallel video-recording of the muscle displacement (Figure 8.5a). For blends synthesized in the presence of large organic anions, the anions remain trapped in the film (Section 1.5.1), resulting in cation exchange for charge balance during oxidation/reduction. Coulodynamic responses from polymers exchanging anions or cations are similar to those presented in Figure 8.3c and c',

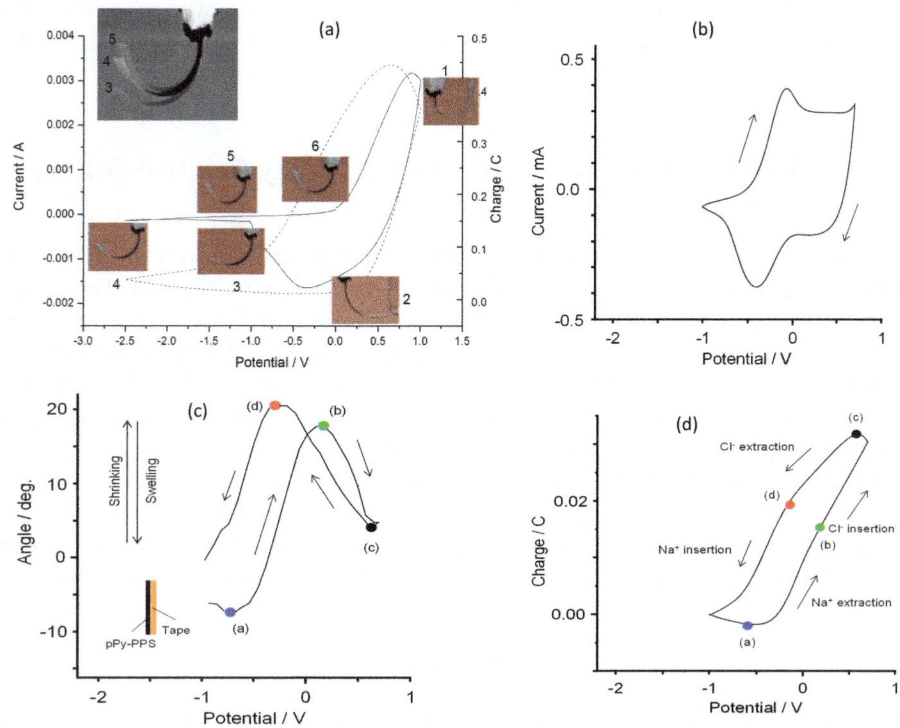

Figure 8.5 (a) Voltammetric (i/E) and coulovoltammetric (Q/E) responses of a Ppy/tape bilayer muscle from a 0.5 M $LiClO_4$ aqueous solution and pictures showing the muscle position at different potentials. Adapted from ref. 78: *Sens. Actuators, B*, 2014, **1990**, 27–30, with the permission of Elsevier. A Ppy-*para*-phenolsulfonic acid (HpPS)/tape artificial muscle was subjected to consecutive potential sweeps from −1.0 V to 0.7 V to the same anodic potential limit of 0.7 V at 10 mV s^{-1} in a 0.5 M NaCl aqueous solution: (b) stationary voltammetric, (c) parallel dynamo-voltammetric (angle–potential) and (d) coulovoltammetric (charge–potential) responses. Reprinted from ref. 105 with the permission of the Royal Society of Chemistry.

respectively. At the moment, only blends of CPs with organic anions (from organic acid salts) with intermediate dimensions present dual anion and cation exchange in different potential ranges, as revealed by the dynamo-voltammetric responses (Figure 8.5c) attained in parallel to voltammetric (Figure 8.5b) and coulovoltammetric (Figure 8.5d) responses.

The existence of different potential ranges, each with a prevalent exchange of cations or anions, indicates the strong variation of the interaction forces (polymer–anion–cation–solvent) inside the film with the film oxidation state. Considering the characteristics that the interaction forces can shift, tuning from mainly anion to mainly cation exchanges can be achieved by changing the solvent or the solved salt, as proved recently.[105,106] In this way, polymeric motors are emerging as unbeatable tools for the future quantification of inter- and intra-molecular force changes during reactions inside dense gels. Attained models could be translated to similar problems in the reactions taking place in the intracellular matrix of living cells that result in life functions and malfunctions.

8.2.8 Artificial Muscles as Tools to Quantify Relative Solvent Exchanges

One question, among so many that are pending with regards to biological processes, is the quantification of the number of water molecules exchanged per biological reaction unit. This key point separates health and illnesses in many cases. Now, biomimetic reactions [eqn (4.1)–(4.4)] and artificial muscles can be used as reaction model and tools, respectively, for the quantification of the number of solvent molecules exchanged per reaction unit (per electron extracted from or injected to the polymer film).[107]

From a chemical point of view, water is not a reactant as both the chemical composition and molecular bonds remain unchanged on both sides of the reaction. A physical process drives the water exchange: the osmotic pressure (Π) follows a linear dependence on the membrane concentration according to the Van't Hoff equation:

$$\Pi = iMRT, \tag{8.4}$$

where i is the dimensionless Van't Hoff factor, M is the molarity, $R = 0.082$ L atm K^{-1} mol^{-1} (the gas constant) and T is the absolute temperature.

As an initial hypothesis it can be assumed that: (a) the volume variation resulting in the muscle movement during the reaction is only that required for lodging the balancing counterions inside the polypyrrole film; (b) strains perpendicular or parallel to the surface are identical (isotropic); and (c) Young's modulus and the stiffness variation during the driving reaction are similar in every electrolyte.

During the CP film oxidation [the forward reaction in eqn (4.1)], the ionic concentration increases inside the membrane according to eqn (4.11). Consequently, eqn (8.4) suggests that the entrance of water follows the reaction-driven ionic concentration rise in the membrane with a linear proportionality. In summary, for each polymer–ion–solvent system, a different proportionality (a different number of solvent molecules per exchanged monovalent ion or per exchanged electron) is expected.

As a consequence, the film volume variation (ΔV) driven by the reaction must lodge n anions [eqn (4.9)] and n' solvent molecules:

$$\Delta V = \Delta V_{\text{ions}} + \Delta V_{\text{solvent}}, \tag{8.5}$$

where ΔV is the volume variation in the film, ΔV_{ions} is the volume variation due to the faradaic exchange of counterions and $\Delta V_{\text{solvent}}$ is the volume variation due to the osmotic exchange of solvent molecules. The volume occupied by the exchanged counterions can be estimated from the ionic radii of every ion:

$$\Delta V_{\text{ions}} = \frac{Q}{zF} N_{\text{A}} \frac{4}{3} \pi r^3, \tag{8.6}$$

where $(Q/Fz)N_{\text{A}}$ is the number of balancing counterions needed to compensate the number of electrons removed from the polymeric chains defined by the consumed oxidation charge Q, where z is the ion charge, F is the Faraday constant (F = 96 485 C mol^{-1}) and N_{A} is Avogadro's number (the number of ions in 1 mole), and $(4/3)\pi r^3$ is the volume of one ion, where r is its crystallographic radii. From eqn (8.5) and (8.6), with α_0 being the initial position of the muscle, the angular displacement of a bilayer muscle by consumption of a constant specific charge in the presence of different ions is described by:

$$\alpha = kq = k \frac{3}{4} \frac{Fz\Delta V_{\text{ions}}}{N_{\text{A}} \pi r^3} \tag{8.7}$$

Eqn (8.7) correlates the volume variation (ΔV_{ions}) with the angle (α), crystallographic radius (r) and the number of charges per anion (z). Experimental results don't follow eqn (8.7),[107] indicating the need to include water exchange in eqn (8.5).

Assuming that among the tested anions the one giving the smaller angular displacement does not trail any water molecule, its ΔV_{ions} can be taken as the reference volume variation ΔV_{ref}, with its α_{ref} being the reference angular displacement of the muscle. Under these conditions, from the angular displacement in the presence of a different anion (α_{x}) for a constant consumed charge (1 mole of electrons), every time correlates with an unknown volume variation (ΔV_{x}) that can be calculated through two different components related to those of the reference ion (α_{ref} and V_{ref}) and considering the ionic molar volume (V_{x}), the water molar volume (V_{H2O}) and the number of exchanged water molecules:

$$\Delta V_{\text{x}} = V_{\text{ref}} \frac{\alpha_{\text{x}}}{\alpha_{\text{ref}}} = \frac{V_{\text{x}}}{z} + nV_{\text{H}_2\text{O}} \tag{8.8}$$

and from the last equality:

$$n = \frac{V_{IO_3^-}\dfrac{\alpha_x}{\alpha_{ref}} - \dfrac{V_x}{z}}{V_{H_2O}}. \tag{8.9}$$

Thus, the number of water molecules exchanged per reaction unit (exchanged electrons) during the reaction shown in eqn (4.1) from aqueous solutions of the anion's sodium salts were 3 for ClO_4^-, 0.6 for ClO_3^-, 0.35 for NO_2^-, 0.5 for CO_3^{2-}, 1.7 for Br^- and 0 for the reference with IO_3^-. Similar studies were performed using bilayer muscles, with polypyrrole (Ppy) films exchanging anions[107,108] or cations.[109] Other methodologies and hypotheses are also being used to obtain the exchanged solvent.[110–125]

8.2.9 Osmotic and Electro-Osmotic Processes During Actuation

Figure 8.6 shows the coulovoltammetric (charge/potential) and coulodynamic (charge/angular displacement) responses from tape/Ppy and tape/Ppy-dodecylbenzenesulfonate (DBS) bilayer muscles to potential cycles in NaBr aqueous solutions. The relative position of the films and the coulodynamic responses indicate that the CP reduction drives the incorporation of cations from this electrolyte in the full potential range. The coulodynamic responses (Figure 8.6b and c) present some deviations from the expected linearity defined by eqn (8.2) and (8.3). After the beginning of the anodic potential sweep, when the film oxidation was initiated during (positive charge increment from the minimum until point 1) the expulsion of cations, the angular displacement still goes on (film swelling) for some time, resulting in

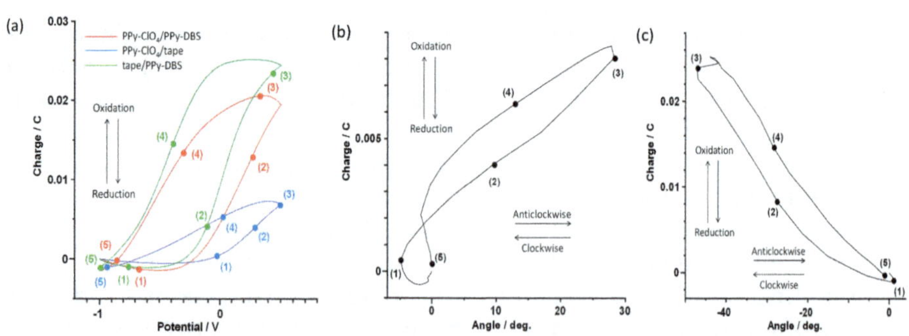

Figure 8.6 (a) Coulovoltammetric (charge/potential) and coulodynamic (charge/angular displacement) responses from (b) tape/Ppy and (c) tape/Ppy-DBS bilayer muscles to a voltammetric cycle between −1.0 V and 0.7 V at 10 mV s^{-1} in a 0.5 M NaBr aqueous solution at room temperature. Reprinted from ref. 126: *Adv. Funct. Mater.*, 2015, **25**, 1535–1541 with the permission of John Wiley and Sons.

some dynamic hysteresis: for the same oxidation state of the CP (the same charge), the angular position is different during the oxidation sweep compared to during the reduction sweep. At any potential, the system composition is outside its equilibrium state: as described in the previous section, the osmotic entrance of solvent is a physical process following the faradaic-driven ionic entrance. The higher concentration of counterions inside the CP membrane is attained at the coulovoltammetric minimum. However, the osmotic equilibrium has not yet been reached due to the delayed entrance of water: the water entrance (film swelling) goes on when the reduction-driven expulsion of anions (film shrinking) is already initiated. The empirical result indicates that, at the beginning of the polymer oxidation, the osmotic volume increase is higher than the decreasing volume due to the expulsion of cations.

As a consequence of the delayed ionic and osmotic exchanges, the muscle's relative position (the angle related to the position) for the same oxidation state of a membrane (same charge) exchanging cations is always a little bit higher during the oxidation sweep than during the reduction sweep. This fact is confirmed by the higher water contents (higher membrane volume and described angle) during membrane oxidation compared to during reduction. In addition, this fact suggests that the water entrance/exit by osmosis, which corresponds to a physical process, always follows the faradaic movement of the counterions. Finally, these facts also indicate that at least a fraction of the exchanged water becomes free and is not directly bonded to the counterions as solvated water: in that case, the swelling–shrinking coulodynamic responses would not present any hysteresis. In summary, the dynamic hysteresis has been attributed to the slowing of the solvent osmotic entrance or exit related to faradaic ionic exchange.[104,126]

For angular displacements arriving up to very high cathodic potential limits, the coulodynamic evolution shows double loop, figure-of-eight shapes (Figure 8.6b). The point where the reduction branch crosses the anodic one is attributed to fast solvent dragging from the film by electro-osmosis when the polymer shrinking approaches the closing potential.[126] The polymer shrinks during its oxidation by the expulsion of cations and the average interchain distance become similar to the counterion unit diameter when a high fraction of the counterions still remains inside the film (Sections 6.13.1 and 6.13.2): the escaping anions drag the solvent in front of them towards the solution (electro-osmotic effect). The number of water molecules expulsed from the membrane per ion rises very fast close to the closing potential with a parallel fast decrease of the membrane volume and increase of the angular displacement: the coulodynamic cycle presents a double loop with a figure-of-eight shape. The electro-osmotic effect has been used for decades during the industrial electrodeposition of paints to get almost dry electrodeposited paint films from very low concentrations of the polyelectrolyte component in the paint bath.[127]

8.3 Smart Membranes Tune Transversal Ionic Flow

The reactions shown in eqn (4.1)–(4.4) drive the swelling/shrinking processes. Free volume is generated between polymeric chains, which is occupied by counterions and solvent molecules coming from the solution (Figures 1.4 or 6.1). The average distance between the chains increases. Being a non-stoichiometric material, the average inter-chain distance can remain constant, or be increased or decreased in a reversible way under control of the reversible CP oxidation charge.

Any self-supported CP film used as a membrane (Figure 8.7A and B) between two solutions including different salts (Figure 8.7C) works as a cation- or anion-driven membrane.[128–135] In addition, we can now move one step ahead, approaching the ion channel behavior of living cells, by tuning our inter-chain distances (pore diameters) by means of the oxidation state

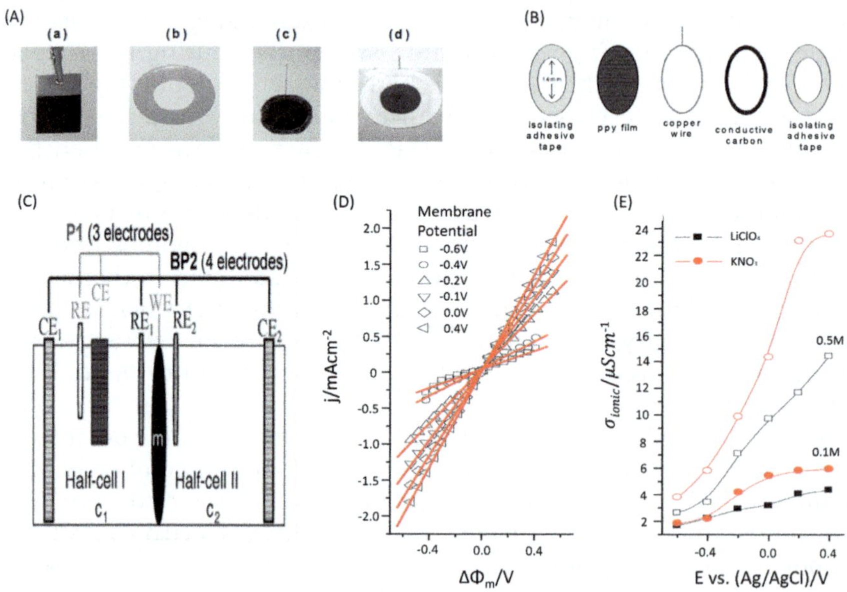

Figure 8.7 Fabrication (a) electrogenerated pPy film on a steel electrode, (b) tape ring (c) copper ring-wire adhered to the pPy film with a carbon tape ring, (d) final membrane: isolating tape ring/pPy membrane/carbon tape ring/isolating tape ring (A) and scheme (B) of a self-supported membrane. (C) Experimental cell for membrane ionic conductivity control: potentiostat 1 (P1) allows control of the membrane oxidation state, then bi-potentiostat 2 (BP2) applies a transmembrane potential gradient and determines the transversal current. (D) Evolution of the stationary ionic current across the membrane (after the membrane oxidation at different potentials indicated on the figure) as a function of the transmembrane applied potential. (E) Evolution of the ionic conductivity as a function of the oxidation state of the membrane in two different electrolytes and for different electrolyte concentrations. Reprinted from ref. 140: *J. Membr. Sci.*, 2007, **290**, 241–249, with the permission of Elsevier.

of the membrane. The membrane now acts simultaneously as a WE. The transversal ionic diffusivity between two different solutions can be followed in the absence of electric fields. The transversal ionic flow can also be measured as a function of the potential gradient applied across the membrane (Figure 8.7C) and this experimental determination can be repeated for different swollen/shrunken, oxidized states of the membrane, as depicted by each line in Figure 8.7D. In this way, the variation of the ionic conductivity as a function of the oxidation state of the membrane and for different concentrations of different salts can be determined (Figure 8.7D).[136–140]

These results indicate that the attractive forces between the polaronic species in the chains and the counterions allow the mobility of the balancing ions, giving a high ionic conductivity across the membrane under chemical or electrochemical potential gradients. In order to maintain the membrane electroneutrality, one counterion penetrates the electrolyte/membrane interface and, simultaneously, another counterion exits through the opposite electrolyte/membrane interface: the charge balance inside the membrane is kept constant with time.

The membrane behaves as a polyelectrolyte with high mobility of the counterions. The cations trying to go through the membrane in the opposite direction, driven by the electric field, feel the repulsion from positive polarons in the chains (for basic CPs exchanging anions), *i.e.*, the membranes present selective (preferential) conductivity of the anions with its transfer number being close to one.

If the solution includes several anionic species, the membrane average inter-chain distance can be tuned through the oxidation state of the membrane with increasing diameters of the different anions. Under a defined oxidation state, the anions with lower diameters than the average inter-chain distance move across the membrane, while those with larger diameters do not pass. By increasing the membrane oxidation state, the average inter-chain distance can be tuned to the larger anions: both the inter-chain average distance and the transversal ionic conductivity can be tuned by the oxidation state of the membrane.

These smart membranes are much closer to biological ones than any present technological membrane. Different ionic separations and desalinations would have immeasurable technological and economic impacts with regards to medical, biochemical, pharmacological and food industry applications following more intense basic exploration of the different CP families in order to obtain optimum anion and cation membranes with tuned ionic conductivities and selectivities.

8.4 Artificial Glands: Smart Chemical Dosage and Smart Drug Delivery

Anions [forwards reaction in eqn (4.1)] and cations [backwards reaction in eqn (4.2)] are stored by the oxidized or reduced state of the different CP families. Moreover, protons can be liberated by oxidation of the polyaniline family of materials. These reactions indicate that ions can be liberated under faradaic control [eqn (4.9) indicates the charge/mole relationship], allowing perfect control of the solution's composition around the polymer film.

When the stored ions have a medical, pharmacological or agricultural (such as in fertilizers) interest, the material presents unexplored possibilities for use as smart chemical or drug delivery devices that work under faradaic control of the ion liberation.[61] At the moment, this is a partial dream: most of the delivered drug occurs under spontaneous diffusive control of ion exchange.[141–143] Difficulties to obtain faradaic liberation control are based on the still limited knowledge (and control) of the involved intra- and intermolecular forces inside the reacting film.

Challenges: The counterion and solvent entrance and expulsion simultaneously involve each of the inter- and intra-molecular, coulombic and van der Waals (polymer–polymer, polymer–ion, polymer–solvent, ion–solvent) interactions involving neutral polymer chains, charged polymer chains, counterions and solvent, as well as the reversible variation of these interactions along with the reversible oxidation/reduction reactions. Thus, for the same salt, the electrochemical responses (chronoamperometric, voltamperometric, coulovoltammetric *etc.*) in the same potential range change in different solvents (Section 6.19) and, for the same solvent, the responses change in solutions with different anions (Section 6.18). These results must allow in the future the quantification, theoretical description and prediction of the variation of the intermolecular forces in the film during reaction as a function of the physical and chemical variables.

Organic amphiphile counterions incorporated into a CP material during synthesis (*i.e.*, by electropolymerization) remain trapped inside the material during its subsequent oxidation/reduction in electrolytes, forcing the exchange of cations for charge balance. Whether this trapping effect is due to the strong polymer–organic anion van der Waals interactions or physical limitations due to the low inter-chain distances produced by the material's reduction–swelling is still an unsolved question. Few efforts have been devoted by the scientific and technological communities in order to determine how the electropolymerization mechanism can be managed and the substituted monomers can be used to generate CP films with larger inter-chain distances by reaction relaxation that can allow the faradaic storage and delivery of large organic and inorganic ions, thus obtaining actual faradaic storage/delivery control of large anions and cations.

On the other hand, once stored, the liberation of ions is also limited by the structural closing of the structure, trapping a high percentage (up to 40% in thick films) of the ions involved in the redox process. Trapped ions can only be delivered very slowly, with high energetic consumption. The positive aspect, from a technological point of view, is that these are permanently stored ions in a packed conformational structure. Concentration gradients with solutions cannot provide the conformational relaxation energy required for the ionic diffusion and storage loss. Whatever the history of the surrounding electrolyte, the stored ions will be there when required. In contraposition, the balancing ions present in the swollen material could be easily exchanged under diffusion control by different ions present in solution, unless the CP includes substituents, such as crown ethers that only accept a specific counterion.

How the trapping effect can be controlled using different plasticizers or blends, or which variables influence the trapping degree, are unsolved points and their solution could help with similar aspects taking place in living cells and for the control of health and illnesses.

Physical limitations, like the maximum inter-chain average distance that the polymer relaxation can provide combined with the counterion dimensions, must also be taken into account. Too large counterions can overcome the average inter-chain relaxation distance, as large counterions present in the electrolyte cannot be stored and the storage reaction [forwards reaction in eqn (4.1) or the backwards reactions in eqn (4.2) and (4.3)] is hindered even at very high overpotentials.

On the other hand, problems also mean opportunities. Life can be considered as the result of a complex interplay between inter- and intra-molecular forces and conformational movements shifting during biochemical reactions and involving reactants and water. In order to clarify this moving interactive scenario and to attain a theoretical description, drug delivery processes, drug delivery limitations, smart membranes, artificial muscles and electrochemical responses from CPs and other electrochemically active organic materials offer an unparalleled platform for the empirical study of the relationships between intermolecular forces and conformational movements driven by reactions. The final development of self-consistent physical–chemical models for reactions driving conformational movements, the generation/destruction of free volume, and the exchange of ions and solvent could allow extension to similar biochemical reactions, biological functions and biochemical malfunctions. The second alternative is expecting that the development of quantum mechanics and super-powerful computers could allow, someday, *ab initio* or semi-empirical treatment and description of these complex systems.

8.5 Chemical Decontamination and Ionic Concentration

In addition to the use of the above-described intelligent membranes for decontamination, the ionic storage reaction suggests the possible elimination of ions from solutions or soils. This unexplored aspect of CPs can be used for the elimination of contaminating or dangerous ions from wastewaters or contaminated soils, with subsequent storage and concentration in the CP.

For analytical purposes, any of the ionic (anion or cation) storage reactions may be used for the pre-concentration of very diluted ions for subsequent analysis.[144–147] Anion or cation storage requires the use of families of CPs exchanging anions or cations, respectively, during the electrochemical reactions. In addition, some CPs can also pre-concentrate organic ions of analytical interest by absorption.[148,149]

8.6 Artificial Chemical Synapse (Man–Computer Interface) and a New Hypothesis for Brain Functions

Eqn (4.1)–(4.4) indicate the transductions of electronic pulses from electronic, conducting circuits to specific ionic pulses in aqueous media: the intermediate CP being the electrochemical (electron–ion) transducer. Figure 8.8a shows a scheme of the two possibilities for the CP families and their reactions to act as electron–anion or electron–cation transducers.

Electronic pulses drive electrical information in electronic devices, such as computers. Ionic pulses drive electrical and chemical information between neurons (bioelectric signals) in nervous pulses. A new way, practically unexplored, is open to develop a direct communication between electronic devices and neurons or the brain using biocompatible, electroactive electrodes that act as electron/ion translators for the transduction of electrical pulses ordered by computers into ionic (electrical and chemical) pulses flowing through an artificial chemical synapse between the CP and neurons (Figure 8.8b).

To date, great effort has been focused on the study of material biocompatibility[150–157] with neuronal cells and tissues and the development of softer electrodes and microelectrodes.[158–163] Comparatively little effort has been devoted to the development of artificial chemical synapses for direct communication between neurons and electronic devices.

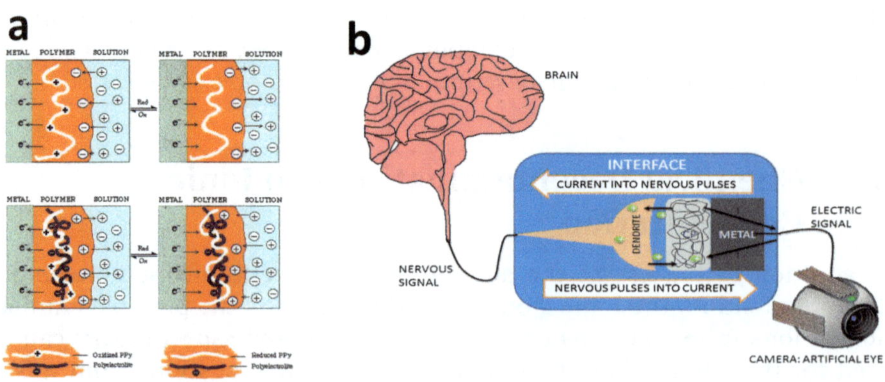

Figure 8.8 (a) Electron–anion [eqn (4.1)] and electron–cation [eqn (4.2)] transducers. Reprinted from ref. 6: *Structural Biological Materials* with the permission of Elsevier. (b) Scheme of the proposed artificial chemical synapse for the connection between electronic equipment and neurons. Reprinted from ref. 8: *Electrochim. Acta*, 2012, **84**, 112–128, with the permission of Elsevier.

Brain function and nervous communication between the brain and organs constitute one of the most stimulating scientific paradoxes: the brain being one of the most explored organs during the last century following Cajal's descriptions, yet brain function still supports high resistance to theoretical description. Current efforts include the two running macro-projects in Europe and the USA, which are devoted to the brain and electronic simulation of brain functions. Here, a new theoretical and experimental (based on reactions from CPs) approach will be presented.

Communication between neighbor neurons includes electrical and chemical exchanges. The full electrical exchange is carried out by ions through ion channels, resulting in electrical currents. These currents give two kinds of external (outside the neuron) signals with different and interconnected natures: electric fields (physical signals) and electrical–chemical currents (bioelectrical or bio-electrochemical signals). Two different methodologies are required to read, collect, understand and use each of the signals generated by the information exchanged between neighbor neurons.

Any current flow, regardless of the origin of the electronic, ionic or polarization current, generates an electric field that expands in a certain space around the carrier-supporting wire, electrolyte or neuron. Ionic flow through neurons and through ion channels results in current peaks with simultaneous peaks of the emitted electric field that can be captured by external *antennae electrodes* (electrocardiograms, encephalograms, electromyograms and so on). Good characterization of these signals from diseased or healthy organs has allowed the development of very useful diagnosis methodologies. In parallel, the biological functions of these organs responding to electric field pulses, such as the heart, have been controlled and regulated by the use of pacemakers connected to very high impedance electrodes (avoiding current flow resulting in water electrolysis).

The second basic component of inter-neuronal electrical communication is flow of the original ionic pulses through the neuron ionic channels. Each electrical pulse (Figure 8.9b and c) generated through an ion channel (*i.e.*, Cl^-, Na^+, K^+ or Ca^{2+}) carries electrical (number of charges), chemical (ion nature) and structural (conformational opening/closing of the channel) information. Reading, collecting, understanding (decoding), reproducing and using these pulses, each including three different packages of information, to solve health problems is a practically unexplored field.

Related to the electrical component of the information transfer between consecutive neurons, each specific ion channel driving the same pulse charge (Q, in C) per channel includes carriers (ions), each driving a different charge (−1, +1 or +2), giving different anionic (−) or cationic (+) electrical currents by the flow of a different number of carriers, n for positively or negatively charged ions and $n/2$ for ions bearing two positive charges each.

Related to the chemical information, the same positive charge resulting from the flow of carriers driving the same charge per carrier unit transfers different chemical information if the carriers have a different chemical nature (Na^+ or K^+).

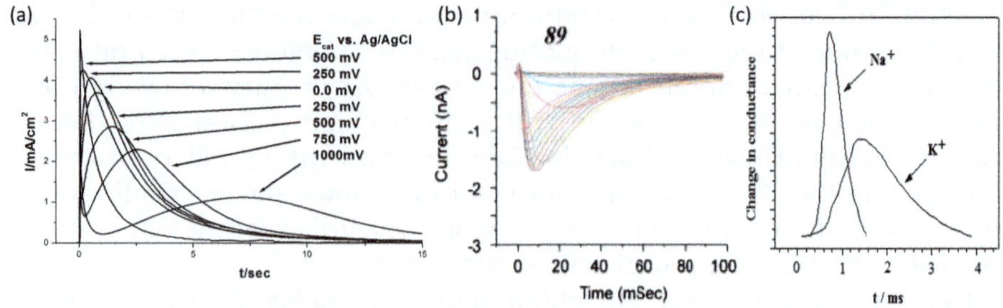

Figure 8.9 (a) Chronoamperometric responses of a polythiophene-coated Pt electrode in a 0.1 M $LiClO_4$ acetonitrile solution subjected to potential steps from different cathodic potentials indicated on the figure, maintained for 30 s, to 950 mV *vs.* Ag/AgCl in 3 M Cl^-. Reprinted from ref. 166: *Electrochim. Acta*, 2008, **53**, 3166–3174, with the permission of Elsevier. (b) Ion channel pulses: voltage clamp chronoamperometric responses across a calcium channel activated by transversal potential steps: initial potential, −100 mV; final potential, ranging between −50 and −10 mV. (c) For the same potential step, conductance variation across potassium or sodium channels. Reprinted from ref. 167: *Proteins: Struct., Funct., Bioinf.*, 2006, **64**, 320–342, with the permission of John Wiley and Sons.

Finally, different ionic channels driving the same ion, thus constituted by the same channel protein, with a different initial conformational state (different intra-chain forces and different conformational energy) will give two ionic pulses with a different peak shape: for the same electrical and chemical information, each current channel peak includes a different piece of structural information. Every ion channel protein constitutes a different multivalent (electrical, chemical, structural) memory read by the ionic flow during every opening/closing process. The energetic conditions of the current flow can modify the conformational energy while closing, storing in this way new structural information.

Apparently, these expensive attempts to simulate brain functions through electronic components seem a quite difficult and out of focus task.

It is suspected that, as a new hypothesis, brain functions, such as memory, thinking, proprioception or consciousness, could be based on this multivalent information unit constituted by ion channels as chemo-ionic-conformational memories and communicated to the surrounding neurons by each of the generated ionic pulses. This is the equivalent of saying that each brain ionic channel works as a multivalent memory unit able to store several values of ionic (electrical and chemical) and energetic conformational states. In addition, every neuron includes different families of basic memory units, each of which can only be read by the flow of its specific ion carrier (Cl^-, Na^+, K^+, Ca^{2+}). Electronic computers work as one-way highways, where only a specific motorcycle model (electrons) can circulate. The brain is a network of closed, interconnected neurons with one-way (dendrite to axon) highways, where a fleet of different vehicles, from bicycles to giant trucks, can transport merchandise. It appears clear that the amount and nature of the transported merchandise by the same number of carriers (charge) are many orders of magnitude higher in the second case. The amount of information that can be translated, treated and stored by the flow of the same number of electronic (computers) or ion channel memory units (the brain) through consumption of the same charge (electrons *vs.* ions) is many orders of magnitude higher in the brain.

In this scenario, the first step required to facilitate communication between any electronic equipment, such as computers, and neurons (ionic systems) is the development of electronic/ionic biocompatible transducers. Electrodes produce at the electrode/electrolyte interfaces the quantitative translation of electronic currents to ionic currents through an electrochemical reaction. At the anode, electrons are extracted from molecules, atoms or ions (present in the electrolyte or taking part of the electrode), generating cations or destroying anions, thus giving new chemical compounds. At the cathode, electrons arrive from the generator through the connecting wires and are injected into molecules, atoms or cations (present in the electrolyte or taking part of the electrode), giving new anions or compensating the cation charges. The electrode/electrolyte system must be charge-balanced at any point in time. The current flow generates (from molecules) or destroys (from ions) charges. In order to keep the charge balance, the simultaneous flow of ions occurs through the electrolyte. The charge flowing by the metal wires and electrodes, and carried by electrons moving in one direction, is translated into the same charge (Kirchhoff's circuit laws) flowing through the electrolyte, carried now by each of the ionic species (anions and cations moving in opposite directions) present in the solution. The number of charges (electrons) flowing through a transversal section of a metal wire in the external circuit equals the number of charges crossing any transversal section of the electrolyte. These charges are determined by the number of anions flowing towards the anode, each multiplied by its valence, plus the number of cations moving in the opposite direction towards the cathode, each multiplied by its valence.

When the electrodic reaction occurs at the synaptic volume close to a dendrite, the expected electron–reaction–ion–neuron channel communication could take place if the chemical conditions were kept under biocompatible conditions. The up-until-now unsolved problem is that, in brain physiological aqueous solutions, the prevalent electrodic reaction from metal electrodes is water electrolysis, producing oxygen and protons (at the anode) or hydrogen and ^{-}OH ions (at the cathode). Such strong chemical variations are not compatible with a neuron's life.

In order to avoid water electrolysis, the alternative way is to try to use the fast charge/discharge of the electrode/electrolyte electrical double layer (EDL) by changing in a very fast way the potential inside the potential window of the aqueous physiological solution through impedance methodologies. The problem is that the EDL charge/discharge is not ionically selective: each of the ionic species present in the electrolyte participates to generate the charge-compensating current during the potential oscillation.

Nowadays, new electrodic materials, such as CPs, are available that can store cations [eqn (4.2) or (4.3)] or anions [eqn (4.1)]. The storage/delivery process occurs (Figure 4.8a and b) inside the potential window of the water discharge, far away from the bio-incompatibility promoted by water electrolysis. Being a faradaic process, these reactions and eqn (4.9) indicate how the charge controls the number of stored or liberated ions: there is only one ion species per polymeric electrode. In addition, most of the CPs doped with these ions that are involved in neuronal pulses present good biocompatibility with neurons. Membranes of intrinsically CPs, and other redox organic films and carbonaceous structures, are presently the champion candidates to obtain good electron/ion transducers for communication between electronic devices and neurons. On this foundation, the long road to the construction of computer/neuron interfaces that allow direct communication between computers and neurons can begin. Then, a new alphabet must be stated to produce words, phrases and book packages that can be interchanged. As an intermediate step, some physical disabilities, such as blindness, could be solved by the permanent connection of video cameras to the brain, with the artificial chemical synaptic interface constituted by an array of specific ionic electrodes that could translate the electronic signals from the video camera into the right nervous pulses understood by the brain.

8.7 Chemo-Ionic-Conformational Memories as Possible Brain Memory Models

The genetic code from each chromosome constitutes a memory unit with the highest information density compared to current natural or artificial memory devices. The stored chemical and conformational information is transferred to proteins or enzymes by biochemical reactions resulting in each individual and personal life. It also transfers to the future generations most of the physical, biochemical, functional and psychological characteristics. Genetic memory and genetic information are carried through chemical information (the amino acid sequence) linked to the structural energetic information (double helix folding) of the macromolecular structure in polymeric nucleic acids. In the previous section, the possibility that brain memory, or a fraction of it, could be stored by the conformational energetic structure of ion channel proteins, which can be read by structural ionic pulses, was mentioned. Eqn (4.1)–(4.4) and Figures 6.2b, 6.8, 6.9 and 7.4f–h reveal that structural and ionic (electrical and chemical) information can be stored by polymeric conformational states.[164,165] In biomimetic reactive gels of CPs, from individual chains to macroscopic volumetric structures, each packed conformational state constitutes a chemo-ionic-conformational energetic memory. Thus, memory units can be constructed from macroscopic to nanoscopic or molecular polymeric electrodes. The molecular memory unit will only be developed when chemists provide the synthesis and manipulation of individual polymeric chains. Each of these memory units could store tens of different values of chemo-ionic-conformational energetic states (outperforming any digital or Qubit memory).

The stored conformational energetic states (mechanical component of memory) resemble the pre-Columbian quipu libraries from the Incas and in biology they approach chromosomes and genetic code storage. Fully reproducible values of conformationally packed energetic states are available, as described in Chapter 6 and quantified in Section 7.2.3 (Figure 7.4f and h).

After conformational compaction, the dried polymer–counterions state constitutes a permanent chemical memory. The value of the stored permanent memory can be read by analytical methods, determining the concentration of trapped counterions or the concentration and energetic state of the conformationally packed polaronic states. Each memory value is linked to both a specific concentration of the trapped ion and a packing conformational energy.

Otherwise, the memory value can be read and erased during its oxidation–relaxation–swelling in a liquid electrolyte with exchange of counterions and water. The structural electrochemical response (Figure 8.9a)[166] includes electrical information (charge flow), chemical information (flow of specific ions) and structural information (conformational energetic changes). The consecutive events and the electro-chemo-ionic-conformational response mimic those responses taking place at the transmembrane ion channels from neurons (Figure 8.9b)[167] influenced by the nature of the ion (Figures 8.9c and 6.15a).

Any chemical or electrochemical way to produce controlled cross-linking degrees of the conformational polymeric structure induces permanent changes in the electrochemical responses, giving permanent memories that can be read but not erased by consecutive oxidation/reduction (reading/writing) processes.

The construction of new computers based on these electro-chemo-conformational memories from dense gels working in the presence of liquid electrolytes, thus providing possibilities to develop three-dimensional structures (mimicking neurons in brain), can be envisaged. The same methodology can improve the construction of artificial chemical synapses for deciphering interneuron communication in the brain, brain memory and the thinking mechanism. As a direct consequence, the hypothesis that our memory, or a fraction of it, can be linked to conformational energetic states in neuron channel proteins could be envisaged for direct exploration.

8.8 Smart Surfaces (Wettability and Self-Cleaning Control)

During the polymeric oxidations/reductions indicated by eqn (4.1)–(4.4), each of the intermolecular (coulombic and van der Waals) interactions between the species taking part in the reaction (neutral polymer chains, oxidized or reduced polymer chains, anions, cations and solvent) change. With the progressive change of the material nature, the molecular interactions with molecules in solvent drops or particles at the polymer surface also change.

The more reduced the polymer film, the closer its nature is to an organic material: low interactions with dipolar molecules, such as water, forming a hydrophobic surface, and strong interactions with organic solvents, forming an oleophilic surface.

The more oxidized the polymer film is the more charges are present at the surface and below the surface: *i.e.*, stronger interactions with water, forming increasingly hydrophilic surfaces, and lower interactions with organic solvents, forming increasingly oleophobic surfaces.

In summary, the surface wettability can be tuned by the oxidation state of the material.[168–175]

8.9 Electrochromic Devices: Smart Windows, Glasses and Mirrors

Eqn (4.1)–(4.3) transform the electronic structures of the highest occupied molecular orbital (HOMO) and lowest unoccupied molecular orbital (LUMO) of the neutral polymer chain, generating new HOMO and LUMO levels, *i.e.*, charged polaronic and polaronic pair (or bipolaronic) π-conjugated structures in the oxidized (p-doped) or reduced (n-doped) polymeric films (Section 5.3, Figure 5.1). Each of these HOMO/LUMO transitions corresponds to a different energy in the ultraviolet (UV), visible (vis) and IR regions: they are chromophores (Figure 8.10c for different CPs). Being faradaic processes, the color change can be stopped at any time or controlled to maintain a constant luminosity inside a car, room or closed space (Figure 8.10a).

The reactions allow perfect control of the concentration of each chromophore inside the film [eqn (5.6) and (5.8)]. Each chromophore can absorb or reflect photons from a region of the UV-vis spectrum. The light absorbance is proportional to the chromophore concentration in the CP film: eqn (5.3), (5.7) and (5.9) indicate that the absorbance can be tuned under control to obtain smart electro-chemo-chromic devices, like smart windows, smart mirrors, smart optical filters or electrochromic screens in both the visible and IR regions. Applications require maximum transparency of one of the species (neutral chain, polaron or polaron pair). The other two species must get maximum absorbance of the fully oxidized or fully reduced film in order to get the maximum contrast and the maximum variation range for the film absorbance and transmittance (Figure 5.1a). Moreover, a long lifetime (maximum number of bleaching/color cycles) can be attained by selecting the optimum working conditions to avoid polymeric degradation reactions or by adding suitable stabilizers (Section 3.15).[176–181]

As with any of the previous electrochemical devices, smart windows, mirrors, goggles, filters or screens are constituted by laminar and transparent electrochemical cells (Figure 2.8 and Figure 8.10b).[182] A transparent plate

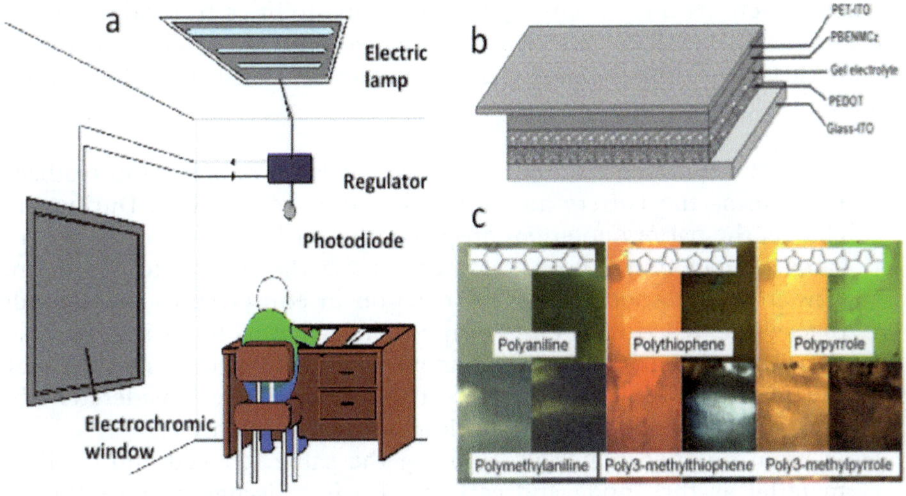

Figure 8.10 (a) Constant luminosity control inside a room, car or closed space. Reprinted from ref. 6: *Structural Biological Materials*, with the permission of Elsevier. (b) Multilayer structure of an electrochromic window. Reprinted from ref. 182: *Org. Electron.*, 2007, **8**, 367–381, with the permission of Elsevier. (c) Color of the oxidized and reduced films from different CPs. Reprinted from ref. 9, with the permission of the Royal Society of Chemistry.

(glass, plastic, graphene) coated with a transparent electronic conductor (indium tin oxide, ITO; graphene) film supports a thin and uniform CP film, produced by electropolymerization or by casting from oligomeric solutions, followed by a transparent ionic conducting and electronic insulator thin film membrane: the electrolyte. Then, the second electrode, which may or may not be electrochromic, completes the electrochemical cell. If electrochromic, it must be color complementary related to the first electrode: when the first electrode is transparent in the oxidized state, the second electrode must be transparent in its reduced state. Thus, the flow of a cathodic current through the first electrode promotes its reduction–darkening, while the second electrode is the anode and its oxidation gives its simultaneous darkening. By reversing the sense of the current flow, both electrode reactions reverse with its simultaneous bleaching.

8.10 All-Organic Batteries and Supercapacitors

Eqn (4.1) and (4.2) indicate the storage of positive charges (p-doping) in the polymer chain. Eqn (4.3) denotes the storage of negative charges (n-doping) in chains. Eqn (4.5)–(4.8) indicate that the oxidation (p-doping) of a polymer chain occurs through n consecutive equilibrium states in a large potential range, as depicted by Figure 4.3. Similar consecutive equilibriums could be stated for the reduction (n-doping) of chains.

These reactions open a large panoply of possibilities to produce different families of batteries, whereby the reduced material always constitutes the anode (− pole) and the oxidized material always constitutes the cathode (+ pole):

- *A*: From a p-doped material by using the same reduced and oxidized material as the battery anode and cathode, respectively. During discharge, the battery potential drops continuously to zero, achieving the same composition of both electrodes. Inside the electrolyte, the anions flow from the anode [forwards reaction in eqn (4.1)] to the cathode [backwards reaction in eqn (4.1)]. When materials following the reaction shown in eqn (4.2) take part in batteries, cations must flow through the electrolyte between electrodes. These batteries are considered in the literature as supercapacitors working at anodic potentials.
- *B*: From n-doped materials by using the same reduced and oxidized material as the anode and cathode. During discharge, the potential drops continuously to zero and cations are exchanged between the electrodes through the electrolyte.
- *C*: The anode (− pole) is an n-doped material and the cathode (+ pole) is the same p-doped material. The potential of the charged battery can attain several volts (Figure 4.3) and it drops slowly during discharge up to the potential of the electrochemical gap.
- *D*: A, B and C using different polymers as the anode and cathode in order to get higher battery potentials. In order to obtain rocking chair batteries both polymers must be selected from materials exchanging the same ion (anion or cation) during charge/discharge. Otherwise, salts will concentrate in the electrolyte, generating new problems.
- *E*: Batteries can be constructed using only one polymeric electrode, the second being a different material, such as a metal (*e.g.*, Li), metal oxide or metal intercalation compound.

The main advantage, differentiating polymeric batteries from any other previous battery, is the three-dimensional nature of the electrode at the chain level: the full electrode material is available at any time to participate in the charge/discharge process. The charge process becomes very fast. The three-dimensional nature of the charge storage constitutes, simultaneously, the main disadvantage in obtaining fast charge/discharge batteries because this is considered by engineers to be the structure of a capacitor and most efforts have been focused on the design and development of supercapacitors[183–188] instead of the new batteries.

By comparison with current batteries, only the electrode material located at the two-dimensional double layer interface is available for charge/discharge processes. In order to increase the battery power and the charge/discharge reaction rates, increasingly porous or powdered materials are used. In this way, rising EDL areas are obtained. By contraposition, in three-dimensional CP electrodes the full material is available at any charge/discharge time.

Unsolved points: The main problem comes from industrial pressure to produce fully solid state batteries. The absence of solvent pushes the system to work under permanent conformationally compacted states, and Figure 7.4f and g illustrate how rising conformational states increase the reaction activation energy and decrease (exponentially) the reaction coefficient and the concomitant reaction rate. Solid state polymeric batteries almost eliminate the advantages of these materials related to the traditional metal, metal compounds, intercalation compounds and so on.

Even when working in the presence of some solvent, the different polymeric materials close and pack their conformational structure, trapping a significant fraction of the counterions. From that point on, the specific current provided by the electrode drops, as does the battery power. In the future, the influence of different solvents, plasticizers and additives must be studied to avoid the closing potential and to increase the specific charge/discharge of the same material.

8.11 Sensors, Biosensors and Sensing Devices

CPs interact in a reversible way with gases and other compounds, changing the electronic conductivity and allowing the development of sensors.[189–194] In this way they behave akin to most classical metal oxide sensors.

Otherwise, the oxidation/reduction of a polymeric chain occurs through *n* consecutive equilibrium steps, as indicated by eqn (4.5)–(4.8). Each of the equilibrium states is described by the Nernst equation [eqn (5.12)], giving an equilibrium potential of E_{i0}, where i denotes the equilibrium step: different oxidation states of the same material can give different sensors.

8.11.1 Le Châtelier's Principle, Chemical Equilibrium and Sensors

Each of the equilibrium steps must follow Le Châtelier's principle, which states that a system at equilibrium, when subjected to a perturbation, responds in a way that tends to minimize its effect.[7,8] Most potentiometric, chemical, biochemical, thermal or mechanical sensors are based on this principle, applied through the Nernst equation: any chemical, biochemical, thermal or mechanical change in the presence of a CP film electrode under any of its chemical equilibrium conditions [eqn (4.5)–(4.8)] shifts the material equilibrium potential. The plethora of available material families from any basic CP and, moreover, the *n* different equilibrium states from every material, are providing and will provide in the future many possibilities to develop, in a very easy way, sensor arrays for any application, such as artificial noses or tongues.[195–202]

The easy interaction between enzymes and CPs allows their incorporation into the growing polymer film during electropolymerization or adsorption once electrogenerated to produce specific biochemical sensors.[203–211]

8.11.2 Biochemical Sensors: The New Conformational (Allosteric) Ways

Nature has solved the problem of signal transduction against complex backgrounds by coupling molecular recognition to a second, equally important feature: binding-induced switching. That is to say that the biomolecules that perform chemoperception *in vivo* invariably couple molecular recognition to a binding-induced change in the conformation or oligomerization state. These switching events, in turn, trigger specific output signals, such as the opening of an ion channel or the activation of an enzyme.[212–215]

Inspired by the high speed, specificity and versatility of these biological sensors, recent years have witnessed significant efforts to create artificial biosensors based on these principles. These efforts will represent a promising route towards versatile, real-time molecular monitoring *in vivo*.

8.11.3 Mechanical Sensors

Among the different possibilities (thermal, chemical, mechanical *etc.*) above-indicated to develop classical sensors from the different oxidation states of CPs, this book is more concerned with mechanical sensors and, in particular, with cantilever mechanical sensors. Bilayer or tri-layer structures (cantilevers), such as those used for artificial muscles, present, whatever the oxidation state of the CP, one of the equilibrium states from eqn (4.5) to (4.8). According to Le Châtelier's principle (Section 8.11.1), the material will respond to any mechanical perturbation (stress, strain, pressure) to minimize the perturbation effect. Oxidized and reduced states of any CP occupy a different volume: when the external pressure increases, the equilibrium shifts (and the potential shifts) to decrease the material volume. For decreasing pressures both the material composition and the material potential shift in opposite directions. The material potential, or the cantilever (bilayer or tri-layer) potential, senses under equilibrium conditions any mechanical perturbation.[216–221]

8.12 Challenges

Each of the composition-dependent properties presented in Chapter 5, or those that will emerge in the future, results in an electro-chemo-biomimetic device driven by the electrochemical reactions. Thus, a new technological world may be announced as each of these laboratory devices will result in a plethora of industrial products and new start-up companies, involving increasing percentages of the economy from advanced countries. Advanced here means open to the new technological border based on reactive products. These new products and companies are waiting for motivated graduates and doctors with solid theoretical and experimental knowledge, and it is hope that they are ready to take economical and industrial risks to advance the current industrial and labor limitations.

References

1. T. F. Otero, in *Modern Aspects of Electrochemistry*, ed. R. E. White, J. O. Bockris and B. E. Conway, Springer, US, New York, 1999, pp. 307–434.
2. T. F. Otero and J. Rodriguez, in *Intrinsically Conducting Polymers: An Emerging Technology*, ed. M. Aldissi, Kluwer Academic Publishers, Dordrecht, 1993, vol. 246, pp. 179–190.
3. T. F. Otero and J. Arias-Pardilla, in *Electropolymerization: Concepts, Materials and Applications*, ed. S. Cosnier and A. Karyakin, Wiley-VCH Verlag GmbH & Co. KGaA, Weinheim, 2010, pp. 241–272.
4. T. F. Otero, in *From Non-Covalent Assemblies to Molecular Machines*, ed. J.-P. Sauvage and P. Gaspard, Wiley-VCH, Weinheim, 2011, pp. 443–452.
5. J. G. Martinez, J. Arias-Pardilla and T. F. Otero, in *Smart Actuation and Sensing Systems – Recent Advances and Future Challenges*, ed. G. Berselli, InTech, Rijeka, 2012.
6. *Structural Biological Materials: Design and Structure-Property Relationships*, ed. M. Elices, Pergamon, Oxford, 2000.
7. T. F. Otero, *Polym. Rev.*, 2013, **53**, 311–351.
8. T. F. Otero, J. G. Martinez and J. Arias-Pardilla, *Electrochim. Acta*, 2012, **84**, 112–128.
9. T. F. Otero and J. G. Martinez, *J. Mater. Chem. B*, 2013, **1**, 26–38.
10. T. F. Otero, S. Villanueva, M. T. Cortes, S. A. Cheng, A. Vazquez, I. Boyano, D. Alonso and R. Camargo, *Synth. Met.*, 2001, **119**, 419–420.
11. E. Bundgaard and F. C. Krebs, *Sol. Energy Mater. Sol. Cells*, 2007, **91**, 954–985.
12. M. Angelopoulos, *IBM J. Res. Dev.*, 2001, **45**, 57–75.
13. M. Gratzel, *Nature*, 2001, **414**, 338–344.
14. G. Gustafsson, Y. Cao, G. Treacy, F. Klavetter, N. Colaneri and A. Heeger, *Nature*, 1992, **357**, 477–479.
15. C. J. Brabec, N. S. Sariciftci and J. C. Hummelen, *Adv. Funct. Mater.*, 2001, **11**, 15–26.
16. F. Garnier, R. Hajlaoui, A. Yassar and P. Srivastava, *Science*, 1994, **265**, 1684–1686.
17. S. R. Forrest, *Nature*, 2004, **428**, 911–918.
18. F. Hide, M. A. DiazGarcia, B. J. Schwartz, M. R. Andersson, Q. B. Pei and A. J. Heeger, *Science*, 1996, **273**, 1833–1836.
19. A. Huxley and R. Simmons, *Nature*, 1971, **233**, 533–538.
20. J. Spudich and S. Watt, *J. Biol. Chem.*, 1971, **246**, 4866–4871.
21. R. H. Baughman, *Synth. Met.*, 1996, **78**, 339–353.
22. M. Shahinpoor, Y. Bar-Cohen, J. O. Simpson and J. Smith, *Smart Mater. Struct.*, 1998, **7**, R15–R30.
23. M. Shahinpoor and K. J. Kim, *Smart Mater. Struct.*, 2004, **13**, 1362–1388.
24. P. Brochu and Q. Pei, *Macromol. Rapid Commun.*, 2010, **31**, 10–36.
25. K. Kaneto, M. Kaneko, Y. Min and A. Macdiarmid, *Synth. Met.*, 1995, **71**, 2211–2212.

26. Y. Liu, A. H. Flood, P. A. Bonvallett, S. A. Vignon, B. H. Northrop, H. R. Tseng, J. O. Jeppesen, T. J. Huang, B. Brough, M. Baller, S. Magonov, S. D. Solares, W. A. Goddard, C. M. Ho and J. F. Stoddart, *J. Am. Chem. Soc.*, 2005, **127**, 9745–9759.
27. K. Uchino, *Am. Ceram. Soc. Bull.*, 1986, **65**, 647–652.
28. Q. M. Zhang, H. F. Li, M. Poh, F. Xia, Z. Y. Cheng, H. S. Xu and C. Huang, *Nature*, 2002, **419**, 284–287.
29. *Electroactive Polymer (EAP) Actuators as Artificial Muscles: Reality, Potential, and Challenges*, ed. Y. Bar-Cohen, Spie Press Book, Bellingham, 2004.
30. Y. Osada, H. Okuzaki and H. Hori, *Nature*, 1992, **355**, 242–244.
31. Y. Osada and M. Hasebe, *Chem. Lett.*, 1985, 1285–1288.
32. T. F. Otero, in *Polymer Sensors and Actuators*, ed. P. Y. Osada and P. D. E. D. Rossi, Springer Berlin Heidelberg, Berlin, 2000, pp. 295–323.
33. K. Kaneto and M. Kaneko, *Appl. Biochem. Biotechnol.*, 2001, **96**, 13–23.
34. R. Kiefer, P. A. Kilmartin, G. A. Bowmaker and J. Travas-Sejdic, in *Electroactive Polymer Actuators and Devices (EAPAD) 2008*, ed. Y. Bar-Cohen, Spie-Int Soc Optical Engineering, Bellingham, 2008, vol. 6927.
35. T. Sendai, H. Suematsu and K. Kaneto, *Jpn. J. Appl. Phys.*, 2009, **48**, 051506.
36. R. H. Baughman, *Synth. Met.*, 1996, **78**, 339–353.
37. E. Smela, *Adv. Mater.*, 2003, **15**, 481–494.
38. D. Derossi and P. Chiarelli, in *Macro-Ion Characterization: From Dilute Solutions to Complex Fluids*, ed. K. S. Schmitz, Amer Chemical Soc, Washington, 1994, vol. 548, pp. 517–530.
39. E. W. H. Jager, E. Smela and O. Inganas, *Science*, 2000, **290**, 1540–1545.
40. G. W. Wang, Y. N. Lu, L. P. Wang, H. J. Wang and J. Y. Wang, *J. Nanosci. Nanotechnol.*, 2014, **14**, 596–612.
41. T. Mirfakhrai, J. D. W. Madden and R. H. Baughman, *Mater. Today*, 2007, **10**, 30–38.
42. L. Valero Conzuelo, J. Arias-Pardilla, J. V. Cauich-Rodriguez, M. Afra Smit and T. Fernandez Otero, *Sensors*, 2010, **10**, 2638–2674.
43. I. Y. Galaev and B. Mattiasson, *Trends Biotechnol.*, 1999, **17**, 335–340.
44. T. F. Otero, *J. Mater. Chem.*, 2009, **19**, 681–689.
45. G. Alici, P. Metz and G. M. Spinks, *Smart Mater. Struct.*, 2006, **15**, 243–252.
46. P. Brochu and Q. Pei, *Macromol. Rapid Commun.*, 2010, **31**, 10–36.
47. A. Huxley and R. Simmons, *Nature*, 1971, **233**, 533–538.
48. P. Metz, G. Alici and G. M. Spinks, *Sens. Actuators, A*, 2006, **130–131**, 1–11.
49. Q. Pei and O. Inganas, *Adv. Mater.*, 1992, **4**, 277–278.
50. Q. Pei and O. Inganas, *J. Phys. Chem.*, 1992, **96**, 10507–10514.
51. T. Otero, E. Angulo, J. Rodriguez and C. Santamaria, *J. Electroanal. Chem.*, 1992, **341**, 369–375.
52. A. DellaSanta, D. DeRossi and A. Mazzoldi, *Synth. Met.*, 1997, **90**, 93–100.
53. A. DellaSanta, D. DeRossi and A. Mazzoldi, *Synth. Met.*, 1997, **90**, 93–100.

54. T. Otero and J. Rodriguez, Electrochemomechanical and Electrochemopositioning Devices – Artificial Muscles, in *Intrinsically Conducting Polymers: An Emerging Technology*, ed. M. Aldissi, Klluwer Academic Press. Dordrecht, 1993, vol. 246.
55. L. Bay, K. West, P. Sommer-Larsen, S. Skaarup and M. Benslimane, *Adv. Mater.*, 2003, **15**, 310–313.
56. F. Vidal, C. Plesse, G. Palaprat, A. Kheddar, J. Citerin, D. Teyssie and C. Chevrot, *Synth. Met.*, 2006, **156**, 1299–1304.
57. T. F. Otero, M. T. Cortes and G. V. Arenas, *Electrochim. Acta*, 2007, **53**, 1252–1258.
58. Q. Pei and O. Inganas, *Adv. Mater.*, 1992, **4**, 277–278.
59. Q. Pei and O. Inganas, *J. Phys. Chem.*, 1992, **96**, 10507–10514.
60. H. Okuzaki, T. Kuwabara, K. Funasaka and T. Saido, *Adv. Funct. Mater.*, 2013, **23**, 4400–4407.
61. T. F. Otero, J. G. Martinez and J. Arias-Pardilla, *Electrochim. Acta*, 2012, **84**, 112–128.
62. R. H. Baughman, *Synth. Met.*, 1996, **78**, 339–353.
63. E. Smela, *Adv. Mater.*, 2003, **15**, 481–494.
64. D. Derossi and P. Chiarelli, in *Macro-Ion Characterization: From Dilute Solutions to Complex Fluids*, ed. K. S. Schmitz, American Chemical Society, Washington, 1994, vol. 548, pp. 517–530.
65. E. W. H. Jager, E. Smela and O. Inganas, *Science*, 2000, **290**, 1540–1545.
66. G. W. Wang, Y. N. Lu, L. P. Wang, H. J. Wang and J. Y. Wang, *J. Nanosci. Nanotechnol.*, 2014, **14**, 596–612.
67. A. DellaSanta, D. DeRossi and A. Mazzoldi, *Synth. Met.*, 1997, **90**, 93–100.
68. E. Smela, W. Lu and B. R. Mattes, *Synth. Met.*, 2005, **151**, 25–42.
69. Y.-Z. Long, M.-M. Li, C. Gu, M. Wan, J.-L. Duvail, Z. Liu and Z. Fan, *Prog. Polym. Sci.*, 2011, **36**, 1415–1442.
70. T. K. Das and S. Prusty, *Polym.-Plast. Technol. Eng.*, 2012, **51**, 1487–1500.
71. I. Y. Galaev and B. Mattiasson, *Trends Biotechnol.*, 1999, **17**, 335–340.
72. T. F. Otero, *J. Mater. Chem.*, 2009, **19**, 681–689.
73. S. Y. Yang, S. K. Mahadeva and J. Kim, *Smart Mater. Struct.*, 2010, **19**, 105026.
74. E. Smela, O. Inganas, Q. Pei and I. Lundstrom, *Adv. Mater.*, 1993, **5**, 630–632.
75. F. Vidal, C. Plesse, D. Teyssie and C. Chevrot, *Synth. Met.*, 2004, **142**, 287–291.
76. T. F. Otero and J. M. Sansinena, *Bioelectrochem. Bioenerg.*, 1997, **42**, 117–122.
77. T. F. Otero and M. T. Cortes, *Chem. Commun.*, 2004, 284–285.
78. T. F. Otero and J. G. Martinez, *Sens. Actuators, B*, 2014, **199**, 27–30.
79. T. F. Otero and J. M. Sansinena, *Bioelectrochem. Bioenerg.*, 1997, **42**, 117–122.
80. T. F. Otero and J. M. Sansinena, *Adv. Mater.*, 1998, **10**, 491–494.
81. T. F. Otero and M. T. Cortes, *Chem. Commun.*, 2004, 284–285.

82. T. F. Otero and M. T. Cortes, in *Smart Structures and Materials 2000: Electroactive Polymer Actuators and Devices (EAPAD)*, ed. Y. BarCohen, Spie-Int Soc Optical Engineering, Bellingham, 2000, vol. 3987, pp. 252–260.
83. F. Garcia-Cordova, L. Valero, Y. A. Ismail and T. Fernandez Otero, *J. Mater. Chem.*, 2011, **21**, 17265–17272.
84. L. Valero, J. Arias-Pardilla, J. Cauich-Rodriguez, M. A. Smit and T. F. Otero, *Electrochim. Acta*, 2011, **56**, 3721–3726.
85. T. F. Otero, J. G. Martinez, L. Valero and J. Arias-Pardilla, *Slow Movements in Conducting Polymer Actuators are Electrochemical Motors*, Proceeding from Actuators. 2010.
86. G. Alici, B. Mui and C. Cook, *Sens. Actuators, A*, 2006, **126**, 396–404.
87. G. Alici, *Sens. Actuators, B*, 2009, **141**, 284–292.
88. A. DellaSanta, D. DeRossi and A. Mazzoldi, *Smart Mater. Struct.*, 1997, **6**, 23–34.
89. G. Di Pasquale, L. Fortuna, S. Graziani, M. La Rosa, D. Nicolosi, G. Sicurella and E. Umana, *IEEE Trans. Instrum. Meas.*, 2009, **58**, 3731–3738.
90. P. Du, X. Lin and X. Zhang, *Sens. Actuators, A*, 2010, **163**, 240–246.
91. Y. Fang, X. Tan, Y. Shen, N. Xi and G. Alici, *Mater. Sci. Eng., C*, 2008, **28**, 421–428.
92. Y. Fang and X. Tan, *Sens. Actuators, A*, 2010, **158**, 121–131.
93. F. Greco, V. Domenici, A. Desii, E. Sinibaldi, B. Zupancic, B. Zalar, B. Mazzolai and V. Mattoli, *Soft Matter*, 2013, **9**, 11405–11416.
94. J. D. W. Madden, P. G. A. Madden and I. W. Hunter, in *Smart Structures and Materials 2001: Electroactive Polymer Actuators and Devices*, ed. Y. BarCohen, Spie-Int Soc Optical Engineering, Bellingham, 2001, vol. 4329, pp. 72–83.
95. J. D. W. Madden, N. A. Vandesteeg, P. A. Anquetil, P. G. A. Madden, A. Takshi, R. Z. Pytel, S. R. Lafontaine, P. A. Wieringa and I. W. Hunter, *IEEE J. Oceanic Eng.*, 2004, **29**, 706–728.
96. A. A. A. Moghadam, K. Torabi, M. Moavenian and R. Davoodi, *J. Intell. Mater. Syst. Struct.*, 2013, **24**, 484–498.
97. C. H. Nguyen, G. Alici and G. Wallace, *IEEE/ASME Trans. Mechatron.*, 2014, **19**, 1279–1288.
98. A. D. Price and H. E. Naguib, *J. Intell. Mater. Syst. Struct.*, 2013, **24**, 548–558.
99. S. Ruiz, B. Mead, V. Palmre, K. J. Kim and W. Yim, *Smart Mater. Struct.*, 2015, **24**, 015007.
100. T. Shoa, J. D. W. Madden, N. R. Munce and V. Yang, *Polym. Int.*, 2010, **59**, 343–351.
101. T. Shoa, D. S. Yoo, K. Walus and J. D. W. Madden, *IEEE/ASME Trans. Mechatron.*, 2011, **16**, 42–49.
102. V. Venugopal, H. Zhang, R. Northcutt and V. B. Sundaresan, *Sens. Actuators, B*, 2014, **201**, 293–299.
103. X. Wang, G. Alici and X. Tan, *Smart Mater. Struct.*, 2014, **23**, 025015.
104. L. Valero, J. G. Martinez and T. F. Otero, *J. Solid State Electrochem.*, 2015, **19**, 2683–2689.

105. M. Fuchiwaki and T. F. Otero, *J. Mater. Chem. B*, 2014, **2**, 1954–1965.
106. R. Kiefer, J. G. Martinez, T. F. Otero, A. Kesküla, F. Kaasik, M. Harjo, R. Valner, V. Vaddepally, A.-L. Peikolainen and A. Aabloo, Solvent and electrolyte effects in PPyDBS free standing films, in *SPIE Smart Structures and Materials+ Nondestructive Evaluation and Health Monitoring*, International Society for Optics and Photonics, 2015, vol. 9430, pp. 94301C–94301C-8.
107. T. F. Otero and J. G. Martinez, *Chem. Mater.*, 2012, **24**, 4093–4099.
108. T. F. Otero, J. G. Martinez and B. Zaifoglu, *Smart Mater. Struct.*, 2013, **22**, 104019.
109. L. Valero, T. F. Otero and J. G. Martinez, *ChemPhysChem*, 2014, **15**, 293–301.
110. L. Bay, T. Jacobsen, S. Skaarup and K. West, *J. Phys. Chem. B*, 2001, **105**, 8492–8497.
111. S. Bruckenstein and A. Hillman, *J. Phys. Chem.*, 1988, **92**, 4837–4839.
112. S. Bruckenstein and A. Hillman, *J. Phys. Chem.*, 1991, **95**, 10748–10752.
113. R. Meier, M. Schindler, P. Mueller-Buschbaum and B. Watts, *Phys. Rev. B*, 2011, **84**, 174205.
114. D. Bagchi and R. Menon, *Chem. Phys. Lett.*, 2006, **425**, 114–117.
115. S. K. M. Jonsson, J. Birgerson, X. Crispin, G. Greczynski, W. Osikowicz, A. W. D. van der Gon, W. R. Salaneck and M. Fahlman, *Synth. Met.*, 2003, **139**, 1–10.
116. J. B. Spires, H. Peng, D. E. Williams, C. Soeller and J. Travas-Sejdic, *Electrochim. Acta*, 2010, **55**, 3061–3067.
117. C. Gabrielli, M. Keddam, N. Nadi and H. Perrot, *J. Electroanal. Chem.*, 2000, **485**, 101–113.
118. A. R. Hillman and M. A. Mohamoud, *Electrochim. Acta*, 2006, **51**, 6018–6024.
119. M. J. M. Jafeen, M. A. Careem and S. Skaarup, *J. Solid State Electrochem.*, 2012, **16**, 1753–1759.
120. W. Plieth, A. Bund, U. Rammelt, S. Neudeck and L. M. Duc, *Electrochim. Acta*, 2006, **51**, 2366–2372.
121. S. Skaarup, M. J. M. Jafeen and M. A. Careem, *Solid State Ion.*, 2010, **181**, 1245–1250.
122. C. Visy, C. Janaky and E. Krivan, *J. Solid State Electrochem.*, 2005, **9**, 330–336.
123. J. Frommer, *J. Macromol. Sci., Chem.*, 1987, **A24**, 449–454.
124. J. J. L. Cascales, A. J. Fernandez and T. F. Otero, *J. Phys. Chem. B*, 2003, **107**, 9339–9343.
125. J. J. Lopez Cascales and T. F. Otero, *J. Chem. Phys.*, 2004, **120**, 1951–1957.
126. M. Fuchiwaki, J. G. Martinez and T. F. Otero, *Adv. Funct. Mater.*, 2015, **25**, 1535–1541.
127. W. Machu, *Handbook of Electropainting Technology*, Electrochemical Publications Limited, Glasgow, 1978.
128. C. Deslouis, M. Musiani and B. Tribollet, *J. Phys. Chem.*, 1994, **98**, 2936–2940.

129. T. Sata, Y. Ishii, K. Kawamura and K. Matsusaki, *J. Electrochem. Soc.*, 1999, **146**, 585–591.
130. H. Zhao, W. Price and G. Wallace, *J. Membr. Sci.*, 1994, **87**, 47–56.
131. A. C. Partridge, C. Milestone, C. O. Too and G. G. Wallace, *J. Membr. Sci.*, 1997, **132**, 245–253.
132. H. Zhao, W. E. Price, C. O. Too, G. G. Wallace and D. Zhou, *J. Membr. Sci.*, 1996, **119**, 199–212.
133. A. C. Partridge, C. B. Milestone, C. O. Too and G. G. Wallace, *J. Membr. Sci.*, 1999, **152**, 61–70.
134. V. Misoska, J. Ding, J. M. Davey, W. E. Price, S. F. Ralph and G. G. Wallace, *Polymer*, 2001, **42**, 8571–8579.
135. J. Pellegrino, in *Advanced Membrane Technology*, ed. N. N. Li, E. Drioli, W. S. W. Ho and G. G. Lipscomb, New York Acad Sciences, New York, 2003, vol. 984, pp. 289–305.
136. C. Ehrenbeck and K. Juttner, *Electrochim. Acta*, 1996, **41**, 511–518.
137. K. Juttner and C. Ehrenbeck, *J. Solid State Electrochem.*, 1998, **2**, 60–66.
138. C. Ehrenbeck and K. Juttner, *Electrochim. Acta*, 1996, **41**, 1815–1823.
139. M. J. Ariza and T. F. Otero, *Colloids Surf., A*, 2005, **270**, 226–231.
140. M. J. Ariza and T. F. Otero, *J. Membr. Sci.*, 2007, **290**, 241–249.
141. V. Pillay, T.-S. Tsai, Y. E. Choonara, L. C. du Toit, P. Kumar, G. Modi, D. Naidoo, L. K. Tomar, C. Tyagi and V. M. K. Ndesendo, *J. Biomed. Mater. Res. A*, 2014, **102**, 2039–2054.
142. S. Geetha, C. R. K. Rao, M. Vijayan and D. C. Trivedi, *Anal. Chim. Acta*, 2006, **568**, 119–125.
143. D. Svirskis, J. Travas-Sejdic, A. Rodgers and S. Garg, *J. Controlled Release*, 2010, **146**, 6–15.
144. C. Hu, M. He, B. Chen and B. Hu, *J. Chromatogr. A*, 2013, **1275**, 25–31.
145. J. C. Wu, W. M. Mullett and J. Pawliszyn, *Anal. Chem.*, 2002, **74**, 4855–4859.
146. G. Liljegren, N. Forsgard, C. Zettersten, J. Pettersson, M. Svedberg, M. Herranen and L. Nyholm, *Analyst*, 2005, **130**, 1358–1368.
147. M. Kaykhaii, G. W. Dicinoski and P. R. Haddad, *Anal. Lett.*, 2010, **43**, 1546–1555.
148. J. Li, Z. Shao, C. Chen and X. Wang, *RSC Adv.*, 2014, **4**, 38192–38198.
149. H. Bagheri, A. Roostaie and M. Allahdadlalouni, *Microchim. Acta*, 2015, **182**, 617–624.
150. P. Fattahi, G. Yang, G. Kim and M. R. Abidian, *Adv. Mater.*, 2014, **26**, 1846–1885.
151. A. R. Harris, S. J. Morgan, J. Chen, R. M. I. Kapsa, G. G. Wallace and A. G. Paolini, *J. Neural Eng.*, 2013, **10**, 016004.
152. P. M. George, A. W. Lyckman, D. A. LaVan, A. Hegde, Y. Leung, R. Avasare, C. Testa, P. M. Alexander, R. Langer and M. Sur, *Biomaterials*, 2005, **26**, 3511–3519.
153. N. Gomez, J. Y. Lee, J. D. Nickels and C. E. Schmidt, *Adv. Funct. Mater.*, 2007, **17**, 1645–1653.
154. S. M. Richardson-Burns, J. L. Hendricks, B. Foster, L. K. Povlich, D.-H. Kim and D. C. Martin, *Biomaterials*, 2007, **28**, 1539–1552.

155. W. R. Stauffer and X. T. Cui, *Biomaterials*, 2006, **27**, 2405–2413.
156. A. S. Widge, M. Jeffries-El, X. Cui, C. F. Lagenaur and Y. Matsuoka, *Biosens. Bioelectron.*, 2007, **22**, 1723–1732.
157. M. Asplund, E. Thaning, J. Lundberg, A. C. Sandberg-Nordqvist, B. Kostyszyn, O. Inganas and H. von Holst, *Biomed. Mater.*, 2009, **4**, 045009.
158. J. E. Collazos-Castro, J. L. Polo, G. R. Hernandez-Labrado, V. Padial-Canete and C. Garcia-Rama, *Biomaterials*, 2010, **31**, 9244–9255.
159. D.-H. Kim, J. A. Wiler, D. J. Anderson, D. R. Kipke and D. C. Martin, *Acta Biomater.*, 2010, **6**, 57–62.
160. C.-H. Lu, Y.-S. Hsiao, C.-W. Kuo and P. Chen, *Biochim. Biophys. Acta*, 2013, **1830**, 4321–4328.
161. B. C. Thompson, S. E. Moulton, J. Ding, R. Richardson, A. Cameron, S. O'Leary, G. G. Wallace and G. M. Clark, *J. Controlled Release*, 2006, **116**, 285–294.
162. D. T. Simon, S. Kurup, K. C. Larsson, R. Hori, K. Tybrandt, M. Goiny, E. H. Jager, M. Berggren, B. Canlon and A. Richter-Dahlfors, *Nat. Mater.*, 2009, **8**, 742–746.
163. W. R. Stauffer, P.-M. Lau, G.-Q. Bi and X. T. Cui, *J. Neural Eng.*, 2011, **8**, 044001.
164. T. F. Otero and J. G. Martinez, *Adv. Funct. Mater.*, 2013, **23**, 404–416.
165. T. F. Otero, *J. Mater. Chem. B*, 2013, **1**, 3754–3767.
166. T. F. Otero and F. Santos, *Electrochim. Acta*, 2008, **53**, 3166–3174.
167. M. C. Emerick, R. Stein, R. Kunze, M. M. McNulty, M. R. Regan, D. A. Hanck and W. S. Agnew, *Proteins: Struct., Funct., Bioinf.*, 2006, **64**, 320–342.
168. L. B. Xu, W. Chen, A. Mulchandani and Y. S. Yan, *Angew. Chem.,Int. Ed.*, 2005, **44**, 6009–6012.
169. S. Taleb, T. Darmanin and F. Guittard, *ACS Appl. Mater. Interfaces*, 2014, **6**, 7953–7960.
170. M. Liu, F.-Q. Nie, Z. Wei, Y. Song and L. Jiang, *Langmuir*, 2010, **26**, 3993–3997.
171. B. Yameen and A. Farrukh, *Chem.–Asian J.*, 2013, **8**, 1736–1753.
172. Y. Pei, J. Travas-Sejdic and D. E. Williams, *Langmuir*, 2012, **28**, 8072–8083.
173. M. Liu, X. Liu, C. Ding, Z. Wei, Y. Zhu and L. Jiang, *Soft Matter*, 2011, **7**, 4163–4165.
174. S. Taleb, T. Darmanin and F. Guittard, *RSC Adv.*, 2014, **4**, 3550–3555.
175. T. Darmanin and F. Guittard, *Prog. Polym. Sci.*, 2014, **39**, 656–682.
176. P. M. Beaujuge and J. R. Reynolds, *Chem. Rev.*, 2010, **110**, 268–320.
177. R. J. Mortimer, *Chem. Soc. Rev.*, 1997, **26**, 147–156.
178. Q. Pei, G. Zuccarello, M. Ahlskog and O. Inganas, *Polymer*, 1994, **35**, 1347–1351.
179. C. M. Amb, A. L. Dyer and J. R. Reynolds, *Chem. Mater.*, 2011, **23**, 397–415.
180. P. Camurlu, *RSC Adv.*, 2014, **4**, 55832–55845.
181. W. M. Kline, R. G. Lorenzini and G. A. Sotzing, *Color. Technol.*, 2014, **130**, 73–80.

182. V. Seshadri, J. Padilla, H. Bircan, B. Radmard, R. Draper, M. Wood, T. F. Otero and G. A. Sotzing, *Org. Electron.*, 2007, **8**, 367–381.
183. B. Conway, *J. Electrochem. Soc.*, 1991, **138**, 1539–1548.
184. P. Novak, K. Muller, K. S. V. Santhanam and O. Haas, *Chem. Rev.*, 1997, **97**, 207–281.
185. E. Frackowiak, *Phys. Chem. Chem. Phys.*, 2007, **9**, 1774–1785.
186. G. A. Snook, P. Kao and A. S. Best, *J. Power Sources*, 2011, **196**, 1–12.
187. R. Gracia and D. Mecerreyes, *Polym. Chem.*, 2013, **4**, 2206–2214.
188. K. S. Ryu, K. M. Kim, N. G. Park, Y. J. Park and S. H. Chang, *J. Power Sources*, 2002, **103**, 305–309.
189. J. Janata and M. Josowicz, *Nat. Mater.*, 2003, **2**, 19–24.
190. H. Bai and G. Shi, *Sensors*, 2007, **7**, 267–307.
191. D. T. McQuade, A. E. Pullen and T. M. Swager, *Chem. Rev.*, 2000, **100**, 2537–2574.
192. B. Ding, M. Wang, J. Yu and G. Sun, *Sensors*, 2009, **9**, 1609–1624.
193. D. S. Correa, E. S. Medeiros, J. E. Oliveira, L. G. Paterno and L. H. C. Mattoso, *J. Nanosci. Nanotechnol.*, 2014, **14**, 6509–6527.
194. D. Nicolas-Debarnot and F. Poncin-Epaillard, *Anal. Chim. Acta*, 2003, **475**, 1–15.
195. U. Yogeswaran and S.-M. Chen, *Sensors*, 2008, **8**, 290–313.
196. K. Maksymiuk, *Electroanalysis*, 2006, **18**, 1537–1551.
197. D. James, S. M. Scott, Z. Ali and W. T. O'Hare, *Microchim. Acta*, 2005, **149**, 1–17.
198. U. Lange, N. V. Roznyatouskaya and V. M. Mirsky, *Anal. Chim. Acta*, 2008, **614**, 1–26.
199. S. Nambiar and J. T. W. Yeow, *Biosens. Bioelectron.*, 2011, **26**, 1825–1832.
200. A. Ramanavicius, A. Ramanaviciene and A. Malinauskas, *Electrochim. Acta*, 2006, **51**, 6025–6037.
201. M. A. Rahman, P. Kumar, D.-S. Park and Y.-B. Shim, *Sensors*, 2008, **8**, 118–141.
202. J. Bobacka, A. Ivaska and A. Lewenstam, *Chem. Rev.*, 2008, **108**, 329–351.
203. H. Sangodkar, S. Sukeerthi, R. S. Srinivasa, R. Lal and A. Q. Contractor, *Anal. Chem.*, 1996, **68**, 779–783.
204. M. Gerard, A. Chaubey and B. D. Malhotra, *Biosens. Bioelectron.*, 2002, **17**, 345–359.
205. M. Ates, *Mater. Sci. Eng., C*, 2013, **33**, 1853–1859.
206. N. K. Guimard, N. Gomez and C. E. Schmidt, *Prog. Polym. Sci.*, 2007, **32**, 876–921.
207. S. Nambiar and J. T. W. Yeow, *Biosens. Bioelectron.*, 2011, **26**, 1825–1832.
208. K. Ghanbari, S. Z. Bathaie and M. F. Mousavi, *Biosens. Bioelectron.*, 2008, **23**, 1825–1831.
209. S. Cosnier and M. Holzinger, *Chem. Soc. Rev.*, 2011, **40**, 2146–2156.
210. M. J. Higgins, P. J. Molino, Z. Yue and G. G. Wallace, *Chem. Mater.*, 2012, **24**, 828–839.
211. C. Dhand, M. Das, M. Datta and B. D. Malhotra, *Biosens. Bioelectron.*, 2011, **26**, 2811–2821.

212. J. L. Vinkenborg, N. Karnowski and M. Famulok, *Nat. Chem. Biol.*, 2011, **7**, 519–527.
213. K. W. Plaxco and H. T. Soh, *Trends Biotechnol.*, 2011, **29**, 1–5.
214. A. Vallee-Belisle and K. W. Plaxco, *Curr. Opin. Struct. Biol.*, 2010, **20**, 518–526.
215. R. Penchovsky, *Biotechnol. Adv.*, 2014, **32**, 1015–1027.
216. W. Takashima, T. Uesugi, M. Fukui, M. Kaneko and K. Kaneto, *Synth. Met.*, 1997, **85**, 1395–1396.
217. W. Takashima, K. Hayasi and K. Kaneto, *Electrochem. Commun.*, 2007, **9**, 2056–2061.
218. T. Shoa, J. D. W. Madden, T. Mirfakhrai, G. Alici, G. M. Spinks and G. G. Wallace, *Sens. Actuators, A*, 2010, **161**, 127–133.
219. Y. Wu, G. Alici, J. D. W. Madden, G. M. Spinks and G. G. Wallace, *Adv. Funct. Mater.*, 2007, **17**, 3216–3222.
220. N. Festin, C. Plesse, P. Pirim, C. Chevrot and F. Vidal, *Sens. Actuators, B*, 2014, **193**, 82–88.
221. S. W. John, G. Alici, G. M. Spinks, J. D. Madden and G. G. Wallace, *Smart Mater. Struct.*, 2009, **18**, 085007.

CHAPTER 9

Multi-Tool Devices Mimicking Brain–Organ Intercommunication

9.1 Introduction

Table 5.1 collects some of the composition-dependent properties of polymers. As a matter of fact, all those properties could change simultaneously in the same material driven by the same reaction. The key point is optimizing the synthesis of the material (Chapter 4) to get good, simultaneous variation of the different selected properties. Chapter 8 was devoted to some initial exploration of the current, univocal technological procedure: each of the composition-dependent properties gives, under control, a new electro-chemo-biomimetic device. Now, the exploration of a new scientific and technological border for multifunctional and multi-tool devices based on that table can be undertaken: *i.e.*, several tools working simultaneously in a physically uniform device driven by the device reaction.

9.2 Electrochemically and Chemically Driven Multifunctionality

The same electrochemical reaction shown in eqn (4.1) promotes the simultaneous variation of the different composition-dependent properties of the materials depicted in Table 5.1. A new technological field can be announced whereby the simultaneous variation of two, three or four material properties driven by the same reaction can be optimized and theoretically described,

RSC Smart Materials No. 19
Conducting Polymers: Bioinspired Intelligent Materials and Devices
By Toribio Fernández Otero

Published by the Royal Society of Chemistry, www.rsc.org

resulting in chemo-multifunctionality, electrochemo-multifunctionality and intelligent multi-tool materials and devices. The variation of each property should be a faradaic function of the material composition (Chapter 5).

9.3 Multi-Tool Devices

Two, three or four of the chemo-biomimetic properties from Table 5.1 can be selected. The polymer synthesis conditions and the working conditions can be optimized in order to warrant the simultaneous variation of each property in a large range of values. In any physically uniform device constructed with this polymer, two, three or four tools will work simultaneously, driven by the same electrochemical reaction. The new concept of *simultaneous multifunctionality* emerges from Table 5.1, announcing the development of an unparalleled technological world of multi-tool devices whereby different tools work simultaneously in a physically uniform device driven by the reaction of its constitutive material.

At the moment, only dual-sensing muscles (artificial muscles and sensors from Table 5.1) have been developed and described by physico-chemical models, including artificial muscles sensing while working under thermal, mechanical, electrical or chemical conditions. This means that one actuator and several sensors are working simultaneously in one physically uniform device.[1–6] Thus, among the multiple possibilities to develop multi-tool devices offered by Table 5.1, this chapter will be focused on the development of dual and simultaneous sensing actuators, sensing muscles or sensing motors from conducting polymers.

Artificial muscles were presented in Section 8.2 as reaction-driven motors. Sensors and biosensors were presented in Section 8.11 as chemical equilibrium-based devices. Here, in order to integrate actuators and sensors, it is necessary to make a preliminary exploration of the sensing abilities of a chemical reaction (here, the polymeric reaction) outside the chemical equilibrium, which means outside the theoretical support of Le Châtelier's principle.

Great scientific and technological motivations push us: a parallel, dual device does not exist in our technological world at present. Only biological evolution has developed some bifunctional or multifunctional devices. Haptic muscles have attracted attention because they work simultaneously as chemical-driven motors and sensors of the mechanical, thermal or chemical working conditions.

This multifunctionality from haptic muscles results in mechanical proprioception in animals and humans, which is considered a psychological brain function.[7] One of the oldest dreams of engineers and robot designers is the availability of motors that can sense, by themselves, working conditions, thus providing dual-sensing motors. Inspired by haptic muscles and taking artificial proprioception as a goal, Table 5.1 offers the possibility to develop an artificial muscle following the evolution of the muscle potential simultaneously with the muscular movement driven by the electrochemical reaction.

The reaction-driven volume variation (Section 8.2) results in faradaic artificial muscles: the flowing current controls the rate of the movement and the consumed charge controls the muscle position, covering two basic proprioception properties. Artificial muscles work under reaction control and outside equilibrium conditions. In this context, a parallel principle to those of Le Châtelier and Nernst that can be applied to reacting systems and that can support and describe the development of dual-sensing muscles does not exist.

9.4 Otero's Sensing Principle During Reaction

Polymeric electrodes working under reaction control, *i.e.*, outside of equilibrium conditions, can take part of any of the electrochemical devices described in Chapter 8. Le Châtelier's principle was reformulated due to empirical evidence that, going from the same initial to the same final oxidation state of the polymeric electrode each time, the evolution of either the material potential, the material's consumed electrical energy, the muscle potential or the muscle's consumed energy (Section 9.5) changes when the experimental working conditions (mechanical, thermal, chemical or electrical) are modified.[3–5,8–19]

The new sensing principle for reactions states that:[20–23] reacting systems (materials or devices here), when subjected to energetic perturbations (thermal, chemical, mechanical or electrical here), respond by adapting the reaction energy to the new energetic conditions.

Here, perturbation is understood as working under different values of the variables acting on the reaction rate: temperature (thermal energy); electrolyte concentration (chemical energy); trailed mass, strain, stress or pressure (mechanical energy); driving current or applied potential (electrical and electrochemical energy) and so on; or when the variable is changed while the device is working.

For any electrochemical device (Chapter 8), the electrical energy (U) consumed by the driving reaction is given by:

$$U = I(\mathrm{A})E(\mathrm{V}) = \frac{Q(\mathrm{C})}{t(\mathrm{s})}E(\mathrm{V}), \tag{9.1}$$

where I (in A) represents the current flowing through the device at time t (in s), Q (in C) is the consumed charge and E (in V) is the material potential. Thus, according to eqn (9.1), the new principle supposes that:

1. For electrochemical devices driven by the flow of a constant current (I), any change of the working energetic (U) conditions (mechanical, thermal, chemical, electrical) will be sensed by the instantaneous adaptation of the device potential (E) evolution to the new working energetic conditions.
2. For electrochemical devices driven by a potential step (working under constant potential), any energetic (U) perturbation (mechanical, thermal, chemical, electrical) will be detected by the instantaneous adaptation of the current (I) flowing through the device to the new working energetic conditions.

Thus, when any electrochemical device works under the flow of a constant current, the sensing magnitudes will be the device potential and the electrical energy consumed by the device after a constant time of current flow [eqn (9.1)]. When the electrochemical reaction driving the device actuation occurs under polarization at a constant potential, the sensing magnitudes will be the flowing current or the consumed electrical energy after a constant polarization time.

9.5 Sensing Materials: Reactive Mechanical, Chemical, Thermal or Electrical Sensors

The simplest reactive system is a conducting polymer membrane (self-supported or coating a metal) electrode subjected to electrochemical reactions by application of different electrochemical methodologies, such as those mentioned previously. After electrogeneration, the material is subjected to consecutive potential sweeps (Section 2.5.1) inside the oxidation–swelling and reduction–shrinking potential range to get stationary voltammetric responses. By its integration, the resulting coulovoltammetric (Q/E) response gives the reversible charge consumed by the membrane oxidation/reduction, detecting the possible presence of irreversible reactions (Section 4.7.2). Then, after reduction at a constant potential inside the reduction–shrinking region, the material is subjected to consecutive square current waves (Section 2.6) until stationary chronovoltammetric responses are attained. Current and time are selected to obtain an oxidation/reduction charge lower than the reversible Q/E charge: the material will stay inside the reversible region during the consecutive square current waves. Starting each time from the reduced material, the procedure is repeated for different experimental values of the chemical, thermal, electrical or mechanical variables.

Normalized oxidation and reduction chronopotentiometric responses (taking the current at $t = 0$ to be $I = 0$) under the influence of each studied energetic condition will be presented to check the adaptation of the material potential evolution (the sensing ability) to the different values of the studied variable (to the different working energetic conditions).

9.5.1 Reaction-Driven Mechanical Sensors

Films of conducting polymers under different oxidation states act as mechanical sensors when subjected to different stress/strain cycles. The sensing characteristics under the simultaneous reaction of the conducting polymer film, as far as this author knows, still have not been investigated.

9.5.2 Reaction-Driven Thermal Sensors

Figure 9.1a shows the chronopotentiometric responses under different working temperatures.[16] Figure 9.1b shows the normalized anodic (oxidation) and cathodic (reduction) cyclic chronopotentiometric responses from a polypyrrole/dodecylbenzenesulfonate (Ppy/DBS) blend film under different temperatures. As expected, the material potential evolves at lower values for rising temperatures: under rising thermal energetic conditions the material potential evolves at lower values, consuming a lower electrical energy for oxidization between the same initial and final oxidation state each time. After integration of the potential responses ($\int E\mathrm{d}t$), the electrical energy ($U = I\int E\mathrm{d}t$) consumed during every oxidation or reduction process at each studied temperature is attained. The consumed energy decreases linearly for rising working temperatures (Figure 9.1c). Oxidation- or reduction-consumed electrical energy by the material under cycling senses the working temperature and Figure 9.1c shows the anodic and cathodic sensor calibration lines.

Similar results under different temperatures for both the material potential or the consumed electrical energy after a constant time of current flow were attained for different conducting polymers,[14,15,17,24,25] carbon nanotubes[13] and graphenes.[26]

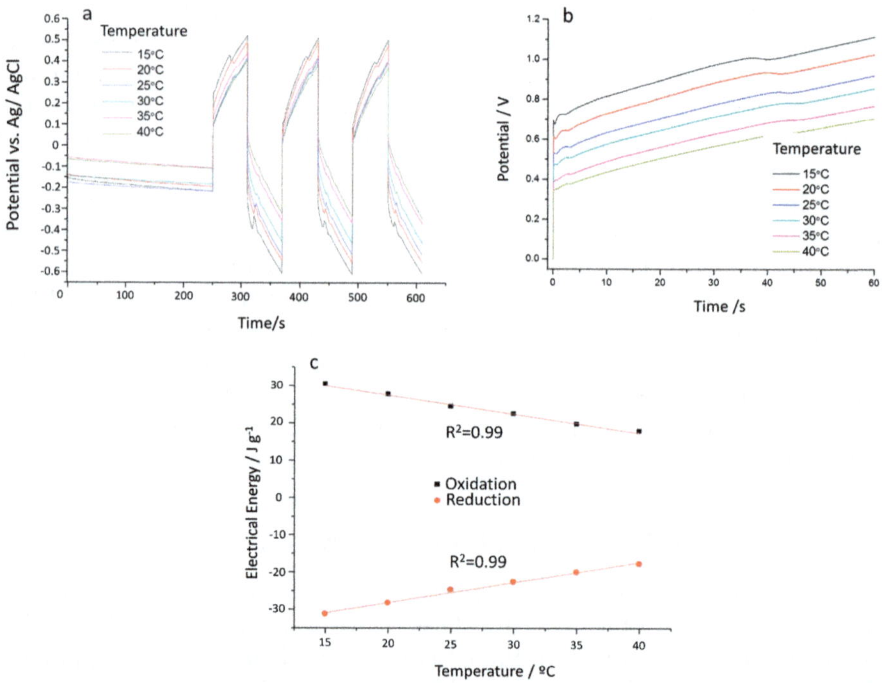

Figure 9.1 (a) Chronopotentiometric responses (following the working scheme in Figure 9.6A) at different temperatures by the flow of a ±4 mA current with 60 s per step through a self-supported polypyrrole/dodecylbenzenesulfonate (Ppy/DBS) film electrode in a 0.1 M $LiClO_4$ aqueous solution. (b) Normalized consecutive oxidation chronopotentiograms. (c) Variation of the electrical energy consumed during oxidation or reduction under the above conditions as a function of the working temperature. Reproduced from ref. 16: *Sensors*, 2010, **10**, 2638–2674.

9.5.3 Reaction-Driven Chemical Sensors

Figure 9.2a and b show the normalized anodic (oxidation) and cathodic (reduction) chronopotentiometric responses when Ppy-coated polyurethane (PU) fibrous mats were subjected to square current waves in different concentrations of electrolyte.[15] As expected, the Ppy-coated PU potential evolves during its oxidation or reduction at lower values for rising electrolyte concentrations: under rising chemical energetic conditions (a rising chemical potential of the reactants) the material requires a lower electrical energy to be oxidized or reduced between the same initial and final oxidation state each time. The consumed electrical energy, attained by integration of the chronopotentiometric responses ($U = I\int E\mathrm{d}t$), decreases logarithmically (Figure 9.2c) for rising working electrolyte concentrations. Under these conditions, oxidation or reduction electrical energies consumed by the material sense the working chemical conditions and Figure 9.2c presents the anodic and cathodic sensor calibration lines.

9.5.4 Reaction-Driven Electrical Sensors

Similar results to those depicted by Figures 9.1 and 9.2 were attained by the flow of rising constant anodic and cathodic currents through the material electrode under a constant electrolyte concentration, working temperature, mechanical conditions and consumed oxidation and reduction charges. The material potential evolves at rising values under the flow of increasing currents and both the material potential and the consumed electrical energy after a constant time of the current flow follow a linear dependence on the flowing current for different conducting polymers,[14,15,17,24,25] carbon nanotubes[13] and graphene-based[26] electrodes.

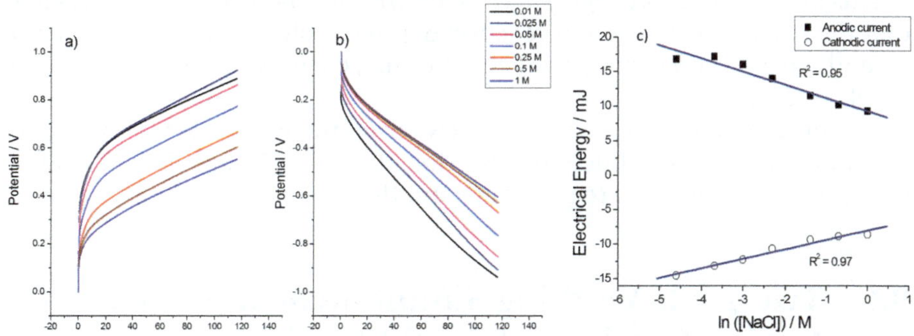

Figure 9.2 Normalized anodic (oxidation) (a) and cathodic (reduction) (b) chronopotentiograms obtained from polyurethane/polypyrrole (PU/Ppy) microfibrous mats in aqueous solutions with different concentrations, indicated in the figure, of NaCl under the flow of (a) +0.2 mA and (b) −0.2 mA currents for 116 s each time. (c) Semilogarithmic evolution of the consumed electrical energy as a function of the NaCl concentration (calibration lines of the chemical sensors). Reprinted from ref. 15: *J. Electroanal. Chem.*, 2014, **719**, 47–53, with the permission of Elsevier.

9.5.5 New Aspects of Three-Dimensional Structural Reactions: Reacting Material, Consumed Charge and Working Energetic Conditions

By combining the structural energy (*i.e.*, conformational), the reaction sensing principle (Section 9.4) and the variations of the consumed electrical energy under different energetic conditions it should be expected that using other electrochemical methodologies, such as the potential cycle, would result in higher electrochemical energies being consumed under rising available working energetic conditions (higher temperature, higher electrolyte concentrations or low demand for mechanical energy). Figure 9.3 shows the voltammetric and coulovoltammetric responses from silk fibroin mats coated with polyaniline (PANI) in different concentrations of aqueous methane sulfonic acid (MSA).[25] Higher available chemical energy (higher chemical potentials of the involved reactants) allows deeper conformational movements of the polymeric chains with deeper oxidation/reduction of the material, consuming higher oxidation/reduction charges. Similar results were found for higher available thermal energies with different materials.[14,15,17,25]

A new principle can be announced for three-dimensional (3D) reacting materials and electrodes, for which the 3D reactions involve structural transformations: the amount of reacting material and the concomitant electrical charge and electrical energy consumed by the driving reaction under constant conditions of the used electrochemical methodology are a function (linear or semilogarithmic) of the local working energetic (thermal, chemical, mechanical or electrical) conditions. A shorter version of the new *energetic principle for 3D structural reactions* (electrochemical, chemical or biochemical) is that the working energetic conditions (thermal, chemical, mechanical or electrical) determine both the extent of the structural reaction as it occurs, or not.

This new *energetic principle for 3D structural electrochemical reactions* states that the reaction extension or reaction depth (involved charge) is a function of, and senses, the working energetic thermal, chemical, mechanical or electrical conditions.

In biology and biochemistry, this new structural energetic principle plays many important roles, illustrated (Section 10.4.4) by the activity of the Ebola virus, which has adapted to ambient conditions.

9.6 Two Tools Working Simultaneously in One Device: Sensing Artificial Muscles

Working under consecutive square current waves, any artificial muscles (as bending bilayer or tri-layer muscles) must move between the same initial and final position (angle) by consumption of the same oxidation and reduction charge per cycle. By changing one of the working variables each time, whilst keeping the rest constant, the energy consumed here by bilayer muscles working under different mechanical, thermal, chemical or electrical energetic conditions will be presented.

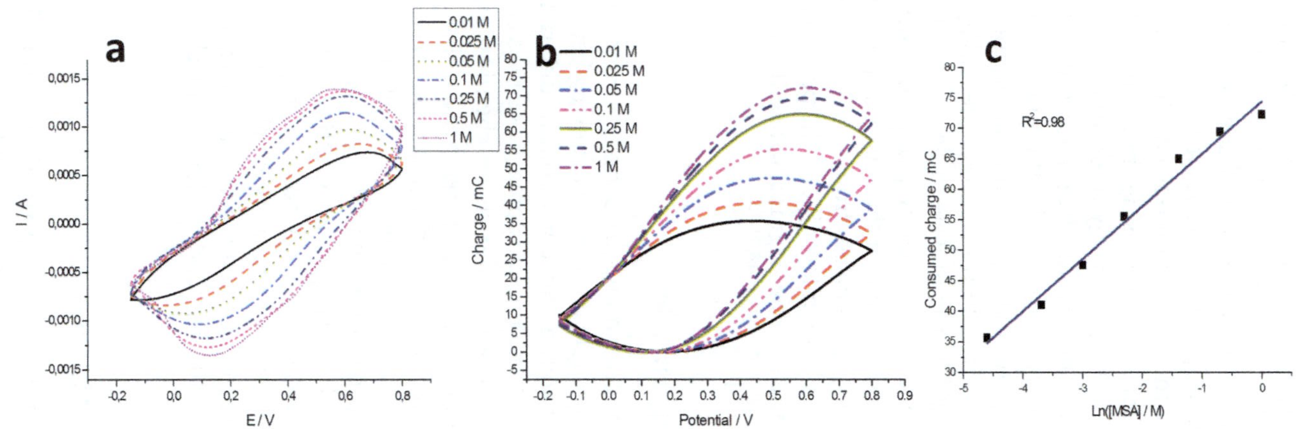

Figure 9.3 (a) Voltammograms obtained from a silk fibroin/PANI microfibrous mat when it was subjected to potential sweeps between −0.15 and 0.8 V *versus* Ag/AgCl, at 10 mV s^{-1}, in different concentrations of an MSA aqueous solution, as indicated in the figure. (b) Coulovoltammograms obtained by integration of the voltammograms presented in (a) for the reaction of PANI at different concentrations of the electrolyte. (c) Variation of the consumed electrical charge obtained from the coulovoltammograms of PANI in (b) for the reaction of PANI with the log of the MSA concentration. Reprinted from ref. 25: *Electrochim. Acta*, 2014, **123**, 501–510, with the permission of Elsevier.

9.6.1 Mechanical Sensing Muscles

Figure 9.4d shows the normalized anodic (oxidation) chronopotentiometric responses from a fully polymeric triple-layer Ppy/tape/Ppy muscle moving free or trailing different steel masses attached to its bottom between the same initial and final position under the flow of square current waves. As expected, the muscle potential (potential gradient between the Ppy working electrode, WE, layer and the Ppy counter-electrode, CE, layer; Figures 2.1c and 8.2c) evolves at higher values for rising trailed masses: the muscle material consumes a higher electrical energy to describe a constant angular displacement while trailing increasingly heavy steel masses attached to its bottom. The electrical energy consumed by the muscle during the WE oxidation, attained after integration of the chronopotentiometric responses ($U = I\int E\mathrm{d}t$), increases linearly with the trailed steel mass (Figure 9.6h).[27,28] The electrical energy consumed by the muscle to describe a constant angle driven by the Ppy WE film oxidation is a linear sensor of the working mechanical conditions, being the linear equation of the sensor calibration.

Similar mechanical sensing responses from working muscles were attained from different materials, and bilayer (here, the potential evolution is followed related to a reference electrode; Figures 8.2a and 2.1d) and tri-layer muscles in different electrolytes.[3,4,8,9,11,16,18,29–32]

9.6.2 Chemical Sensing Muscles

Figure 9.4a shows the normalized anodic (oxidation) chronopotentiometric responses from a Ppy/tape/Ppy triple layer muscle moving between the same initial and final angle under square current waves in different concentrations of electrolyte. As expected, the muscle potential evolves at lower values for rising electrolyte concentrations: under rising chemical energetic conditions (a higher chemical potential of the reactant) the muscle adapts to the ambient, consuming a lower electrical energy to move between the same initial and final position. The electrical energy (U) consumed by the muscle, attained after integration of the chronoamperometric responses, follows a semilogarithmic decrease for increasing concentrations of the electrolyte. The oxidation or reduction electrical energy consumed by the muscle is a semilogarithmic sensor (Figure 9.4e) of the working chemical conditions.

Similar chemical sensing responses from working muscles were attained for different materials, and bilayer and tri-layer muscles in different electrolytes.[3,4,8,9,11,16,18,29–32]

9.6.3 Thermal Sensing Muscles

Figure 9.4b shows the normalized anodic (oxidation) and cathodic (reduction) chronopotentiometric responses from a Ppy/tape/Ppy triple layer muscle moving between the same initial and final angular position under different temperatures driven by square current waves. As expected, the muscle potential evolves at lower values for rising available thermal energy at

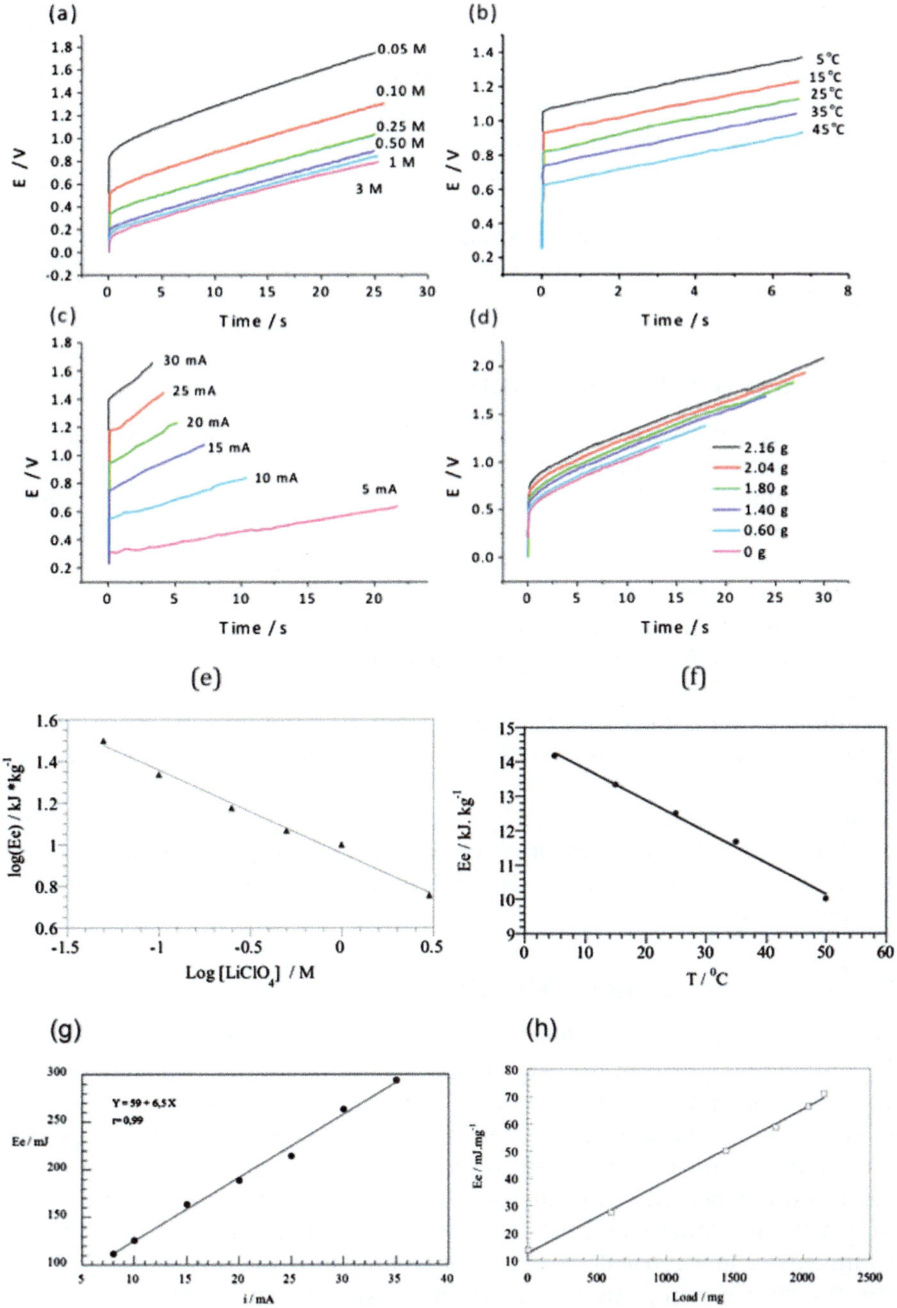

Figure 9.4 Chronopotentiograms obtained when a triple Ppy/tape/Ppy layer ($2 \times 1.5\ cm^2$, 12 mg Ppy) gives a 90° angle (a) in aqueous solutions of $LiClO_4$ (3, 1, 0.5, 0.25, 0.1 and 0.05 M) under a constant current of 10 mA, (b) in 0.1 M $LiClO_4$ at different temperatures: 5, 15, 25, 35 and 45 °C, (c) under the flow of different currents: 5, 10, 15, 20, 25 and 30 mA, and (d) shifting different attached steel weights: 0, 0.6, 1.4, 1.8, 2.04 and 2.16 g (adapted from ref. 28 and 29). Electrical energy consumed by the device as a function of the different studied variables: (e) electrolyte concentration, (f) temperature, (g) current and (h) shifted weight. Adapted from ref. 19, with the permission of the Royal Society of Chemistry.

rising working temperatures: under rising thermal energetic conditions the muscle adapts, consuming lower electrical energy to move between the same initial and final angle. The consumed energy, attained after integration of the chronopotentiometric responses ($U = I\int E\mathrm{d}t$), is a linear sensor (Figure 9.4f) of the working temperature.

Similar thermal sensing responses from working muscles were attained for different materials, and bilayer and tri-layer muscles in different electrolytes.[3,4,8,9,11,16,18,29–32]

9.6.4 Electrical Sensing Muscles

Figure 9.4c shows the normalized anodic (oxidation) chronopotentiometric responses from a bilayer Ppy/tape muscle moving between the same initial and final angle driven by square current waves of rising and constant charge. As expected for a system having constant impedance, the muscle potential evolves at higher values for rising driven currents: the muscle reaction is equivalent to resistance following Ohm's law and the muscle potential evolves at rising potentials under rising driving currents. The consumed energy, attained by integration of the response ($U = I\int E\mathrm{d}t$), is a linear sensor (Figure 9.4g) of the driving current.

Similar electrical sensing responses from working muscles were attained from different materials, and bilayer and tri-layer muscles in different electrolytes.[3,4,8,9,11,16,18,29–32]

Artificial muscles constituted by conducting polymers exchanging cations or anions during actuation produce reverse bending movements but both of them sense, while working, thermal, mechanical, chemical and electrical working conditions.

9.6.5 Tactile Artificial Muscles

When an artificial muscle describes a bending movement between the same initial angular position and the same final angle under constant thermal, chemical, mechanical and electrical conditions, a reproducible evolution of the muscle potential is attained. By placing an obstacle at any intermediate position of the muscle movement (Figure 9.5a), the muscle potential follows the expected reproducible evolution until the contact time. There, just at the touching time, the muscle potential steps (Figure 9.5b) in order to increase the electrical energy to push and shift the obstacle.[10,11,16,29,33] If the procedure is repeated, positioning obstacles with increasing mass at the same intermediate position each time, then the muscle potential always steps at the same point (Figure 9.5b) and the stepping potential gradient is proportional to the resistance opposed by the obstacle to be shifted:[10,11] this is a tactile muscle. When the obstacle resistance overcomes the capacity of the muscle to produce mechanical energy, a potential step of several volts is produced at the muscle–obstacle contact point: the sensing potential jump indicates the inability of the muscle to move the obstacle. Artificial muscles based on conducting polymers, or other electrochemically active materials,

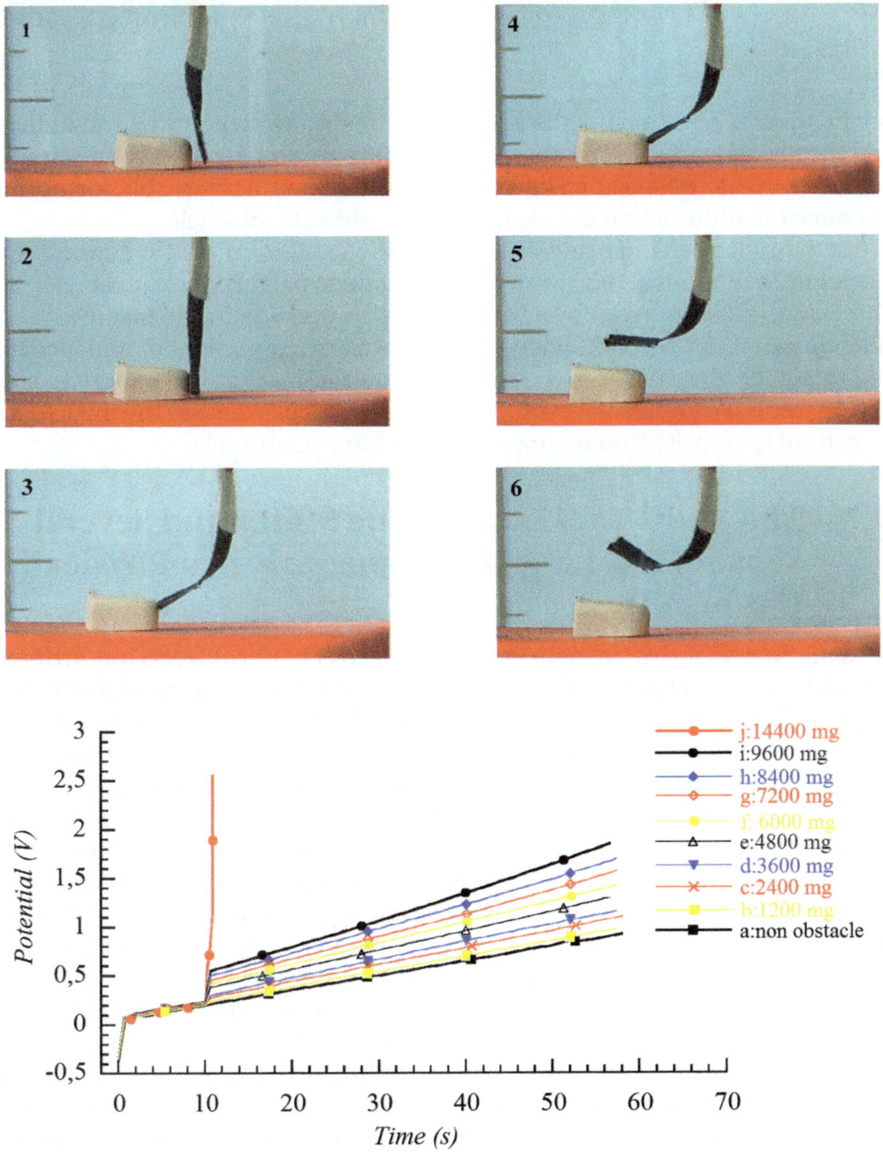

Figure 9.5 (a) The triple-layer muscle initiates its movement under a constant current of 5 mA; the black Ppy film on the right (1) acts as the anode, while the one on the left acts as the cathode in a 1 M aqueous $LiClO_4$ solution. (2) Ten seconds later, the muscle meets an obstacle weighing 6000 mg. (3) and (4) The obstacle is pushed. (5) The angular movement allows the muscle to overcome the border of the obstacle. (6) The free movement continues until the current stops. (b) Chronopotentiograms (potential increment across the device) obtained from a triple-layer macroscopic muscle containing two Ppy films (2 cm × 1.5 cm × 13 μm), each weighing 6 mg, under a current flow of 5 mA in a 1 M $LiClO_4$ aqueous solution. The muscle moved freely (a) or it met an obstacle after 10 s and pushed it for 3.5 s; it then overcame the obstacle and continued with a full angular movement of 108° (from −18° to +90°). The initial position was recovered by applying a current of −5 mA for 57 s. Obstacles weighing 1200, 2400, 3600, 4800, 6000, 7200, 8400 and 9600 mg were pushed, but the muscle was unable to push and slide an obstacle weighing 14 400 mg. Reprinted from ref. 10: *Adv. Mater.*, 2003, **15**, 279–282, with the permission of John Wiley and Sons.

are tactile muscles sensing the presence of obstacles in their way and indicating through the muscle potential increment both the mechanical resistance opposed by the obstacle to be shifted and if the muscle is able (or not) to produce the required mechanical energy to shift the obstacle.

Tactile muscles constitute a paradigmatic exhibition of the new sensing principle for chemical reactions: any mechanical perturbation (obstacle) of a three-layer muscle driven by a constant current results in the instantaneous adaptation of the reaction energy by stepping the muscle potential, according to eqn (9.1), in order to develop the required mechanical energy to shift the obstacle. The potential evolution also informs us whether the muscle can or cannot (potential step by several volts) shift the obstacle.

9.7 The Multi-Tool Device: One Motor and Several Sensors Working Simultaneously in a Physically Uniform Device

As a summary of the previous sections, electrochemical (reactive) artificial muscles are polymeric dual-sensing motors, for which the movement rates [eqn (8.1)] and relative positions [eqn (8.2)] are controlled by the driving current and consumed charge, respectively. During the muscle's movement between the same initial and final position, the muscle's potential evolution and the consumed electrical energy sense the mechanical, chemical, thermal or electrical working conditions. Motors and sensors work simultaneously, driven by the same reaction in a physically uniform device. Only two wires connect the multi-tool device to the electrical generator (Figure 8.3B). Both motor-driving magnitudes (current and charge) and sensing magnitudes (muscle potential and consumed electrical energy) are simultaneously present at any working time in the two connecting wires. Quite simple software can read and treat this information to make fast decisions.

For the same actuating–sensing functions, present-day technologies will require the presence of one motor with two connecting wires and a thermal sensor, a chemical sensor and several mechanical sensors, each of these sensors requiring two or four connecting wires. A higher complexity is required and much higher energy is consumed for a fast reading and treatment of the magnitudes provided by every device.

Artificial muscles fulfil the old dream of engineers and robot designers: the availability of electrical motors sensing, by themselves, the working conditions, with actuating and sensing information simultaneously present in the only two connecting wires at any actuation time (sensing motors).

9.8 Proprioception: Artificial Proprioception from Sensing Artificial Muscles

Proprioception is the sense of the relative position of neighboring parts of the body and the strength of effort being employed in movements. Since 1826 (Charles Bell), proprioception has been considered a psychological mechanism of our brain.[34] Proprioception is based on haptic muscles (muscles sensing the working conditions) and its permanent communication with the brain, which also controls it (Figure 9.6A). At this point, the emerging question is if sensing and tactile artificial muscles could allow us to construct proprioceptive artificial devices and, in that case, if the attained artificial proprioception could be described by self-consistent physical-chemical equations.

When we catch an object in our hand, proprioception also provides us with a full knowledge of the mechanical conditions: weight of the object, if our muscles are able (or not) to move it, the object's permanent position related to any other body part, and the rate and direction of the movement at any time. Any proprioceptive artificial device must provide information about the relative position at any time of the different parts involved in the movement. This aspect is covered by artificial muscles that provide quantitative control of the movement rate [eqn (8.1)] and relative position [eqn (8.2)] attained at any time of the movement.

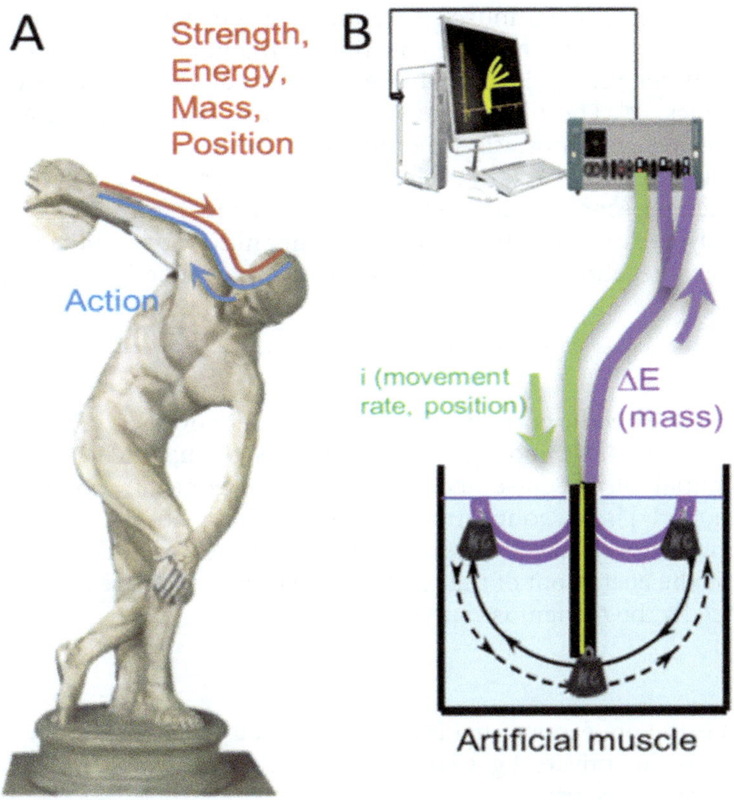

Figure 9.6 (A) Human mechanical awareness and (B) artificial expected mechanical awareness from artificial muscles. Reprinted from ref. 22: *Sens. Actuators, B*, 2014, **195**, 365–372, with the permission of Elsevier.

In addition, artificial muscles also sense the weight of any trailed object (Section 9.5.1), the presence of obstacles in its way (giving back information about the muscle–obstacle contact time) and the possibility to shift, or not, the obstacle (Section 9.5.5).

In natural proprioception, the brain orders the movements and receives back actuating and sensing information from haptic muscles. By processing the generated information, the brain develops proprioception, which allows rapid decision-making with the generation of new orders. The fuel providing the required energy for the brain nerves and muscle functions is given by reactions involving adenosine triphosphate (ATP). Artificial proprioceptive systems now in progress include a computer, an electrical generator and two wires connecting the generator with the artificial sensing muscle (Figure 9.6B). A full integration of the system for fast decision-making requires a single piece of software based on a simple physical–chemical theoretical description of the artificial proprioception.

9.9 Theoretical Description of Artificial Proprioception

Proprioceptive artificial muscles are faradaic sensing motors driven by an electrochemical reaction of the constitutive material. Electrochemical reactions are described by the basic Butler–Volmer equation of the electrochemical kinetics.[35,36] The flow of a constant anodic current (i_a, in A) through the electrode (galvanostatic experiment) defines a constant reaction rate [eqn (2.4)]. Now the muscle's electrochemical reaction involves two reactants, each present in a different media: anions (A^-) from the solution and active polymeric sites (Pol*) from the polymer film. Vetter studied these electrochemical reactions and the basic Butler–Volmer equation became:[36]

$$i_a = FVk_{a0}\left[A^-\right]^d [\text{Pol}*]^e \exp\left(\frac{(1-\alpha)nF(E-E_0)}{RT}\right), \tag{9.2}$$

where i_a is the anodic current driving the muscle movement, F is the Faraday constant ($F = 96\,485$ C mol^{-1}), V is the volume of the film, $E - E_0$ is the muscle overpotential (also represented by η), [A^-] is the concentration of anions in solution, [Pol*] is the concentration of active chain sites in the film, α is the symmetry factor of the electrochemical reaction (1/2 at the initial approach) and k_{a0} is the coefficient of the reaction rate (formerly the reaction constant) described by the Arrhenius expression:

$$K = A\mathrm{e}^{-\frac{E_a}{RT}}, \tag{9.3}$$

where A is the pre-exponential factor, E_a (in J mol^{-1}) is the reaction activation energy, R is the universal gas constant ($R = 8.314$ J K^{-1} mol^{-1}) and T (in K) is the working temperature.

When a constant overpotential is applied to the muscle, eqn (9.2) determines the current flowing through the muscle to keep the imposed potential.

In our case, under the flow of a constant anodic current, eqn (9.2) allows determination of the evolution of the muscle potential during current flow.

The simplest equation of the electrochemical kinetics [eqn (9.2)] links most of the experimental variables required for the development of the expected artificial proprioceptive equation describing the sensing muscle: driving current (i), temperature (T), electrolyte concentration $[A^-]$ and muscle potential (E). It was developed for electrochemical reactions taking place under constant atmospheric pressure, which means under constant mechanical conditions. In order to include changes in the mechanical conditions, such as those above studied empirically, the Arrhenius expression of the reaction coefficient must be reformulated. Taking into account thermodynamic relationships linking the activation energy, the reaction entropy increment (ΔS, in J K^{-1}), the reaction enthalpy increment (ΔH, in J mol^{-1}), the mechanical work increment, [$\Delta(PV)$, in N m] and the Gibbs equation,[37] eqn (9.2) can be rewritten to now include the reaction mechanical work, $\Delta(PV)$:[22]

$$i_a = FVk_{a0}[A^-]^d[Pol*]^e \exp\left(\frac{(1-\alpha)nF(E-E_0)}{RT}\right)\exp\left(\frac{\Delta(PV)}{RT}\right). \quad (9.4)$$

This is the reformulated basic Butler–Volmer equation for an electrochemical reaction producing mechanical work, $\Delta(PV)$, and sensing mechanical working conditions through the reaction-driven volume variation, as it happens with conducting polymers. Eqn (9.4) should be the basic electro-chemo-mechanical equation for the description of dual actuating–sensing devices.

Temporal evolutions are included by the progressive decrease of the concentration of the active chain sites, [Pol*], of the polymer film with the oxidation time to produce charged polaronic structures as a function of the consumed charge (a faradaic process):

$$[Pol*] = [Pol*]_{initial} - [Pol*]_{consumed} = [Pol*]_{initial} - \frac{Q}{FV} = [Pol*]_{initial} - \frac{it}{FV}, \quad (9.5)$$

where $[Pol*]_{initial}$ is the initial concentration of active sites in the conducting polymer film, $[Pol*]_{consumed}$ is the consumed concentration of active sites after a time t of current flow and Q is the consumed charge. Starting from the same initial concentration, flow of the same charge always gives the same final concentration. By substituting eqn (9.5) in eqn (9.4), after including the constant electrochemical cell impedance (Z, in Ω) and taking logarithms, the evolution of the muscle potential with time is obtained:

$$E(t) = E_0 + i_a Z - \frac{\Delta(PV)}{(1-\alpha)nF} + \frac{RT}{(1-\alpha)nF}\left\{\ln\left(\frac{i_a}{FV}\right) - d\ln[A^-] - e\ln\left([Pol*]_{initial} - \frac{i_a}{FV}\right) - \ln k_{a0}\right\} \quad (9.6)$$

Taking into account eqn (8.3), the muscle potential evolution includes the muscle bending movement rate (ω) and the muscle relative position (α) at any time of current flow:

$$E(t) = E_0 + i_a Z - \frac{\Delta(PV)}{(1-\alpha)nF} + \frac{RT}{(1-\alpha)nF} \times \left\{ \ln\left(\frac{\omega}{kFV}\right) - d\ln\left[\mathrm{A}^-\right] - e\ln\left([\mathrm{Pol}*]_{\mathrm{initial}} - \frac{(\alpha-\alpha_0)}{k'FV}\right) - \ln k_{\mathrm{a}0} \right\} \quad (9.7)$$

Eqn (9.7) describes the muscle potential evolution during the time required to extract, by flow of a constant anodic current (i_a), n electrons from each film chain.

Following polymer science, the n electrons extracted from every chain are not energetically equivalent, giving n consecutive equilibrium states defined by eqn (4.5)–(4.8). The n electrons are extracted through n consecutive steps of one electron per step, with E_1, E_2, E_3, ..., E_n being the oxidation potential for the equilibrium states by extraction of the first, second, third, ..., nth consecutive electrons, respectively. Assuming that a constant potential step (ΔE) is produced every time between the extraction, the last $(i-1)$ electron from the chains, the extraction of the first ith electron [$\Delta(E_i - E_{i-1}) = \Delta E(i-1)$] and the potential evolution during extraction of the nth electron from every film chain, $E_n(t)$, are related by:

$$E(t) = E_0 + i_a Z + \Delta E(n-1) - \frac{\Delta(PV)}{(1-\alpha)nF} + \frac{RT}{(1-\alpha)nF} \times \left\{ \ln\left(\frac{\omega}{kFV}\right) - d\ln\left[\mathrm{A}^-\right] - e\ln\left([\mathrm{Pol}*]_{\mathrm{initial}} - \frac{(\alpha-\alpha_0)}{k'FV}\right) - \ln k_{\mathrm{a}0} \right\} \quad (9.8)$$

This is the basic equation describing artificial proprioception [in the second term, time is included by the relative muscle position, α; eqn (8.3)].

Eqn (9.8) describes the muscle potential evolution during the flow of a constant anodic current (i_a) for t seconds, $E_n(t)$, by extraction of the nth electron from each of the film's polymeric chains according to the reaction shown in eqn (4.1), as a function of the working mechanical conditions, $\Delta(PV)$, the working thermal conditions, T, and the working chemical conditions, $[\mathrm{A}^-]$, with the angular rate of the movement at that time being ω and the relative position of the muscle related to the muscle's position at time zero being ($\alpha - \alpha_0$). Thus, the muscle potential while moving through a relative position α at an angular rate ω senses, simultaneously, the mechanical, thermal, chemical and electrical working conditions.

The consumed electrical energy during extraction of the nth electron from each polymer chains is then attained from eqn (9.8), giving:

$$U_n(t) = I\int E(t)\mathrm{d}t = E_0 t + zi^2 t + \Delta e(n-1)i_{\mathrm{a}}t$$
$$+ \frac{2lt\left(Y_{\mathrm{Pol}}x + Y_{\mathrm{tape}}x + mg\right)\sin\left(k_1 Q/2\right)}{(1-\alpha)F}$$
$$+ \frac{RTt}{(1-\alpha)F}\left\{d\ln\left[\mathrm{A}^-\right] - e\ln\left(\left[\mathrm{Pol}*\right]_{\mathrm{initial}} - \frac{i_{\mathrm{a}}t}{FV} - 1\right)\left(\frac{\left[\mathrm{Pol}*\right]_{\mathrm{initial}} - i_{\mathrm{a}}t}{i_{\mathrm{a}}V}\right)\right.$$
$$\left. - \ln\left(\frac{i_{\mathrm{a}}}{FV}\right) + \ln k_{\mathrm{a}0}\right\} \quad (9.9)$$

As expected, the consumed energy is also a function of (*i.e.*, it senses) the mechanical, thermal, chemical and electrical working conditions.

9.9.1 Potential and Consumed Energy Evolution During Actuation: Stair Functions

Eqn (9.8) describes the basic artificial proprioception for any of the electrochemical devices presented in Chapter 8 during extraction of the *n*th electron from each of the constitutive polymeric chains. According to eqn (4.5)–(4.8), the oxidation of the polymer film moves through *n* steps. Eqn (9.8) describes the potential evolution during the time required to extract the first, second, third, *etc.* electron from each of the film's constitutive chains, where *n* is 1, 2, 3, *etc.* After extraction of the last first electron from every chain, the potential steps by ΔE to begin the extraction of a second electron from the first chain. This step is repeated after extraction of each last *i*th electron from the constitutive chains. The potential evolution becomes, for a monodisperse polymer film (all the chains have the same conjugation length), a stair function:[20]

$$E(t) = \Sigma E_n(t)p_n(t) = E_1(t)p_1(t) + E_2(t)p_2(t) + \ldots + E_n(t)p_n(t), \quad (9.10)$$

where $p_n(t)$ is the unitary pulse function, which has a value of 1 during the time required to extract the *i*th electron from each of the film's chains and 0 outside this time; $E_n(t)$ is given by eqn (9.8).

The number of steps equals the number of electrons extracted from each chain and must be adjusted to fit the experimental evolution of the muscle potential during the time of the muscle movement: chronopotentiometric response.

Under the same conditions, the consumed electrical energy $U(t)$ during actuation also follows a stair function:

$$U(t) = \Sigma U_n(t)p_n(t) = U_1(t)p_1(t) + U_2(t)p_2(t) + \ldots + U_n(t)p_n(t), \quad (9.11)$$

where $U_n(t)$ is described by eqn (9.9). Eqn (9.8)–(9.11) describe artificial proprioception.

9.10 Dual Actuator–Sensor Systems

Eqn (9.8)–(9.11) describe the simultaneous mechanical, thermal, chemical and electrical awareness of any electrochemical device based on conducting polymers. Here, we are concerned with artificial muscles; thus, these equations indicate that any artificial muscle is simultaneously a mechanical sensor, a thermal sensor, a chemical sensor and an electrical sensor of the muscle working conditions. One motor and four sensors work simultaneously in the same physically uniform device, all of them driven simultaneously by the reaction of the device's constitutive material.

As an initial step for the exploration of these multi-tool devices, the experimental muscle potential evolution and the concomitant theoretical description of each of the simplest dual systems (actuator–mechanical sensor, actuator–chemical sensor, actuator–thermal sensor and actuator–electrical sensor) will be presented here. For simplicity, only the theoretical description during the device oxidation actuation will be developed. A detailed presentation of the theoretical description during the reduction-driven actuation, if required, can be found in the literature.[20–22]

The same experimental methodology was used for every dual muscle sensor. The muscle is driven between the same initial angular position to the same final position and back to the original initial position each time by the flow of consecutive square current waves consuming constant and equal anodic and cathodic charges. Usually, after two consecutive square waves, steady state chronopotentiometric responses (anodic and cathodic evolutions of the muscle potential) are obtained. The full procedure is repeated for a different value of the studied experimental variable every time, with the rest remaining constant.

9.10.1 Dual Actuator–Mechanical Sensor: Experiments and Model

Eqn (9.8) and (9.10) indicate that, under a constant experimental electrolyte concentration (chemical energy), working temperature (thermal energy) and driving current (electrical energy), the muscle's potential evolution is a sensor of the working mechanical conditions or mechanical perturbations (pressure, strain, stress, trailed weight, obstacles and so on). After a constant time of current flow, all these constant terms from eqn (9.8) are collected by the constant C_1 (Table 9.1) and the sensing equation becomes:[22]

$$E = C_1 - \frac{\Delta(PV)}{(1-\alpha)nF}. \tag{9.12}$$

Table 9.1 Constants for the different sensing equations. Reprinted from *Prog. Polym. Sci.*, 2015, **44**, 62–78.

Mechanical sensor	$C_1 = E_0 + i_a Z + \Delta E(n-1) - \frac{2l(Y_{Pol}x + Y_{tape}x)\sin(k_1 Q/2)}{(1-\alpha)F} + \frac{RT}{(1-\alpha)F}\left\{\ln\left(\frac{i_a}{FV}\right) - d\ln[A^-] - e\ln\left([Pol*]_{intial} - \frac{q}{FV}\right) - \ln k_{a0}\right\}$
	$C_2 = -\frac{2lg\sin(k_1 q/2)}{(1-\alpha)F}$
	$C_3(t) = E_0 t + Zi_a^2 t + \Delta E(n-1)i_a t + \frac{2lt(Y_{Pol}x + Y_{tape}x)\sin(k_1 Q/2)}{(\alpha-1)F} + \frac{RTt}{(\alpha-1)F}\left\{d\ln[A^-] - e\ln\left([Pol*]_{initial} - \frac{i_a t}{FV} - 1\right)\left(\frac{[Pol*]_{initial} - i_a t}{i_a V}\right) - \ln\left(\frac{i_a}{FV}\right) + \ln k_{a0}\right\}$
	$C_4(t) = \frac{2glt\sin(k_1 q/2)}{(\alpha-1)F}$
Thermal sensor	$C_1' = E_0 + (n-1)\Delta E$
	$C_2' = \frac{R}{(1-\alpha)F}\left\{\ln\left(\frac{i_a}{FV}\right) - d\ln[A^-] - e\ln\left([Pol*]_{initial} - \frac{q}{FV}\right) - \ln k_{a0}\right\}$
	$C_3' = i_a\{E_0 + (n-1)\Delta E\}$
	$C_4'(t) = \frac{Ri_a t}{(1-\alpha)F}\left\{\ln\left(\frac{i_a}{FV}\right) - d\ln[A^-] - \ln k_{a0}\right\} + \frac{RVe}{(1-\alpha)}\left\{\ln\left([Pol*] - \frac{i_a t}{FV}\right) - 1\right\}\left\{[Pol*] - \frac{i_a t}{FV}\right\}$

(continued)

Table 9.1 (*continued*)

Chemical sensor	$C_1'' = E_0 + i_a Z + (n-1)\Delta E + \frac{RT}{(1-\alpha)F}\left\{\ln\left(\frac{i_a}{FV}\right) - e\ln\left([\text{Pol}*]_{\text{initial}} - \frac{q}{FV}\right) - \ln k_{a0}\right\}$
	$C_2'' = \frac{RT}{(1-\alpha)F}d$
	$C_3''(t) = i_a\int E\,\mathrm{d}t = E_0 i_a t + Z i_a^2 t + (n-1)\Delta E i_a t + \frac{RTVe}{(1-\alpha)}\left\{\ln\left([\text{Pol}*]_{\text{initial}} - \frac{i_a t}{FV}\right) - 1\right\}$ $\times\left\{[\text{Pol}*]_{\text{initial}} - \frac{i_a t}{FV}\right\} + \frac{RT}{(1-\alpha)F} i_a t\ln\left(\frac{i_a}{FV}\right) - \frac{RT}{(1-\alpha)F} i_a t\ln k_{a0}$
	$C_4''(t) = -\frac{RT}{(1-\alpha)F} d i_a t$
Current sensor	$C_1'' = E_0 + \Delta E(n-1) + \frac{RT}{(1-\alpha)F}\left\{\ln\left(\frac{i_a}{FV}\right) - d\ln\left[A^-\right] - e\ln\left([\text{Pol}*]_{\text{initial}} - \frac{q}{FV}\right) - \ln k_{a0}\right\}$
	$C_3''(t) = i_a\int E\,\mathrm{d}t = E_0 i_a t + (n-1)\Delta E i_a t + \frac{RTVe}{(1-\alpha)}\left\{\ln\left([\text{Pol}*]_{\text{initial}} - \frac{i_a t}{FV}\right) - 1\right\}$ $\times\left\{[\text{Pol}*]_{\text{initial}} - \frac{i_a t}{FV}\right\} + \frac{RT}{(1-\alpha)F} i_a t\ln\left(\frac{i_a}{FV}\right) - \frac{RT}{(1-\alpha)F} i_a t\ln k_{a0} - \frac{RT}{(1-\alpha)F} d i_a t\ln\left[A^-\right]$

Considering a bending bilayer (conducting polymer/tape) muscle as a reference, the term $\Delta(PV)$ includes the different types of mechanical work involved during the muscle actuation and is required to:

(1) bend the conducting polymer and tape films;
(2) tug the mass of those films;
(3) tug any other mass (steel mass) attached to the muscle bottom;
(4) touch and shift an obstacle;
(5) produce or respond to any change of internal or external pressure, strain or stress acting on the device while working.

Considering here the mechanical components 1, 2 and 3, eqn (9.12) becomes:

$$E = C_1 - \frac{2l\left(Y_{\mathrm{Pol}}x + Y_{\mathrm{tape}}x + mg\right)\sin\left(\frac{k'Q}{2}\right)}{(1-\alpha)F}, \tag{9.13}$$

where Y_{Pol} and Y_{tape} are the Young's moduli of the conducting polymer film and tape, respectively; $Y_{\mathrm{Pol}}x$ and $Y_{\mathrm{tape}}x$ are the forces required to bend the polymer layer or tape, respectively, following Hooke's expression; x is the vertical displacement produced by the bending movement; mg is the force required to elevate the mass of the two bending layers plus the mass of any object attached to the bilayer bottom (m) for a vertical distance x, where g is the gravimetric acceleration ($g = 9.81$ m s^{-2}); $k'Q$ is the angle described by the muscle [eqn (8.2)] moving each time from the same initial position to the same final position for every value of the trailed mass during time t by flow of a constant current i and consuming charge $Q = it$, and l is the length of the bilayer bending inside the electrolyte.

Eqn (9.13) describes a muscle mechanical sensor or a mechanical sensor constituted by an electro-chemo-mechanical cantilever. For the same muscle trailing a different steel mass attached to its bottom each time, eqn (9.13) becomes:

$$E = C_1 + C_2 m, \tag{9.14}$$

where constants C_1 and C_2 are collected in Table 9.1.

Experimental responses obtained from bending Ppy/tape bilayer muscles trailing different steel masses attached to the muscle bottom each time and driven by square current waves of ±2 mA flowing for 100 s each are presented in Figure 9.7A and B (solid lines). The bending movement always starts from the same initial position with a constant angular displacement of 70 ± 2°. Constant flowing currents (anodic and cathodic) and constant flowing times determine constant consumed charges (oxidation and reduction), Q, and constant specific charges, $q = Q/w$, consumed per unit of conducting polymer mass (w) immersed in the electrolyte.

The theoretical evolution of the muscle potential when pushing different attached steel masses was simulated from eqn (9.8) and (9.10) for an initial

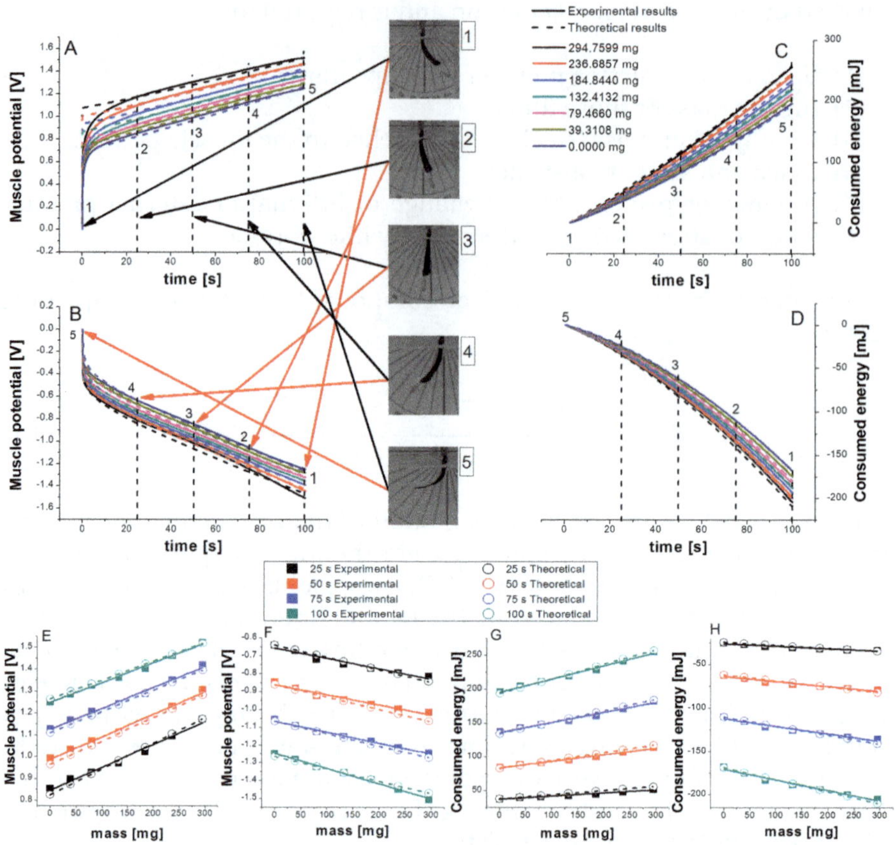

Figure 9.7 Chronopotentiometric response from a bilayer Ppy/tape artificial muscle bending in a 0.5 M $LiClO_4$ aqueous solution by flow of (A) 2 mA (clockwise bending) or (B) −2 mA (anticlockwise bending) trailing different steel masses, indicated in the figure, attached to the muscle's bottom through an angle of 70°. Parallel evolution attained for the consumed electrical energy (C) during the anodic process and (D) during the cathodic process (solid lines correspond to experimental results and dotted lines correspond to theoretical results). Sensing calibration curves: (E) muscle potential from A after different times of anodic current flow (at different constant angles) as a function of the trailed mass. (F) Muscle potential from B after different times of cathodic current flow (at different constant angles) as a function of the trailed mass. (G) Consumed electrical energy from C after different times of anodic current flow as a function of the trailed mass. (H) Consumed electrical energy from D after different times of cathodic current flow as a function of the trailed mass. Reprinted from ref. 22: *Sens. Actuators, B*, 2014, **195**, 365–372, with the permission of Elsevier.

concentration of active centers, [Pol*], in an immersed polymer film of 2 M L^{-1}, considering a constant electrochemical symmetry factor, $\alpha = 0.5$, when the average number of electrons exchanged per polymer changed, n = 190,[38] with the dimensions of the Ppy film inside the solution having length = 2 cm, width = 1 cm, film mass = 1.6 mg and film density = 1540 g L^{-1},[39] when the electrolyte concentration [A^-] was 0.5 M, and working at ambient temperature (25 °C). From electrochemical kinetic studies,[40] either the reaction orders for the muscle driving reaction ($d = 0.7$ and $e = 1.2$) or the rate coefficient $k_{a0} = k_{c0} = 1.0142\ 10^{-10}$ M L^{-1} s^{-1} were obtained. The Young's moduli were measured with a universal test frame machine: Y_{Pol} = 1250 MPa and Y_{tape} = 96 MPa.

A good agreement exists between theoretical (dotted lines) and experimental (solid lines) evolutions of the muscle potential (Figure 9.7A and B), confirming that eqn (9.8) and (9.10) describe the influence of the mechanical working conditions. The mechanical sensing equation [eqn (9.14)] was calibrated from the experimental and theoretical muscle potentials after three different times of current flow, *i.e.*, during the extraction of the same *i*th electron from the film's polymer chains or at the same film oxidation state for every experiment. Figure 9.7E and F confirm that the electrochemical bilayer device works simultaneously as a polymeric motor and a mechanical sensor. In a physically uniform device, the reaction of the conducting polymer drives and powers, simultaneously, the artificial muscle and the mechanical sensor: electrochemical muscles sense the working mechanical conditions and any mechanical perturbation.

Following a similar procedure for the consumed electrical energy from eqn (9.9) under constant chemical and physical conditions, when the muscle trails different masses through the same angle (same time of current flow), the energy consumed up to the same position every time becomes a linear sensing function of the trailed mass:

$$U_n(t) = C_3(t) + C_4(t)m, \tag{9.15}$$

where constants $C_3(t)$ and $C_4(t)$ are collected in Table 9.1. These energetic constants are time dependent, indicating that the muscle mechanical sensitivity increases for rising times of current flow. This becomes a unique characteristic of the reactive film sensors that can be generalized as follows: when the energy consumed by any electrochemical device is used as the mechanical sensing magnitude, the sensitivity of the sensor increases for rising reaction times.

The experimental and theoretical [eqn (9.11)] evolutions of the consumed electrical energy during the muscle actuation trailing different attached masses and driven by the flow of the anodic or cathodic current are presented by Figure 9.7C and D, respectively. Both experimental and theoretical mechanical sensing calibration curves [eqn (9.15)] are presented in Figure 9.7G and H. As described by the theoretical constants, the slope (mechanical sensitivity, in J g^{-1}) increases for rising actuation times.

9.10.2 Dual Actuator–Chemical Sensor: Experiments and Model

Eqn (9.8) and (9.10) indicate that, under a constant working temperature (thermal energy), driving current (electrical energy) and trailing only the two layer masses (mechanical energy), the muscle potential evolution is a sensor of the working chemical conditions or chemical perturbations (electrolyte concentration, ionic strength, solvent). After a constant time of current flow, eqn (9.8) becomes:[21]

$$E = C_3 - \frac{dRT}{\left[\left(1-\alpha\right)nF\right]}\ln\left[A^-\right] \tag{9.16}$$

$$E = C_3 - C_4\ln[A^-], \tag{9.17}$$

where C_3 collects (Table 9.1) all the constant terms from eqn (9.8) under actual empirical conditions and C_4 is also a constant constituted by the constant values of d, R, T, α, n and F identified above. Eqn (9.17) describes the muscle chemical sensor or the electro-chemo-mechanical cantilever acting as a chemical sensor.

Experimental and theoretical muscle potential evolutions while moving between the same initial and final position in a different electrolyte concentration each time are presented in Figure 9.8A and B, showing a good correlation between experiments and the model. The chemical sensing equation, eqn (9.17), was calibrated from the experimental and theoretical muscle potentials after three different times of current flow. Figure 9.8E and F confirm that the electrochemically driven bilayer muscle works simultaneously as a chemical sensor. In a physically uniform device, the reaction of the conducting polymer from the device drives, simultaneously, the artificial muscle and the chemical sensor: the electrochemical muscles sense, while working, the chemical working conditions and any chemical perturbation.

Under the same conditions, eqn (9.9) becomes a semilogarithmic sensing function of the chemical environment:

$$U_n(t) = C'_3(t) + C''_4(t)\ln[A^-]. \tag{9.18}$$

Here, the constants $C'_3(t)$ and $C''_4(t)$ are collected in Table 9.1. These energetic constants are also time dependent, indicating that the chemical sensitivity of the cantilever muscle increases for rising constant times of current flow: rising oxidized states for the anodic chemical sensor or rising reduced states for the cathodic chemical sensor.

Experimental and theoretical [eqn (9.11)] evolutions of the consumed electrical energy during the anodic and cathodic actuation in different electrolyte concentrations are presented in Figure 9.8C and D, respectively. Both, experimental and theoretical [eqn (9.18)] chemical sensing calibration curves are presented in Figure 9.8G and H. As described by the theoretical constants, the slope (mechanical sensitivity, in J M^{-1}) increases for rising actuation times.

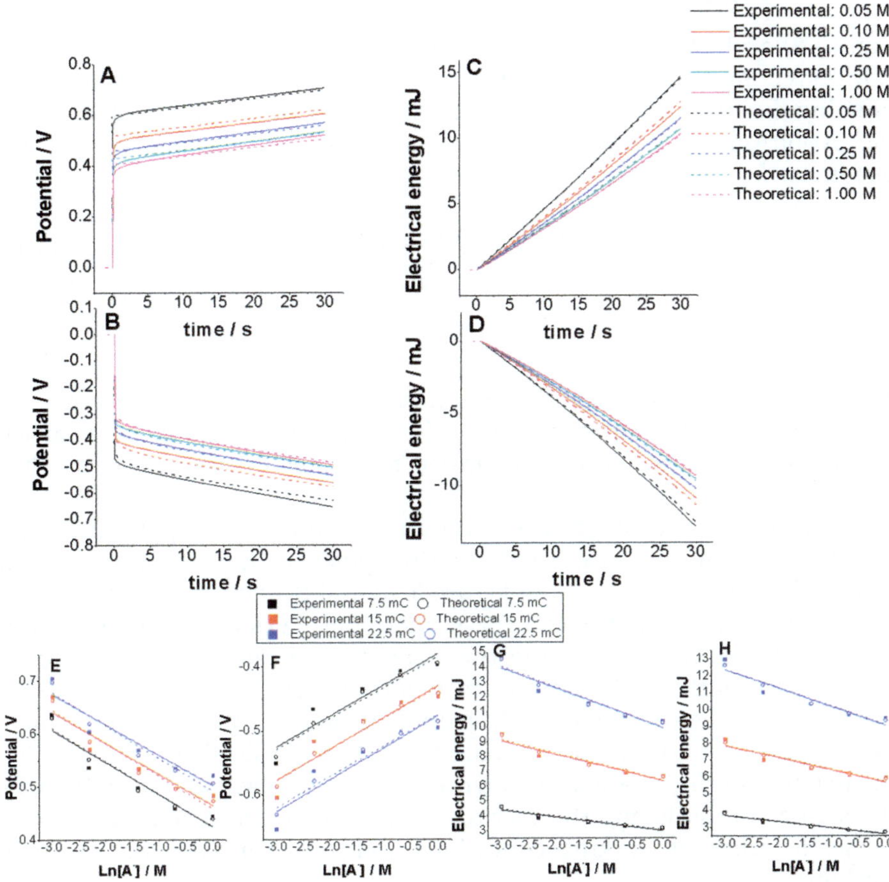

Figure 9.8 Chronopotentiometric responses from a bilayer Ppy/tape artificial muscle bending in a $LiClO_4$ aqueous solution with different salt concentrations, indicated on the figure, by flow of (A) 2 mA (clockwise bending) or (B) −2 mA (anticlockwise bending) through an angle of 70°. Parallel evolution of the consumed electrical energy (C) for the anodic process and (D) for the cathodic process (solid lines correspond to experimental results and dotted lines correspond to theoretical results). Sensing calibration curves: (E) muscle potential from A after different times of anodic current flow (at different constant angles) as a function of the electrolyte concentration. (F) Muscle potential from B after different times of cathodic current flow (at different constant angles) as a function of the electrolyte concentration. (G) Consumed electrical energy from C after different times of anodic current flow as a function of the electrolyte concentration. (H) Consumed electrical energy from D after different times of cathodic current flow as a function of the electrolyte concentration. Adapted with permission from ref. 21: *J. Phys. Chem. B*, 2012, **116**, 9223–9230. Copyright (2012) American Chemical Society.

9.10.3 Dual Actuator–Thermal Sensor: Experiments and Model

Eqn (9.8) and (9.10) indicate that, under a constant electrolyte concentration (chemical conditions), driving current (electrical conditions) and pushing the two layer masses (mechanical conditions), the muscle potential evolution is a sensor of the working temperature. Thus, eqn (9.9) becomes:[20]

$$E = C_5 + C' \frac{dRT}{(1 - \alpha) nF} \quad (9.19)$$

or

$$E = C_5 + C_6 T, \quad (9.20)$$

where C_5 and C_6, which are expression collected in Table 9.1, become constant under actual experimental conditions. Eqn (9.20) describes the muscle thermal sensor or the electro-chemo-mechanical cantilever acting as a thermal sensor.

Experimental and theoretical evolutions of the muscle potential while moving between the same initial and final position at a different temperature each time are presented in Figure 9.9A and B, showing a good correlation between experiments and the model. The thermal sensing equation, eqn (9.20), was calibrated from the experimental and theoretical muscle potentials after three different times of current flow. Figure 9.9E and F confirm that the electrochemically driven bilayer device works simultaneously as a polymeric motor and a thermal sensor. In a physically uniform device, the reaction of the conducting polymer drives and powers, simultaneously, the artificial muscle and the thermal sensor: the electrochemical muscles sense the thermal working conditions and any thermal variation while working.

Under the same conditions, eqn (9.9) becomes a semilogarithmic sensing function of the chemical ambient:

$$U_n(t) = C'_3(t) + C'_4(t)T, \quad (9.21)$$

where the constants $C'_3(t)$ and $C'_4(t)$ are collected in Table 9.1. These energetic constants are also time dependent, indicating that the thermal sensitivity of the cantilever muscle increases for rising constant times of current flow: rising oxidized states for the anodic chemical sensor or rising reduced states for the cathodic chemical sensor.

Experimental and theoretical [eqn (9.11)] evolutions of the consumed electrical energy during the muscle actuation at different temperatures are presented in Figure 9.9C and D. Both experimental and theoretical thermal sensing calibration curves [eqn (9.21)] are presented in Figure 9.9G and H. As described by the theoretical constants, the slope (mechanical sensitivity, in $J\,K^{-1}$) increases for rising actuation times.

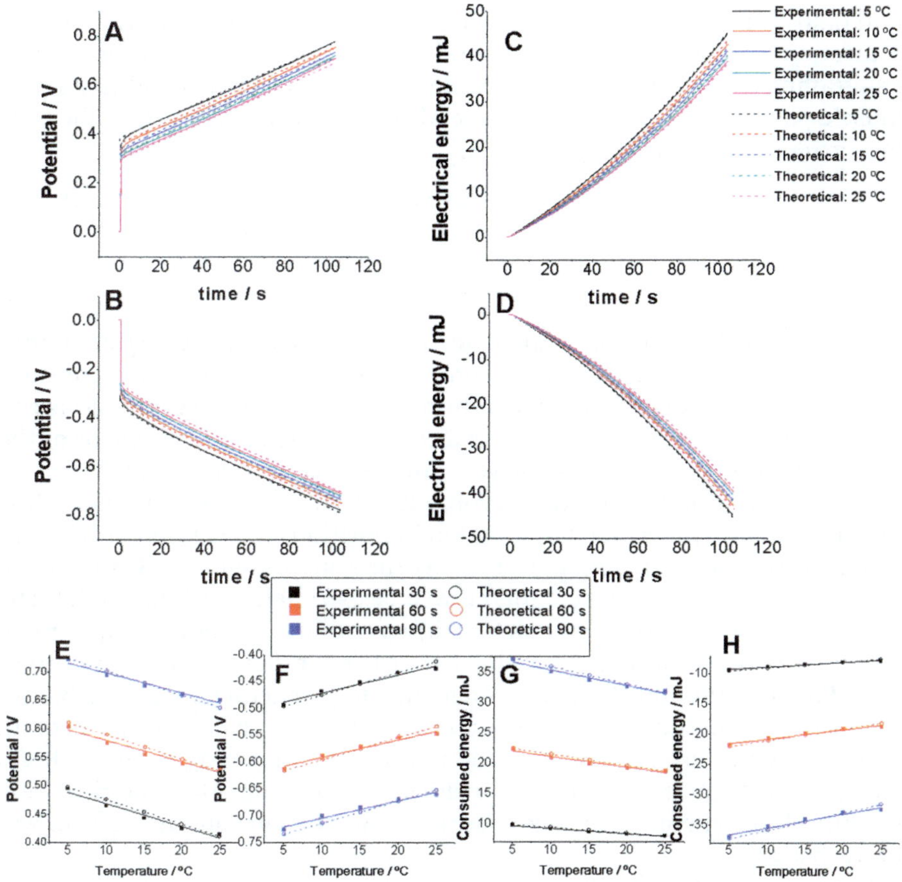

Figure 9.9 Chronopotentiometric responses from a bilayer Ppy/tape artificial muscle bending in a 0.1 $LiClO_4$ aqueous solution at different working temperatures by flow of (A) 2 mA (clockwise bending) or (B) −2 mA (anticlockwise bending) through an angle of 70°. Parallel evolution of the consumed electrical energy (C) for the anodic process and (D) for the cathodic process (solid lines correspond to experimental results and dotted lines correspond to theoretical results). Sensing thermal calibration curves for: (E) the muscle potential from A after different times of anodic current flow (at different constant angles); (F) the muscle potential from B after different times of cathodic current flow (at different constant angles); (G) the consumed electrical energy from C after different times of anodic current flow; and (H) the consumed electrical energy from D after different times of cathodic current flow. Reprinted with permission from ref. 20: *J. Phys. Chem. B*, 2012, **116**, 5279–5290. Copyright (2012) American Chemical Society.

9.10.4 Dual Actuator–Electrical Sensor: Experiments and Model

Eqn (9.8) and (9.10) indicate that, working under a constant electrolyte concentration (chemical conditions), working temperature (thermal conditions) and trailing only the two layer masses (constant mechanical conditions), the muscle potential after a constant time of current flow is a sensor of the driving current. Thus, eqn (9.8) becomes:[21]

$$E = C_7 + Zi_a \tag{9.22a}$$

or

$$E = C_7 + C_8 i_a. \tag{9.22b}$$

Here, C_8 is the constant impedance of the electrochemical system after exclusion of the muscle sensor driving reaction and C_7 became a constant under the studied experimental conditions described by the expressions collected in Table 9.1. Eqn (9.22b) describes the muscle electrical sensor or the electro-chemo-mechanical cantilever acting as an electrical sensor.

Experimental and theoretical evolutions of the muscle potential while moving between the same initial and final positions (consuming a constant charge) driven by a different current every time are presented in Figure 9.10A and B, showing a good correlation between experiments and the model. The electrical sensor described by eqn (9.22b) was calibrated from the experimental and theoretical muscle potentials after three different times of current flow. Figure 9.10E and F confirm that the electrochemically driven bilayer device works simultaneously as polymeric motor and an electrical sensor. In a physically uniform device, the reaction of the conducting polymer drives and powers, simultaneously, the artificial muscle and the electrical sensor: the electrochemical muscles sense the driving current and any current perturbation.

Under the same conditions, the consumed electrical energy, eqn (9.9), becomes a linear sensing function of the driving current:

$$U_n(t) = C''_3(t) + qZi_a, \tag{9.23}$$

where the constant $C''_3(t)$ is collected in Table 9.1. This energetic constant is also time dependent, indicating that the sensitivity to the driving current of the cantilever muscle increases for rising constant times of current flow: rising oxidized states for the anodic chemical sensor or rising reduced states for the cathodic chemical sensor.

Experimental and theoretical [eqn (9.11)] evolutions of the consumed electrical energy during the muscle actuation driven by different current flows are presented in Figure 9.10C and D. Both experimental and theoretical electrical sensing calibration curves [eqn (9.23)] are presented in Figure 9.10G and H. As described by the theoretical constants, the slope (mechanical sensitivity, in J A^{-1}) increases for rising actuation times.

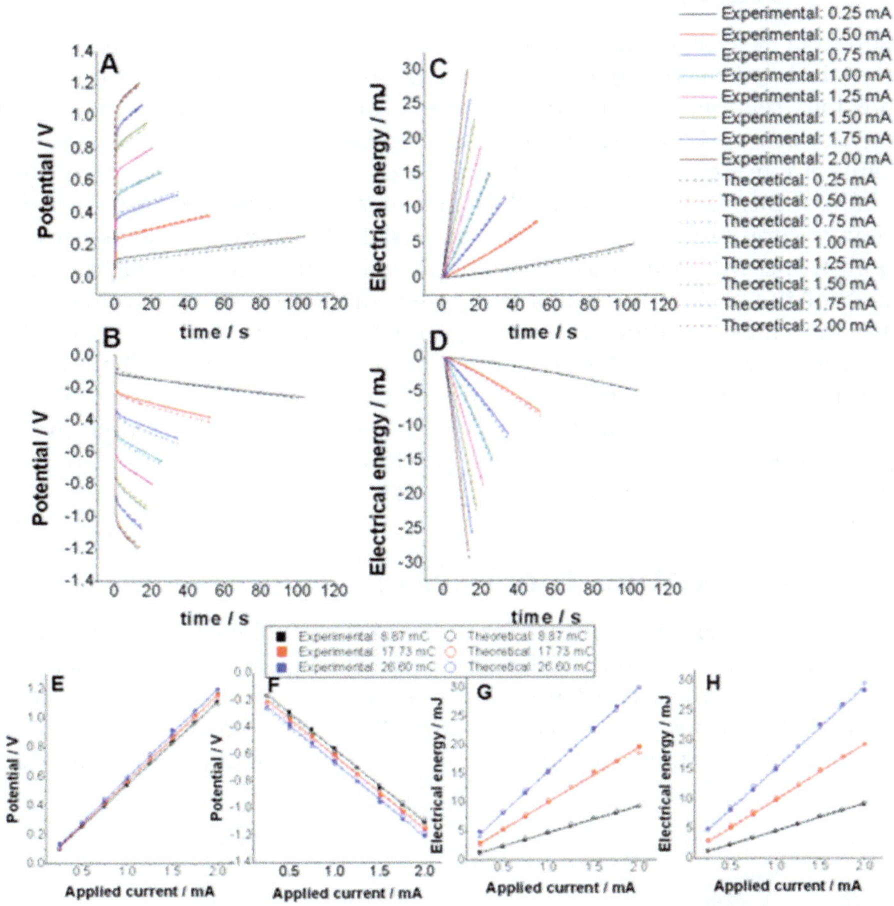

Figure 9.10 Chronopotentiometric responses from a bilayer Ppy/tape artificial muscle bending in a 0.1 $LiClO_4$ aqueous solution driven by different currents (square waves) by consumption of the same anodic (A) and cathodic (B) charges every time through an angle of 70°. Parallel evolution of the consumed electrical energy (C) for the anodic process and (D) for the cathodic process (solid lines correspond to experimental results and dotted lines correspond to theoretical results). Sensing thermal calibration curves for: (E) the muscle potential from A after different times of anodic current flow (at different constant angles); (F) the muscle potential from B for the same oxidation state (at different constant angles); (G) the consumed electrical energy from C at different constant angles; and (H) the consumed electrical energy from D at different constant angles. Adapted with permission from ref. 21: *J. Phys. Chem. B*, 2012, **116**, 9223–9230. Copyright (2012) American Chemical Society.

9.11 One Actuator and Several Simultaneous Sensors in One Device

Eqn (9.8) describes one actuator and several sensors (mechanical, thermal, chemical and electrical) working simultaneously in a uniform device. In the previous sections we have proved one-by-one how the dual actuator–sensor devices work empirically as described by the proprioceptive equation. Table 9.1 presents the sensing constants. Each sensor constant includes different physical and chemical magnitudes. When two or more variables change simultaneously, eqn (9.10) and Table 9.1 should allow de-convolution of the empirical muscle potential evolution to get the magnitude of each of the shifting variables. Actuating (driving current, movement rate and movement position) and sensing (muscle potential and consumed electrical energy) magnitudes are present at any working time in the only two connecting wires: the computer can read this information at any time. The computer–electrical generator–two wires–artificial muscle ensemble and the proprioceptive equations constitute an artificial proprioceptive system mimicking natural proprioception (Figure 9.11).

9.12 Other Multi-Tool Devices

The door is open in the future for the empirical exploration, technological exploitation and theoretical description of chemical- and electrochemical-driven devices where two, three, four or more tools work simultaneously. As described previously for sensing motors, Table 5.1 allows the selection of the properties required to develop any new multi-tool device. An important

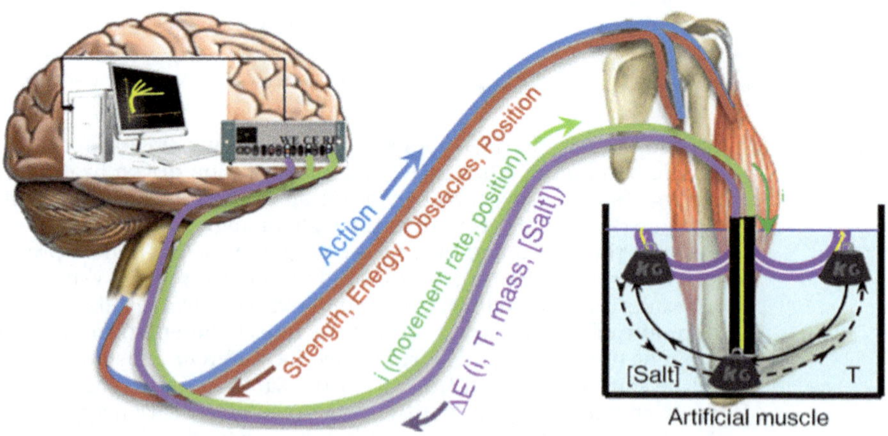

Figure 9.11 The computer–potentiostat–wires–artificial muscle ensemble and the proprioceptive equations mimic brain–nerve–muscle proprioceptive in natural systems. Reprinted with permission from ref. 20: *J. Phys. Chem. B*, 2012, **116**, 5279–5290. Copyright (2012) American Chemical Society.

challenge emerges for chemists, biochemists, physicists, engineers and designers of new tools and robots: some control of the complex material synthesis mechanism (Chapter 3) is required in order to improve the synthesis conditions to get large simultaneous variation of different composition-dependent properties (Chapter 5) to produce different multi-tool devices (two, three or four tools working simultaneously), as described by the extended proprioceptive equations.

9.13 Intelligent Electrochemical Materials for Multi-Tool Devices

Eqn (9.8)–(9.11) describe the mechanical, chemical and thermal awareness of the oxidation reaction involving any reactive material (intrinsically conducting polymer, redox polymer, carbon nanotube, graphene, porphyrins, phthalocyanines and so on). Any electrochemical device (Chapter 8) based on the composition-dependent properties (Chapter 5) of those materials will also sense during actuation any thermal, chemical, mechanical or electrical perturbation.

Artificial muscles, actuators and motors, batteries and supercapacitors, windows, mirrors, glasses, filters and screens, membranes, drug delivery agents, neuron interfaces, electron–ion transducers or contaminant captors, and solvo-phobic/solvo-phillic surfaces, among other electrochemical and reactive devices, will sense mechanical, chemical, thermal and electrical conditions or perturbations while working. All of them are simultaneously reaction-driven actuators and sensors.

In Chapter 1, biological materials that form part of reactive organs were considered to be the highest reference for intelligent, reactive materials that develop three simultaneous functions as actuators, sensors and in self-healing. As soon as conducting polymers (or any other electroactive materials giving electrochemical devices) become self-healing they will attain the same intelligent order of magnitude depicted by biological organs. At the moment, conducting polymers (and any other material forming electroactive films) give, by reaction, dual sensing and actuating materials, as described by eqn (9.8)–(9.11), becoming some of the most intelligent artificial materials. The concomitant sensing–actuating devices here presented may become some of the most intelligent devices in our technological world to date.

References

1. T. F. Otero, *J. Mater. Chem.*, 2009, **19**, 681–689.
2. J. G. Martinez, J. Arias-Pardilla and T. F. Otero, in *Smart Actuation and Sensing Systems – Recent Advances and Future Challenges*, ed. G. Berselli, InTech, Rijeka, 2012.
3. T. F. Otero, in *Artificial Muscle Actuators Using Electroactive Polymers*, ed. P. Vincenzini, Y. BarCohen and F. Carpi, 2009, vol. 61, pp. 112–121.

4. T. F. Otero, M. T. Cortes and G. V. Arenas, in *Smart Sensors, Actuators, and MEMS II*, ed. C. Cane, J. C. Chiao and F. V. Verdu, 2005, vol. 5836, pp. 477–483.
5. T. F. Otero, in *From Non-Covalent Assemblies to Molecular Machines*, ed. J.-P. Sauvage and P. Gaspard, Wiley-VCH, Weinheim, 2011, pp. 443–452.
6. T. Otero, M. Cortes, L. Boyano and G. Vazquez, in *Smart Structures and Materials 2004: Electroactive Polymer Actuators and Devices (EAPAD)*, ed. Y. BarCohen, Spie-Int Soc Optical Engineering, Bellingham, 2004, vol. 5385, pp. 425–432.
7. Wikipedia Free Encycl., 2015, https://en.wikipedia.org/wiki/Proprioception, accessed September 2015.
8. T. Otero and J. Sansinena, *Bioelectrochem. Bioenerg.*, 1995, **38**, 411–414.
9. T. F. Otero and M. T. Cortes, *Sens. Actuators, B*, 2003, **96**, 152–156.
10. T. F. Otero and M. T. Cortes, *Adv. Mater.*, 2003, **15**, 279–282.
11. T. F. Otero, I. Boyano, M. T. Cortes and G. Vazquez, *Electrochim. Acta*, 2004, **49**, 3719–3726.
12. F. G. Cordova, Y. A. Ismail, J. G. Martinez, A. S. Al Harrasi and T. F. Otero, *Electroact. Polym. Actuators Devices*, 2013, **8687**, 868708.
13. J. G. Martinez, T. Sugino, K. Asaka and T. F. Otero, *ChemPhysChem*, 2012, **13**, 2108–2114.
14. Y. A. Ismail, J. G. Martinez, A. S. Al Harrasi, S. J. Kim and T. F. Otero, *Electroact. Polym. Actuators Devices*, 2011, **7976**, 79761L.
15. Y. A. Ismail, J. G. Martinez and T. F. Otero, *J. Electroanal. Chem.*, 2014, **719**, 47–53.
16. L. Valero Conzuelo, J. Arias-Pardilla, J. V. Cauich-Rodriguez, M. Afra Smit and T. Fernandez Otero, *Sensors*, 2010, **10**, 2638–2674.
17. Y. A. Ismail, J. G. Martinez, A. S. Al Harrasi, S. J. Kim and T. F. Otero, *Sens. Actuators, B*, 2011, **160**, 1180–1190.
18. F. G. Cordova, L. Valero, Y. A. Ismail and T. F. Otero, *Electroact. Polym. Actuators Devices*, 2012, **8340**, 83401B.
19. T. F. Otero, *J. Mater. Chem.*, 2009, **19**, 681–689.
20. T. F. Otero, J. J. Sanchez and J. G. Martinez, *J. Phys. Chem. B*, 2012, **116**, 5279–5290.
21. J. G. Martinez and T. F. Otero, *J. Phys. Chem. B*, 2012, **116**, 9223–9230.
22. J. G. Martinez and T. F. Otero, *Sens. Actuators, B*, 2014, **195**, 365–372.
23. T. F. Otero and J. G. Martinez, *Prog. Polym. Sci.*, 2015, **44**, 62–78.
24. M. S. Kang, M. K. Shinb, Y. A. Ismail, S. R. Shin, S. I. Kim, H. Kim, H. Lee and S. J. Kim, *Nanotechnology*, 2009, **20**, 085701.
25. Y. A. Ismail, J. G. Martinez and T. F. Otero, *Electrochim. Acta*, 2014, **123**, 501–510.
26. J. G. Martinez, T. F. Otero, C. Bosch-Navarro, E. Coronado, C. Marti-Gastaldo and H. Prima-Garcia, *Electrochim. Acta*, 2012, **81**, 49–57.
27. T. F. Otero and M. T. Cortes, in *Smart Structures and Materials 2000: Electroactive Polymer Actuators and Devices (EAPAD)*, ed. Y. BarCohen, Spie-Int Soc Optical Engineering, Bellingham, 2000, vol. 3987, pp. 252–260.

28. T. F. Otero and M. T. Cortes, in *Smart Structures and Materials 2001: Electroactive Polymer Actuators and Devices*, ed. Y. BarCohen, 2001, vol. 4329, pp. 93–100.
29. T. F. Otero, M. T. Cortes, L. Boyano and G. Vazquez, in *Smart Structures and Materials 2004: Electroactive Polymer Actuators and Devices (EAPAD)*, ed. Y. BarCohen, 2004, vol. 5385, pp. 425–432.
30. T. F. Otero and J. Arias-Pardilla, in *Electropolymerization: Concepts, Materials and Applications*, ed. S. Cosnier and A. Karyakin, Wiley-VCH Verlag GmbH & Co. KGaA, Weinheim, 2010, pp. 241–272.
31. T. F. Otero and M. Broschart, *J. Appl. Electrochem.*, 2006, **36**, 205–214.
32. T. F. Otero, J. G. Martinez, L. Valero, K. Asaka and Y. A. Ismail, in *Electroactive Polymers: Advances in Materials and Devices*, ed. P. Vincenzini and S. Skaarup, 2013, vol. 79, pp. 16–25.
33. T. F. Otero, *Bioinspiration Biomimetics*, 2008, **3**, 035004.
34. L. Jones, *Can. J. Physiol. Pharmacol.*, 1994, **72**, 484–487.
35. A. J. Bard and L. R. Faulkner, *Electrochemical Methods: Fundamentals and Applications*, Wiley, New York, 1980.
36. K. Vetter, *Electrochemical kinetics: Theoretical and Experimental Aspects*, Academic Press, New York, NY, 1967.
37. A. J. Bard and L. R. Faulkner, *Electrochemical Methods: Fundamentals and Applications*, Wiley, New York, 1980.
38. T. F. Otero, J. J. Sanchez and J. G. Martinez, *J. Phys. Chem. B*, 2012, **116**, 5279–5290.
39. T. F. Otero and J. Rodriguez, in *Intrinsically Conducting Polymers: An Emerging Technology*, ed. M. Aldissi, Kluwer Academic Publishers, Dordrecht, 1993, vol. 246, pp. 179–190.
40. T. F. Otero and J. G. Martinez, *Adv. Funct. Mater.*, 2013, **23**, 404–416.

CHAPTER 10

Final Comments and Challenges

10.1 Introduction

Many experimental evidences have been presented revealing conformational and structural changes driven by reactions involving reactants, polymers and large macromolecules. Reaction-driven shrink–closing and compaction effects result in ion trapping. Basic electrochemical reactions must be re-written after structural considerations in order to get a more precise description.

Reaction-driven structural changes are usual from different materials and ubiquitous in biochemistry. Structural electrochemical responses from other materials, such as bipyridyl monolayers or carbon nanotubes, will be presented here. Biochemical processes, such as muscular action, allosteric effects, enzymatic reactions, Ebola virus activity, aptasensors and aptazymes, or ionic channel actions driving conformational movements, are presented also.

The structural activation energy and the structural reaction coefficient from the structural chemical kinetic (SCK) model presented in Chapter 7 give a qualitative description of both enzymatic reaction rates and hemoglobin allosteric reaction coefficients. Some new experimental results are required to obtain the different constants there included in order to get a quantitative description of the biochemical reactions and predictions for new natural or artificial processes.

Some exciting and accessible (in the mid-term) experimental and theoretical challenges, such as developing electro-chemo-conformational memories, the theoretical description of brain memory and other brain functions, or

RSC Smart Materials No. 19
Conducting Polymers: Bioinspired Intelligent Materials and Devices
By Toribio Fernández Otero

Published by the Royal Society of Chemistry, www.rsc.org

the production of massive materials with high and uniform volumetric electroactivity giving volumetric devices and industrial products, are presented and discussed.

10.2 Reactions and Structures

After description of the reaction-driven conformational processes (Chapter 6) eqn (4.1)–(4.3) result in quite simplified versions of the actual reactions, not including any indication of the trapping effects due to reaction-compaction processes.

Now a most realistic version can be presented and the p-doping reaction in eqn (4.1) becomes:

$$\mathrm{Pol}^{a+}(\mathrm{A}^-)_a(\mathrm{H_2O})_n + n\mathrm{A}^- + m\mathrm{H_2O} \rightleftarrows [\mathrm{Pol}^{(a+n)+}(\mathrm{A}^-)_{n+a}(\mathrm{H_2O})_{m+n}]_{\mathrm{gel}} + ne^-, \qquad (10.1)$$

indicating that any reduced and conformationally packed state still contains positive polaronic species (Pol^+) and anions (A^-), which are required for charge balance.

Related to the polymer blends, the p-doping reaction in eqn (4.3) can be re-written as:

$$[(\mathrm{Pol}^{a+})(\mathrm{MA}^{(a+m)-})(\mathrm{C}^+)_m(\mathrm{H_2O})_o]_{\mathrm{gel}} \rightleftarrows (\mathrm{Pol}^{(a+n)+})(\mathrm{MA}^{(a+m)+})(\mathrm{C}^+)_{m-n} + n\mathrm{C}^+ + o\mathrm{H_2O} + ne^-, \qquad (10.2)$$

indicating that any oxidized and conformationally compacted structure includes the constituent macroions (MA), some positive charges on the chains and balancing cations (C^+). Taking into account that full reduction-swelling is not possible (Figure 6.3) and that the reduced polymer includes charged chains and balancing cations in addition to the constituent macroions, the n-doping reaction in eqn (4.3) becomes:

$$\mathrm{Pol}^{a-}(\mathrm{C}^+)_a(\mathrm{H_2O})_o + n\mathrm{C}^+ + m\mathrm{H_2O} + ne^- \rightleftarrows [(\mathrm{Pol}^{(a+n)-})(\mathrm{C}^+)_{a+n}(\mathrm{H_2O})_{o+m}]_{\mathrm{gel}}, \qquad (10.3)$$

indicating that any oxidized and conformationally compacted state still contains negative polaronic species and cations, which are required for charge balance.

For simplicity, most of the reaction-driven structural effects, properties and devices have been described in this book for materials exchanging anions and following eqn (4.1). The used concepts, theoretical descriptions, properties and devices are common, with some minor remarks, for materials exchanging anions [eqn (4.1)] or for those exchanging cations [eqn (4.2) and (4.3)].

Thus, related to the reaction-driven volume changes, from the families of conducting polymers following eqn (4.1) in a large potential range, the identified reaction-driven basic structural processes are: oxidation–relaxation, oxidation–swelling, reduction–shrinking and reduction–compaction.[1] From the families of conducting polymers following eqn (4.2) and (4.3), the reaction-driven basic structural processes are: reduction–relaxation, reduction–swelling, oxidation–shrinking and oxidation–compaction.[2] In addition, from blends of conducting polymers with polyelectrolytes and amphiphilic ions a new slow reduction process was revealed from the coulovoltammetric responses at cathodic potentials, which in the case of the amphiphilic macroions has been attributed to the formation of vesicles inside the gel membrane.[2]

Structural processes driven by eqn (4.2) are related to p-doping and p-dedoping processes taking place at anodic potentials by exchange of cations and water close to the anodic limit of the electrolyte potential window. Structural processes driven by eqn (4.3) are linked to n-doping and n-dedoping processes also by exchange of cations and water, but taking place now at high cathodic potentials: very stable electrolytes (having an electrolyte potential window with a high cathodic potential limit) are required for their study and applications.

Any conducting polymer synthesized by oxidation–polymerization in the presence of small anions follows eqn (4.1) in the presence of a different small anion, yielding a new polymeric compound. The same polymer in the presence of a large anion (a polyelectrolyte) doesn't present any electrochemical response in the same potential range.[1] It doesn't react due to structural limitations: the free volume that the conformational relaxation could produce is not enough to lodge the large counterion required to compensate for the polymeric charges on the chains inside the polymeric structure. Once returned to the presence of a small anion, the usual voltammetric and chronoamperometric responses are obtained from the conducting polymer.[1] During both p-doping and n-doping reactions, the diameter of the exchange cation can also influence the kinetics of the doping process.[3,4]

Any conducting polymer synthesized by oxidation–polymerization in the presence of *large* anions follows eqn (4.2) in the presence of electrolytes with small ions. Any conducting polymer synthesized by oxidation–polymerization in the presence of *small* anions follows eqn (4.1) in the presence of electrolytes with small ions. The question now is which ionic exchange prevails when anions with intermediate dimensions are used during the polymer synthesis. Using bending bilayer (conducting polymer/tape) muscles as useful tools to translate small ionic exchanges into large bending movements, a mixed exchange was observed:[5] cation exchange at the more cathodic potential range and anion exchange at the more anodic potential range.

Another pending point is related to the dissociation constant of the macroion salt in the polymer film: the cation is exchanged [eqn (4.2)] because the interaction force with the macro-anion is not strong. This force is determined by the dissociation constant, the value of which can be managed by changing the solvent. Recently it has been reported, using artificial muscles as tools, that the system can be switched from cation exchange to anion exchange by moving it back and forth from aqueous to organic electrolytes.[6,7]

Obtaining *ab initio* calculations of the intermolecular force variations during the reactions in eqn (4.1), (4.2) or (4.3) is required to attain quantitative descriptions of the composition-dependent properties and devices, and this is currently outside the possibilities of quantum chemistry. Molecular dynamic calculations can provide useful results by using monodisperse oligomers, solvent, small ions and cations.[8–14] The treatment of complex and partially cross-linked polymer gels is beyond current theoretical possibilities. Under these circumstances, the results from the previous paragraphs, together with new voltammetric and coulovoltammetric results from different materials and supported by dynamo-voltammetric and coulodynamic results from bilayer muscles, can provide the required quantitative information to model polymer–polymer, polymer–anion, polymer–cation, anion–cation, polymer–solvent, anion–solvent and cation–solvent interaction variations inside the dense gel during reaction-driven structural changes.[15] The attained model, including the structural chemical kinetics (Section 7.2.3), should constitute a good platform for the quantitative description and prediction of similar processes driven by biochemical reactions in living cells.

10.3 Other Artificial Materials Giving Reaction-Driven Structural Responses

The literature presents structural chronoamperometric, voltammetric or coulovoltammetric responses, such as those treated in this book for conducting polymers exchanging anions, for conducting polymers exchanging cations during p-doping, p-dedoping, n-doping and n-dedoping, and from a large number of different materials. Of course, most of them are not identified as reaction-driven structural processes by the authors.

10.3.1 Conducting Polymers Exchanging Cations During p-Doping/p-Dedoping

The basic components of the reaction-driven structural changes have been treated in Section 6.4. The formation and growth of lamellar or vesicular structures can be followed during consecutive voltammetric responses (the oxidation depth during each anodic sweep does not arrive to collapse the vesicles; Figure 10.1a). The formation of a deeply reduced–compacted conducting polymer around every vesicle is confirmed by the subsequent relaxation–nucleation maximum on the subsequent chronoamperometric response to an anodic potential step during the film oxidation (Figure 10.1b). The compacted structure of the oxidized film is confirmed by the reduction nucleation–relaxation shape of the subsequent chronoamperometric responses to cathodic potential steps (film reduction), confirming the compacted structure of the oxidized film (Figure 10.1b).

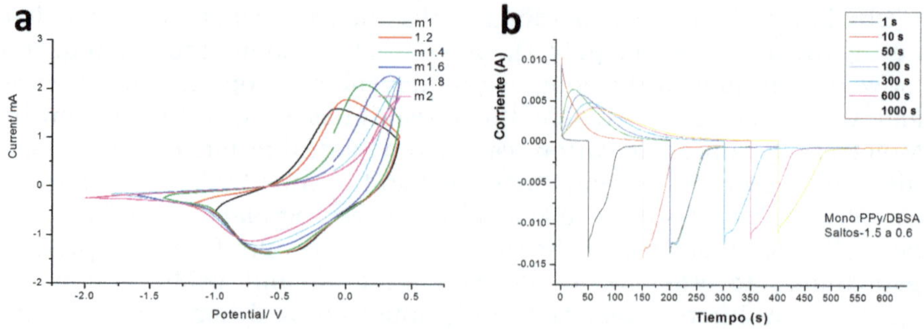

Figure 10.1 (a) Voltammetric response from polypyrrole/dodecylbenzenesulfonate (Ppy/DBS) films cycled from 0.4 V to different cathodic potential limits of −1, −1.2, −1.4, −1.6, −1.8 and −2 V in 0.1 M $LiClO_4$ at 6 mV s^{-1}. (b) Chronoamperometric responses to consecutive square potential steps from −1.5 V, kept for reduction times of 1, 10, 50, 100, 300, 600 and 1000 s for every experiment, to the same anodic potential of 0.6 V, kept for enough time to allow the oxidation current to drop close to zero.

10.3.2 Conducting Polymers Exchanging Cations During n-Doping/n-Dedoping

Increasingly conformationally packed states of the film are attained by oxidation–compaction, giving the expected nucleation–relaxation maximum on the subsequent chronoamperometric response during the film reduction by a cathodic potential step (Figures 6.5 and 10.1b)

10.3.3 Monolayers of Bipyridyl Cations

Structural electrochemical responses are attained also from monolayers or multilayers of relatively small reactive molecules. A monolayer of any viologen, 4,4′-bipyridine (Bpy) or its alkyl derivatives adsorbed on mercury electrodes exchanges iodide ions with the solution during its electrochemical oxidation/reduction, giving up to two consecutive structural reduction chronoamperometric maxima by reorganization of the adsorbed layer.[16] Figure 10.2 shows the chronoamperometric responses obtained in the presence of iodide ions by potential steps from −500 mV to different cathodic potentials (−564, −568, −570 and −572 mV) attributed by the authors to reduction–desorption–nucleation and reduction–reorientation–nucleation structural processes, similar to those happening in Li_xFePO_4.[17] The positive currents in Figure 10.2 indicate reduction reactions.

10.3.4 Very Large Carbon Nanotubes

Self-supported films of “super-growth”, millimeter-long single-walled carbon nanotubes (SG-SWNTs) in the presence of an ionic liquid give similar nucleation–relaxation–oxidation chronoamperometric responses to those observed using conducting polymers exchanging anions (Figure 10.3).[18,19]

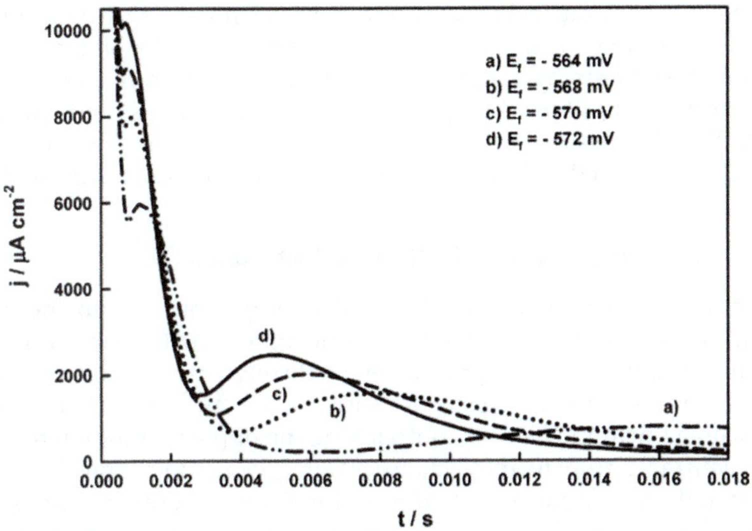

Figure 10.2 Chronoamperometric response from a monolayer of 4,4′-bipyridyl radical cation salts on mercury in the presence of iodide ions. Reprinted with permission from ref. 16: *Langmuir*, 1994, **10**, 723–729. Copyright (1994) American Chemical Society.

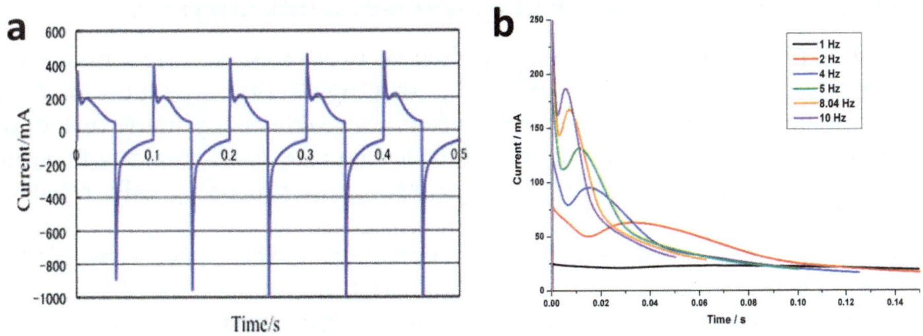

Figure 10.3 (a) Stationary consecutive chronoamperometric responses from super-growth single-walled carbon nanotubes (SG-SWCNT/EMITFSI film thickness, 17.5 mm) in the ±2 V range when applying square wave voltages with a frequency of 10 Hz (EMITFSI, 1-ethyl-3-methylimidazolium-bis(trifluoromethyl-sulfnyl)imide). (b) Stationary anodic chronoamperometric responses at frequencies of 1, 2, 4, 5, 8.04 and 10 Hz. Reprinted from ref. 18: *Electrochim. Acta*, 2012, **60**, 177–183, with the permission of Elsevier.

10.4 Biological Processes and Conformational Structures

Conformations play a central role in most biological reactions and functions from enzymatic reactions to respiratory functions or the transcription and translation of genetic information, signalling and others. They can be re-considered, and are being considered, as resulting from molecular chemo-conformational machines. The energy required for the biological conformational movements have different origins and aims. A full description of biological conformational changes is out of the scope of this book. Here, some of them have been selected just for illustration of the links with the electrochemically stimulated conformational relaxation (ESCR) and SCK models.

10.4.1 Muscular Action in Striated Muscles

As mentioned previously, the chemical energy from adenosine triphosphate (ATP) hydrolysis to adenosine diphosphate (ADP) drives the myosin head conformational bending movement through consecutive steps along the actin chains, each linked to the sarcomere wall.[20] The final result is a macroscopic linear movement approaching the opposite sarcomere wall: the muscle contracts, transforming chemical energy into mechanical energy and heat through reaction-driven conformational movements. Inspired by natural muscles, the artificial muscles treated in this book are based on chain polymeric motors, the reversible electrochemical-driven conformational movements of which translate electrical energy into mechanical energy.

10.4.2 Allosteric Chemical Responses from Enzymes

Enzymes play a key role in many biological processes, accelerating a variety of metabolic reactions, as well as controlling energy transduction. They possess the remarkable capacity to accelerate reactions by many orders of magnitude compared to their non-catalyzed counterparts, making feasible crucial processes that would otherwise not occur on biologically relevant timescales.

Coming from a quasi-static vision of enzymes, an integrated view of protein structure, dynamics and function is emerging mainly based on *in situ* nuclear magnetic resonance (NMR) or atomic force microscopy (AFM) determinations. Proteins are considered as dynamic and active motors. The internal conformational movements (dynamics) of enzymes are closely linked to their functions. The catalytic effect is viewed as a series of correlated conformational changes that occur throughout the enzyme and act along the reaction coordinate to facilitate the chemical reaction.[21,22] Such facilitation may be viewed as active site conformational reorganization, stabilizing interactions as well as increased sampling of chemically productive substrate/active site configurations. Now the SCK model can contribute with a quantitative explanation due to the low reaction conformational relaxation component of the activation energy [eqn (7.15)] for the enzyme–substrate reaction. Enzymes catalyzing the same reactions belong to a protein family: the overall characteristic shape and the allosteric movements linked to the catalytic specificity are similar. Enzymes catalyzing mechanistically similar reactions often belong to the same protein superfamily.

In addition, the flexibility of the conformational movements gives enzymes the possibility to evolve very fast, giving near-by enzymes acting now on new substrates. An increasing number of researchers are considering this flexible and fast enzymatic evolution as one of the forefronts of biological evolution.

The benefits of understanding the enzymatic reaction-driven conformational dynamics better through the structural chemical kinetics include the possibility to obtain a full quantitative theoretical description of the biochemical process, and its allosteric and cooperative effects. A predictive theoretical model should allow improvement of the efficiency of microbial factories by the engineering of enzymes, as well as the design of new enzymes with novel functionalities. Further, there are medical implications of allosteric and cooperative effects for enzyme activity in novel drug design.

10.4.2.1 The Enzymatic Activation Energy May Quantitatively Describe Enzymatic Reaction Rates

From eqn (7.7) and (7.15), the reaction rate (r) of any biochemical reaction involving conformational changes is quantitatively described by:

$$r = k_0\,\mathrm{e}^{-\frac{RT + \Delta H* + z_c\eta_c - z_r\eta}{RT}}[A^-]^a[\mathrm{Pol}*]^b. \tag{10.4}$$

Here, k_0 is the reaction coefficient, R is the gas constant, T is temperature, $\Delta H*$ collects the energetic components of the system, z_c is the conformational compaction coefficient, η_c is the compaction overpotential, z_r is the conformational relaxation coefficient and η is the oxidation overpotential. For enzymatic reactions, the active polymeric sites [Pol*] are the enzymatic active centers [E*]; the anions A^- are now the substrate [S].

Enzymatic reactions occur in the absence of electric fields and/or currents. We assume that the packed conformational state is the most stable state of the enzyme under biological conditions with a constant conformational energy, E_{conf}. The enzymatic chemical reaction drives the molar conformational relaxation (E_{relax}) energy and the term $\Delta H*$ includes all the enzyme–electrolyte interaction variations ($E_{\mathrm{E\text{-}e\text{-}S}}$) when the substrate approaches the enzymatic active site. Thus, the enzymatic reaction rate becomes:

$$r = k_0\,\mathrm{e}^{-\frac{RT + E_{\mathrm{E\text{-}e}} + E_{\mathrm{conf}} - E_{\mathrm{relax}}}{RT}}[\mathrm{S}]^a[\mathrm{E}*]^b. \tag{10.5}$$

The presence of the substrate, and only the substrate, increases the reaction-driven (enzyme–substrate) conformational relaxation energy (E_{relax}) with a likely simultaneous decrease of the enzyme–water and enzyme–ions intermolecular forces ($E_{\mathrm{E\text{-}e}}$). Both terms take part in the enzymatic activation energy: any small increase of the molar conformational relaxation energy or any small decrease of the enzyme–electrolyte intermolecular forces promotes the exponential increase of the enzymatic reaction rate [eqn (10.5)]. The enzymatic reaction is accelerated by many orders of magnitude (exponential dependence) compared to their uncatalyzed counterparts.

10.4.3 Allosteric and Cooperative Effects from Hemoglobin and Other Proteins

Heme groups constitute the base for many biochemical processes, one of the most important being breathing. The human hemoglobin protein includes four heme groups.[23] The chemical reaction rate of hemoglobin with oxygen constitutes one of the allosteric paradigms. The reaction constant of the fully reduced protein is very low. Once the first heme group has reacted with oxygen, the reaction rate of the second heme group with oxygen is faster than that of the first one. The evolution goes on increasing the reaction rate for the oxygen reaction with the third heme group of the protein and a higher value is obtained of the reaction rate for the oxygen reaction with the fourth heme group of the protein. The surprising feature of this is that the affinity for the fourth oxygen that binds is between 100 and 1000 times higher than for the first oxygen.[24] In biochemical reactions, these are named *cooperative effects*, giving sigmoidal (Figure 10.4, curve a) shapes of the reaction rate evolution, similar to reactions with initial induction times.[25] Figure 6.1, showing the oxidation kinetics of conducting polymers from open and compacted initial states, illustrates the parallel fast hemoglobin oxidation starting from free solved heme groups (Figure 10.4, curves a and b) or from the fully reduced and conformationally compacted protein.

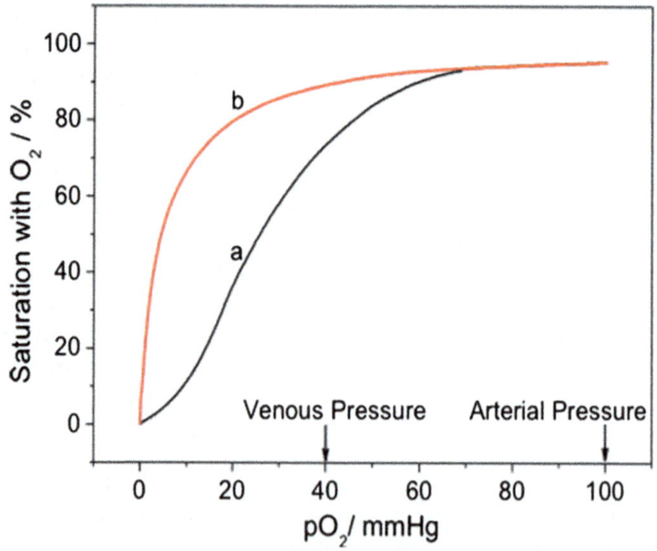

Figure 10.4 The four heme groups in hemoglobin react with oxygen through four consecutive steps of increasing reaction rates, curve (a). The free heme group reacts with oxygen in a similar way to that of the initial open conformational states during oxidation, curve (b). Adapted from ref. 25: *J. Mol. Biol.*, 1965, **12**, 88–118, with the permission of Elsevier.

10.4.3.1 Structural Reaction Coefficients can Describe Allosteric Reactions

Now the structural chemical kinetics offer a good theoretical description of these allosteric cooperative biochemical effects. For a constant concentration of hemoglobin and a constant oxygen concentration, the concentration of each of the four different heme groups is the same. For the same concentration of the reactants, different reaction rates mean rising reaction coefficients. From the reaction shown in eqn (7.19), now in the absence of electric fields:

$$\ln k = \ln k_0 + \left(\frac{E_{\text{relax}} - \Delta H_{\text{H-e}}}{RT} \right) - \left(\frac{E_{\text{comp}}}{RT} \right), \tag{10.6}$$

where $\Delta H_{\text{H-e}}$ represents the variation of the hemoglobin–electrolyte interactions along the reaction. Any fully reduced hemoglobin protein presents a highly compacted conformational structure, *i.e.*, high E_{comp}. At the constant temperature of the human body, the subsequent oxidation–relaxation rate is very slow because the high compaction energy of this conformationally packed structure gives [eqn (10.6)] a very low reaction coefficient. The reaction of the first heme group results in a more relaxed conformational structure, which constitutes the reactant of the second heme group. The most relaxed structure increases the oxygen accessibility in relation to that of the first heme group. From a quantitative point of view, the molar conformationally packed energy, E_{comp}, [eqn (10.6)] has promoted, under constant temperature, the exponential increase of the reaction coefficient of the oxygen molecule with the second heme group to a lesser extent in relation to that of the first heme group. In the same way, consecutive exponential increases of the reaction coefficients for the third and fourth heme groups are attained from eqn (10.6). The SCK model quantitatively explains how, under constant chemical and thermal conditions, the heme–oxygen reaction rate coefficient must exponentially increase for increasingly conformationally relaxed initial states attained by oxidation of the first, second or third heme groups of the initial reduced and compacted protein.[24]

10.4.4 Molecular and Viral Activity and Conformational Structure: The Ebola Virus

In this context, it becomes clear why many biological enzymes, proteins or viruses change their activity according to the energetic (thermal, chemical, electrical, electromagnetic, magnetic) conditions. Electromagnetic radiation, magnetic or electrical fields can force the open state of membrane proteins and pore channels in cells, giving electroporation. Some viruses, such as the Ebola virus, are inactive outside the human body or at low temperatures in some hosts due to its conformationally packed structure. The virus structure relaxes and opens at the temperature of the human body, thus becoming active.[26]

10.4.5 Allosteric Effects from Nucleic Acids

Aptamers are useful for allosteric regulation because they are nucleic acid-based structures in which ligand binding induces conformational changes that may alter the function of a connected oligonucleotide at a distant site.[27] Through this approach, a specific input is efficiently converted into an altered output. This property makes these biomolecules ideally suited to function as sensors or switches in biochemical assays or inside living cells. The ability to select oligonucleotide-based recognition elements *in vitro* in combination with the availability of nucleic acids with enzymatic activity has led to the development of a wide range of engineered allosteric aptasensors and aptazymes.[28,29]

10.4.6 Conformational Movements of Ion Channel Proteins

Figure 10.5 presents the ionic flow through Ca^{2+} and K^+ ion channels driven by increasing potential gradients across the cell membrane stimulated through internal and external clamps. The ion-compacted protein of the closed pore opens under influence of the applied potential step, resulting in rising ionic currents that go through a maximum and then decrease by closing of the channel protein (Figure 8.9b).[30] The peak position (polarization time) and intensity is a function of the potential step gradient. Similar results were attained (Figure 6.8) from thick films of conducting polymers.

In summary, reaction-driven conformational and allosteric changes are ubiquitous in biology, resulting in life functions.

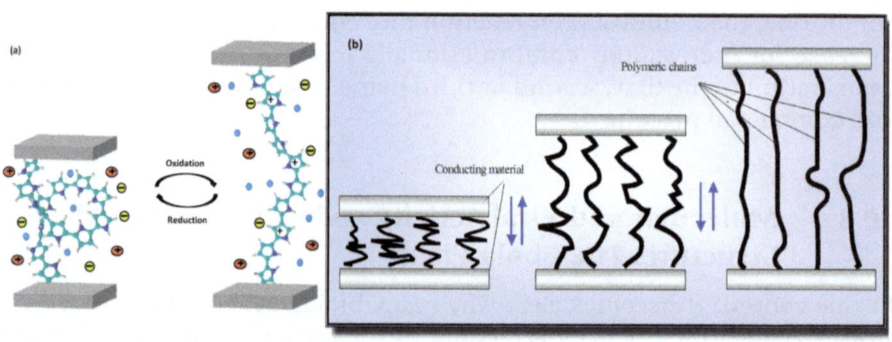

Figure 10.5 (a) Basic element and (b) ideal mono-dimensional and fully anisotropic artificial molecular muscle, with mono-disperse polymeric chains, the ends of which are linked to two parallel microelectrodes. Adapted from ref. 34, with the permission of the Royal Society of Chemistry.

10.5 Challenges

Chemical kinetic models from chemical, biochemical and biological textbooks are stingy when describing and quantifying biochemical reactions, life functions, health and disease. The SCK model presented in this book has included reaction-driven conformational energies in chemical models. A new way has been opened to initiate a quantitative description of life functions based on biochemical reactions driving conformational, allosteric or bending movements.

At the moment, only a few structural and conformational chemical-driven aspects have been described by the SCK model, using a few reactions and involving a few materials, allowing only a couple of families (among the eight different families) of conducting polymers to be identified (see Section 1.5.1). Similar reactive gels can be attained from the different material families resulting from fullerenes, carbon nanotubes, graphenes, redox conducting polymers, porphyrins, phthalocyanines and any redox organic molecule or structure that can be embedded in a polymer to give electronic conducting and electroactive films. It should be expected that new reaction-driven structural aspects will emerge, being detected by coulovoltammetric, chronoamperometric and voltammetric (among other) responses during the chemo-structural exploration of these materials. Structural and kinetic implications will require a careful description and quantification from the empirical chemical kinetics using the methodology described in Section 7.2.3. This unexplored scientific border will require great efforts, open minds, some chemical, biochemical, electrochemical and polymeric background, and close collaboration between chemists, engineers, tool and robot designers, physicists, and medical doctors, among others, to obtain fast medical and technological applications.

Here, we have advanced a model and described in a quantitative way, for the first time, the conformational energy required to go from the initial conformational state of the reactant to a different conformational state of the product, resulting in a conformational movement. Figure 7.4f and h reveal that the value of the conformational energy can surpass the value of the chemical activation energy of the driving reaction. This fact opens great technological expectancies for the technological applications based on the conformational energy, such as artificial muscles, intelligent membranes or new artificial allosteric biosensors.

Here, we have learned that this conformational energy is one of the components of the reaction activation energy [eqn (7.15)]. The reaction rate depends exponentially on the activation energy [eqn (7.7)] and consequently on the conformationally compacted energy of any macromolecule acting as a reactant [eqn (10.5)]: small variations of the conformational energy will change the reaction rate by several orders of magnitude, such as enzymatic reactions do. We now know how to manage small or large variations of the initial conformational energy in conducting polymers under the control of the compaction overpotential [eqn (7.3)]. In addition, we have also learned that any change of the intermolecular forces acting during the reaction will also have an exponential influence on the reaction rate through the activation energy component $\Delta H*$ [eqn (10.5)].

In order to illustrate the unexplored potential of the conformational and intermolecular components of the activation energy [eqn (10.2) and (10.4)], some attention must be paid to the only parallel chemical field where the activation energy can be managed under control: electrochemistry. The experimental electrode reaction overpotential, $\eta = E - E_0$, which can be controlled from the electrical generator, takes a part of the activation energy of any electrochemical reaction [eqn (9.2)]. Metals are obtained by chemical reduction in ovens from their metal salts from mines. The reduction reaction only occurs when the temperature of the oven attains several hundreds or thousands of degrees centigrade. Nevertheless, the same reaction takes place under ambient, soft laboratory conditions in electrochemical cells from aqueous or organic solutions of the salts only requiring a few tens of millivolts of cathodic overpotential. In addition, the metal deposition reduction rate increases exponentially with the applied overpotential due to the exponential decrease [eqn (9.2)] of the activation energy. Just the same exponential influence now funds [eqn (10.4) and (10.5)] the conformational, structural and molecular interaction components of the activation energy.

Coming back to the conformational component of the activation energy, at the moment we still ignore how this powerful conformational reaction lever can be used and how it is quantitatively influenced by small variations of the polymer structure (using materials from the substituted monomer family), substitution of the anion or cation from the electrolyte in different solvents and so on. We have advanced the experimental methodology required to get the final quantitative expressions. We can imagine, but at the moment we also ignore due to the lack of experimental results, the medical, scientific and technological implications this energy, its fluctuations and variations could provide.

Among a plethora of subsequent possible biochemical implications of the conformational compaction and conformational relaxation components of the biochemical reaction-driven conformational, allosteric or bending processes, we have illustrated the catalytic specific activity of enzymes and the allosteric effects in hemoglobin. At present, we still ignore how and in which biological or medical conditions these conformational effects on the reaction activation energy or on the reaction coefficient could be used to inhibit the activity of proteins or nucleic acids that cause diseases. We also ignore how the model can be used to improve the activity of those for which inhibition results in new diseases. At the moment, empirical biosensors coupling molecular recognition to a binding-induced change in the conformation state constitutes the technological forefront of this unexplored technological universe.

Finally, the SCK model has also succeeded in quantifying the unexpected variation of the reaction orders as a function of the initial conformationally packed state of the reactant. At the moment, both scientific and technological expectancies from this fact constitute an unexplored world. As an initial approach, we can remember that, for polymeric and biochemical reactions, such as those treated in this book, the reaction order accounts for the number of active polymeric sites interacted by the approaching second reactant (counterion, enzymatic substrate and so on). More conformationally compacted initial states mean shorter average distances between neighbor polymeric active sites, giving higher empirical reaction orders. In parallel, these active site/reactant interactions must be under the quantitative influence of every intra- and inter-molecular interaction force present in the local arena (solvent, every chemical group of the macromolecule, and each of the anions, cations and molecules there present). Under these circumstances, the reaction order should contribute to some of the most complex aspects of allosteric, folding/unfolding and conformational processes driven by biological reactions: the variation of the local intermolecular interactions during reaction. It appears clear that a deep consideration of the structural reaction order's physical, chemical and biochemical meaning is required before advancing its scientific, medical and technological implications.

Chemistry's heart, chemical kinetics, has been through a long slumber since Marcus theory.[31–33] New analytical methodologies have been trying to substitute for the lack of improved chemical kinetic models by a detailed detection of each biochemical reaction product and reaction intermediate. During the last 40 years, biological and medical advances have been based on the great success of analytical methodologies for the discovery of new smaller and very diluted proteins or nucleic acids as reaction products or as reaction intermediates. Increasingly complex and expensive analytical methodologies were consequently required. The lack of supporting chemical kinetic models was dodged by treating the attained results from each studied biochemical event with increasingly complex specific mathematical or physical treatments. As a result, sentences equivalent to: *the unexpected, unsolved or unexplained results could be related to molecular conformations, folding or unfolding effects* became among the most frequently used in the highest impact journals during the last 30 years. This book has been written to address the urgent need for improved and predictive chemical kinetic models, including every reaction-driven structural change from chemistry and biology. The emerging SCK model here presented announces a new way to describe, quantify and predict, in a closer and simpler way, biochemical reactions and biochemical functions.

The reaction-driven conformational movements are linked to changes, by several orders of magnitude, of the material ionic composition. So, large concentration variations are only available in biological environments during biological reactions. A deeper empirical and theoretical study of the composition-dependent biomimetic properties will provide new information to advance the SCK model in parallel to the development of the most efficient chemo-biomimetic devices and products. As useful feedback, some new ideas and concepts from the biomimetic devices can be used for new explorations of natural, mimicked devices.

A special challenge concerns the use of conformational energetic states (each trapping a different concentration of counterions) as electro-chemo-ionic-conformational memories. Electrochemical reading/writing pulses from these memories mimic electrical, chemical and conformational energetic information components of neuronal pulses. So, the high information content per pulse may allow the storage and treatment of a huge amount of information by the consumption of very small amounts of electrical energy. After development of some basic memory units, the attained basic information may allow the development of new logic systems and computers with new primitive brain-mimicking functions. As feedback, the theoretical treatment could be applied to neuron channel pulses to try to decipher some initial aspects of brain memory code. In this way, a plethora of electronic/neuronal (ionic–chemical) pulse translators could be developed to implant artificial devices (such as video cameras, arms, legs) directly connected through the nervous system to the brain and thus working under direct brain orders, as natural organs do. A great deal of exciting experimental and theoretical work is expected in the coming years.

At the moment, most of the studied chemo-biomimetic devices were constructed using films thinner than 100–400 μm. A problem comes from the polymerization mechanism (Chapter 3). Behind the electropolymerization front, a polymeric degradation–cross-linking starts, resulting in asymmetric bilayers with a composition gradient: most are electroactive (most charge stored per unit of polymer weight) close to the fresh electrogenerated surface with low electroactive material close to the polymer/metal interface. As a result, the final, composition gradient and thick self-supported films behave during oxidation–reduction as bending artificial muscles. For thicker films, the average electroactivity is very low. Some electrochemical properties, such as electroactivity, are very sensitive to this partial degradation: in fact, films thicker than a few micrometers lose their electrochromic properties despite keeping a quite high electroactivity. The exploration of new reactive dense gels from the different materials previously mentioned and from the most unexplored families of conducting polymer (blends, hybrids and composites) will open the possibility to obtain larger materials with a high volumetric electroactivity. They will support the development of three-dimensional electrochemical devices, such as artificial muscles with similar shapes and dimensions to natural ones.

On the other hand, control of the molecular structure will give monodimensional systems, such as mimetic molecular sarcomeres for artificial muscles. For present electro-chemo-mechanical actuators, the statistical distributions of the polymer chains cover all directions of the three-dimensional space. The electrochemical reaction promotes quasi-three-dimensional swelling–shrinking processes (Figure 6.1c). Both bending and linear muscles use only dimensional variations along one of the three space directions: the maximum theoretical energetic efficiency can only attain one-third of the applied electrical energy. The construction of mimetic sarcomeres with polymeric chains going between two electronic conducting micro-plates (Figure 10.5) acting as the working electrode should produce prevalent one-direction dimensional changes under reversible oxidation/reduction, yielding much higher energetic efficiencies.[34] At the moment, different laboratories have been fabricating molecular brushes. The remaining unsolved step is how the end of the molecular brushes can be connected to a second micro-plate.

Despite so many pending challenges, the development of the new technological biomimetic world constituted by devices where two tools, three tools or four tools work simultaneously in one physically uniform device will emerge. Both materials and devices will work simultaneously as actuators and sensors, becoming the closest ones to intelligent materials (Chapter 1) among artificial materials. Using multi-tool devices, a new world of intelligent and proprioceptive zoomorphic and anthropomorphic tools, functions and robots is envisaged, with closer and closer textures to the natural ones.

Finally, as a consequence of the construction and theoretical description of the first dual-sensing muscles, the hitherto considered physiological and behavioral concept, proprioception, has been quantitatively expressed by a physicochemical equation attained from basic electrochemical, polymeric and mechanical principles applied to the dual sensing and actuating properties of artificial muscles. The open question is which other hitherto considered brain functions and physiological or behavioral concepts (memory, thinking, consciousness) can be expressed by physicochemical equations? A clear answer has now emerged: all those involving electric (ionic or electronic) pulses and chemical reactions involving biopolymers as reactants, reaction-driven conformational movements, ions and water. Now, a plethora of dense reactive gel materials are available to be used as models for artificial empirical bio-mimicry and theoretical description. These are exciting tasks for clever, open-minded scientists, with a solid multidisciplinary basic background and being ready to address unexpected experimental results with complex theoretical and experimental obstacles.

References

1. T. F. Otero, M. Alfaro, V. Martinez, M. A. Perez and J. G. Martinez, *Adv. Funct. Mater.*, 2013, **23**, 3929–3940.
2. T. F. Otero, J. G. Martinez, M. Fuchiwaki and L. Valero, *Adv. Funct. Mater.*, 2014, **24**, 1265–1274.
3. C. Weidlich, K. M. Mangold and K. Juttner, *Electrochim. Acta*, 2005, **50**, 1547–1552.
4. G. Zotti, S. Zecchin, G. Schiavon and A. Berlin, *Macromolecules*, 2001, **34**, 3889–3895.
5. M. Fuchiwaki and T. F. Otero, *J. Mater. Chem. B*, 2014, **2**, 1954–1965.
6. N. Aydemir, P. A. Kilmartin, J. Travas-Sejdic, A. Kesküla, A.-L. Peikolainen, J. Parcell, M. Harjo, A. Aabloo and R. Kiefer, *Sens. Actuators, B*, 2015, **216**, 24–32.
7. R. Kiefer, J. G. Martinez, T. F. Otero, A. Kesküla, F. Kaasik, M. Harjo, R. Valner, V. Vaddepally, A.-L. Peikolainen and A. Aabloo, in *SPIE Smart Structures and Materials & Nondestructive Evaluation and Health Monitoring*, International Society for Optics and Photonics, 2015, p. 94301C.

8. M. M. Ostwal, M. Sahimi and T. T. Tsotsis, *Phys. Rev. E*, 2009, **79**, 061801.
9. J. J. Lopez Cascales and T. F. Otero, *J. Chem. Phys.*, 2004, **120**, 1951–1957.
10. S. D. Oliveira Costa and J. J. Lopez Cascales, *J. Electroanal. Chem.*, 2010, **644**, 13–19.
11. T. Bjornholm, W. Brostow, K. Schaumburg, P. V. Shibaev, V. Sinani and R. Vinokur, *Macromol. Symp.*, 1999, **148**, 31–45.
12. J. J. L. Cascales, A. J. Fernandez and T. F. Otero, *J. Phys. Chem. B*, 2003, **107**, 9339–9343.
13. J. M. Fonner, C. E. Schmidt and P. Ren, *Polymer*, 2010, **51**, 4985–4993.
14. M. M. Ostwal, T. T. Tsotsis and M. Sahimi, *J. Chem. Phys.*, 2007, **126**, 124903.
15. T. F. Otero, I. Cantero and H. Grande, *Electrochim. Acta*, 1999, **44**, 2053–2059.
16. M. Maestre, R. Rodriguezamaro, E. Munoz, J. Ruiz and L. Camacho, *Langmuir*, 1994, **10**, 723–729.
17. G. Oyama, Y. Yamada, R. Natsui, S. Nishimura and A. Yamada, *J. Phys. Chem. C*, 2012, **116**, 7306–7311.
18. P. Gimenez, K. Mukai, K. Asaka, K. Hata, H. Oike and T. F. Otero, *Electrochim. Acta*, 2012, **60**, 177–183.
19. K. Asaka, K. Mukai, T. Sugino, H. Randriamahazaka and T. Fernandez Otero, *Electroact. Polym. Actuators Devices EAPAD 2013*, 2013, **8687**, 86871U.
20. A. Huxley and R. Simmons, *Nature*, 1971, **233**, 533–538.
21. S. C. L. Kamerlin and A. Warshel, *Proteins: Struct., Funct., Bioinf.*, 2010, **78**, 1339–1375.
22. B. Ma and R. Nussinov, *Curr. Opin. Chem. Biol.*, 2010, **14**, 652–659.
23. M. F. Perutz, A. J. Wilkinson, M. Paoli and G. G. Dodson, *Annu. Rev. Biophys. Biomol. Struct.*, 1998, **27**, 1–34.
24. A. Ferst, *Structure and Mechanism Protein Science. A Guide to Enzyme Catalysis and Protein Folding*, W.H. Freeman and Co., New York, 1999.
25. J. Monod, J. Wyman and J. Changeux, *J. Mol. Biol.*, 1965, **12**, 88–118.
26. G. G. Weber, J. Kortmann, F. Narberhaus and K. E. Klose, *Proc. Natl. Acad. Sci. U. S. A.*, 2014, **111**, 14241–14246.
27. J. L. Vinkenborg, N. Karnowski and M. Famulok, *Nat. Chem. Biol.*, 2011, **7**, 519–527.
28. J. L. Vinkenborg, N. Karnowski and M. Famulok, *Nat. Chem. Biol.*, 2011, **7**, 519–527.
29. E. Luzi, M. Minunni, S. Tombelli and M. Mascini, *TrAC, Trends Anal. Chem.*, 2003, **22**, 810–818.
30. M. C. Emerick, R. Stein, R. Kunze, M. M. McNulty, M. R. Regan, D. A. Hanck and W. S. Agnew, *Proteins: Struct., Funct., Bioinf.*, 2006, **64**, 320–342.
31. R. Marcus, *J. Chem. Phys.*, 1956, **24**, 966–978.
32. R. Marcus and N. Sutin, *Biochim. Biophys. Acta*, 1985, **811**, 265–322.
33. R. Marcus, *Annu. Rev. Phys. Chem.*, 1964, **15**, 155–&.
34. T. F. Otero, *J. Mater. Chem.*, 2009, **19**, 681–689.

Subject Index